Performing Security Analyses Of Information Systems

By

Charles L. Smith, Sr., PhD, CISSP

ISBN: 1-4033-1475-6 (Electronic)
ISBN: 1-4033-1477-2 (Softcover)
ISBN: 1-4033-1476-4 (Rocket)

Library of Congress Control Number: 2002091063

This book is printed on acid free paper.

Printed in the United States of America
Bloomington, IN

1stBooks - rev. 5/22/02

ACKNOWLEDGEMENTS

I wish to express appreciation to several of my former employers who allowed me the opportunities to acquire the expertise and knowledge in computer network security without which it would have been impossible to write this book.

Charles L. Smith, Sr.

FOREWORD

One of the reasons for this book is that the technology of information systems network security is advancing so rapidly, and yet there seems to be no single reference book that provides an explanation and compilation of the myriad aspects of security. One can quickly get completely inundated with the gamut of reference information available in the technical literature, on the Internet at the Web pages of the various security vendors, and in reference books published by Government agencies and by commercial institutions. I have endeavored to present a complete treatise of the issue of performing a security analysis of an information system while taking care to limit the presentation to only those issues that pertain directly to understanding this vast and complex issue. That certainly is a subjective assessment on my part, and many readers may disagree with me, but hopefully more will be happy with this book.

There is a lot of reference information on this subject so one of the tasks that must be performed is to review these references and glean the essential relevant information and then to concentrate this information into a single whole. To achieve that objective, this book is divided into five parts:

I. Basic Concepts
II. Defining the Problem
III. Finding a Solution
IV. Implementing and Testing the Preferred Solution
V. Conclusions and Recommendations

Each of these parts is further decomposed into one or more chapters:

- Part I (five chapters) contains a presentation of a security analysis process and explains the various concepts required for understanding the process.
- Part II (two chapters) contains a presentation of the process for defining the hypothesis that needs to be resolved.
- Part III (one chapter) contains a presentation of the process for determining the countermeasures needed to protect the system.
- Part IV (one chapter) contains a presentation of the process for migrating from the current legacy system to the target system.
- Part V (one chapter) contains a presentation of the conclusions and recommendations that result from the other nine chapters.

The new Itanium chips from Intel will contain encryption and decryption capabilities not present on the older chips. This will greatly speed up the implementation of encoding capabilities required to provide confidentiality and other Public Key Infrastructure (PKI) needs. Intel, Sun Microsystems, and IBM, as well as other chip manufacturers, are competing to provide the very best cost-effective chips for future needs that include security capabilities. This means that the computation overhead for many of the security needs of the future, such as confidentiality, will be much less than in previous systems.

The approach taken here is an attempt to simplify a complex subject that few people seem to wish to understand. Most industry leaders, whether in government or the commercial area, have opted to outsource this problem and then have proceeded to bury their heads in the sand (so to speak) assuming that what they don't know (about their information system security) can't hurt them, i.e., "ignorance is bliss." At the present time, security analyses are more art than science, but with the advent of structured analyses, this should change. As Benjamin Barber has stated in "Jihad vs McWorld," the airlines, with respect to airport security, "… have taken a cost effectiveness approach rather than a 'safety' approach with disastrous consequences." And similarly, large and small organizations, government and commercial, with respect to information system security, have taken a "simplicity" approach (by outsourcing) rather than an "understanding (of a complex issue)" approach, sometimes with disastrous consequences.

Being primarily a mathematician, I feel a need to offer definitions of pertinent terms, and these appear in Appendix A, Glossary. Appendix B contains most of the acronyms used in the area of security. Appendix C contains a sample questionnaire for obtaining needed information. Appendix D, actually two appendices, contains samples of a security policy and a security rules base. A table of security threats is contained in Appendix E and a table of potential vulnerabilities is contained in Appendix F. Some security mechanisms are contained in Appendix G. A set of security mechanisms designed to counter certain threats is shown in Appendix H. The so-called C2 requirements for controlled access protection are presented in Appendix I. A minimum set of security requirements is contained in Appendix J. Rules for architecture model development are shown in Appendix K and a mathematical analysis of the answer to the question "Why is information system security needed?" can be found in Appendix L.

PREFACE

This book is primarily for commercial and government organizations (owners, managers, system administrators, and users) who must deal with sensitive but unclassified and nonsensitive unclassified information, namely, data that is not classified (i.e., it is not Confidential, Secret, Top Secret, nor any of the higher classifications), yet needs to be protected from viewing by the public or those who do not have a need-to-know. This is the situation for most commercial organizations as well as most civilian government organizations, or at least most of the information these types of organizations must handle.

The security information presented here is for information system owners (or their named representatives) and security analysts who operate at an operational and analytical level rather than at a detailed program level. The information and processes presented here will enable the owner or security analyst to minimize the various security risks to any extent desired, but there is no way to totally eradicate the risks completely. This is a characteristic of security for information systems (i.e., networks of computing and associated devices). In addition, this book can be a supplemental guide for novices or competents or even some experts who need additional information and analytical processes in order to perform their required duties as security analysts.

The approach to the subject of information system security taken here is to limit to a minimum the educational and knowledge background required for reading and understanding the presented material. Thus, the text here will not contain theorems and proofs that may be contained in many papers and books on the subject.

Rather than hiring security professionals or learning about security themselves, it is becoming common practice for many organizations to outsource their security problems. This means that the organizations' owners (i.e., information system stakeholders) are totally dependent upon their selected security vendor to ensure that the owner's system is secure. If a security breach should occur, then the security vendor is expected to respond to the problem, whatever this may involve. However, some owners are reluctant to respond to some malicious intrusions because this may require publicizing a particular vulnerability in their system. However, if certain information is stolen and the vendor has promised that this specific event would not occur, then it may be the case that the owner may not hear about the loss of the information from the vendor. So the owner is somewhat at the vendor's mercy and may be sorry for this trust if the owner should somehow discover that this trust was inappropriate.

To preclude even the possibility of such an event, some security vendors allow for the owner's organization to co-manage the security system by being privy to the auditing or intrusion detection system results, that is, the auditing results are communicated and displayed at an owner's designated site in addition to the vendor's site. However, this privilege will only be as worthy as the owner's organization's abilities to both monitor and understand monitoring results. This places a significant onus upon the owner's organization since any intrusion should be identified by their organization whether or not the vendor detects and reports the incident.

Unfortunately, because malicious incursions may occur infrequently, owners and vendors may be lulled into a sense of complacency assuming that these (malicious) actions are not going to happen (to their system). It is very important for the owners to understand what threats affect their information system and what their system's vulnerabilities are. Some owners feel that a firewall will somehow magically block all malicious activities launched against their system and that this firewall will offer all the protection that they need. This simply is not so.

Moreover, all information systems require a periodic upgrading of the protection being provided because security requirements, threats, asset values, and technologies are constantly changing and today's security mechanisms that provide sufficient protection cannot be guaranteed to be sufficient for the life of the system nor to be cost-effective always.

When security is considered, each owner's information system is different from every other information system so that there is no single silver bullet that will provide universal protection. Even a particular firewall may require a set of rules that will make it behave differently from all other firewalls. Each owner has an information system that may not consist of a unique set of devices, but its software and purposes will make it different in how these devices are used. And the larger the information system, the more complex it is and the higher the likelihood that this system is unique among all other existing information systems and hence most likely has unique security requirements.

This means that when considering the security requirements for a specific information system, a security analysis must be performed for this specific information system, and no existing security document or prior analysis will suffice for specifying its security requirements. Any security vendor that delivers a standard analysis to the customer is shortchanging the client and the client should be aware of this.

The most important product of any security analysis is the definition of the various security issues or problems affecting the information system. That is, a definition of the security problem for this specific system is required before attempting to discover a solution that will include a specific set of security devices, software, and methods. Years ago there was a term for those who were always ready to go out and begin creating a system to satisfy a problem they did not even understand. They were called "tin benders." In this book, a prescriptive model is offered. This model ensures that an appropriate issue definition is required before attempting to define the security mechanisms needed to protect the information system. This book is not for tin benders.

This book contains an explanation of the various phases of performing a security analysis of an information system. In its simplest terms, *security* for an information system consists of those mechanisms and methods that, when installed or implemented, offer improved protection of the system and its information from damage or theft. Thus an *information system security analysis* is the determination of those specific security mechanisms and methods required for cost-effectively improving the protection of a particular system and its information from damage or theft. This analysis should also contain the rationale for why a particular set of mechanisms and methods was chosen from among the various alternatives available. The reader will find that the figures used in this book provide a simple and readily understandable graphical explanation of the various concepts presented here so that, for most ideas presented here, there is both a verbal and a pictorial explanation of the concepts involved.

Security mechanisms are the products or services that provide system protection, and *security methods* are those processes that ensure that system users or managers are knowledgeable of the behaviors required to ensure system protection. An example of a security mechanism is a "firewall" and an example of a security method is a rule stating, "No internal user should ever divulge a password to a caller, no matter who the caller says he or she is." This method of acquiring internal information by a malicious hacker about some organization is called *social engineering* and will be discussed later in this book.

To create a security architecture for some reference system, it is necessary for the analyst to understand the architecture of the reference system. This is so since the security architecture must be embedded in the reference architecture and the creation of this protective subsystem requires an understanding of the reference system into which this subsystem must be embedded. The reference system plus the security system will be referred to as the "overall system" which can be described with an overall system architecture. In most cases, the reference architecture (or legacy system) will have some security mechanisms, but will require some improvements to meet the security requirements.

There is a saying that "the deeper you get, the deeper you get," which means that as one delves more deeply into some subject to better understand and explain the situation, the more complex the explanation gets. The approach here has been to avoid, as far as is possible, getting too complex. Another saying is "keep it simple," which is what I have tried to do. However, the success or failure of these attempts is quite subjective and will depend upon the individual reader's background. Unless they are absolutely essential, acronyms will not be used in this book.

Dr. Charles L. Smith, Sr., CISSP
TLA Associates
6412 Beulah Street, Alexandria, VA 22310
csmith@tla.com   **703-254-2040**

TABLE OF CONTENTS

PART III.   FINDING A SOLUTION

PART IV.   IMPLEMENTING AND TESTING THE PREFERRED SOLUTION

APPENDICES

PART I: BASIC CONCEPTS

This portion of the book consists of five introductory chapters that provide a presentation and explanation of the fundamental concepts for a security analysis of an information system. The approach is to start out with simple and easy to understand explanations and to get progressively more complex. In some cases, this causes some redundancy; however I believe that the result is a book that is easier to understand than one that has no redundancy.

Charles L. Smith, Sr.

CHAPTER ONE

AN INTRODUCTION TO SECURITY ANALYSIS

1.1 Introduction

The terrorist events of September 11, 2001, in New York City, Washington, DC[1], and Pennsylvania, have stimulated a national interest in all types of security by leaders and citizens of the United States of America, as well as various leaders worldwide. Although the results of this increased awareness of U.S. security laxness may include the diminution of personal privacy, a large majority of Americans are for improving security, in all its manifestations. Some of the lack of security precautions that could have contributed to the successful attacks on the World Trade Center and the Pentagon are the ready access the terrorists had to Internet information and the use of the Internet for sending and receiving information using encrypted electronic mail messages.

Given the current status of interest in information system network security, this book is primarily for commercial and government organizations that deal with nonsensitive unclassified information and sensitive but unclassified information (sometimes referred to as SBU information). This is information that is *not classified* (i.e., it is data that *is not* Confidential, Secret, Top Secret, nor any of the higher classifications) but is sensitive and needs to be protected from unauthorized observation, destruction, or modification. This is the situation for most commercial organizations as well as most civilian government organizations, or at least most of the information these types of organizations must handle.

Throughout this book the word "organization" will be used to denote a commercial or government operation such as a company, corporation, business, enterprise, organization, agency, or the like. "Owners" of an organization's information system may be the actual persons who own the system but may also be the leading stakeholders of the organization's system such as a Chief Executive Officer (CEO), Chief Information Officer (CIO), or other high-level manager or government official.

"According to federal intelligence and national security officials, no generally accepted methodology for strategic analysis of cyber threats to the nation's infrastructures has been developed. ... The intelligence community officials we met with said that developing such a methodology would require an intense interagency effort and a dedication of significant resources (GAO 2001)." The prescriptive model for performing a comprehensive security analysis that is presented in this book can be used by any government or commercial organization to identify the cost-effective security mechanisms and methods required to offer the protection they need for their specific information system.

When organizations decide that they want a quality security system for protecting their information technology assets (often in response to a striking publicized event), they usually want it quickly. The best way to ensure that cost-effective countermeasures are defined accurately and completely is to first perform a security analysis that includes a formal risk assessment. Charging forward by defining solutions (i.e., security methods or products) for issues that are poorly understood (if at all) and that can be quickly acquired and installed may seem like a good idea at first, but when these methods and products do not work properly and/or are too expensive, owners will be displeased and disappointed. Hopefully, the prescriptive model for performing a comprehensive security analysis provided in this book will be used to avoid making these costly and time-consuming mistakes.

---

[1] This is in reference to the aircraft attack on the Pentagon, which actually is located in Northern Virgina, not Washington, DC, which is commonly referred to as the location of the Pentagon.

1.1.1 Purpose

A primary objective of this book is to present a formal prescriptive process for performing a security analysis, which, if properly followed, should provide a comprehensive end-to-end method and greatly diminish the possibility of an organization purchasing and/or implementing security mechanisms or methods that are not appropriate for meeting the organization's security requirements. The focus of this book is on *providing directions for performing a formal comprehensive security analysis*. Because the definition of what constitutes an acceptable security analysis varies from organization to organization and from week to week within an organization, it is not recommended that a computer program for supporting security analyses be built unless it incorporates the capability to easily modify the embedded comprehensive prescriptive analysis process.

Information system owners (even if the system consists of a single computer) have three options (Bernstein 1996):

1) Do not plug your organizational computer network into a public network (such as the Internet) by keeping it isolated (often referred to as a "stand alone" or "private" system);
2) Place security software on each client in your system; and/or
3) Place security subsystems (platform and software) at a network perimeter[2] location (e.g., a firewall).

The guidance offered by this book is for those who select option 2 and/or option 3. Some owners may wish to employ option 1 (stand-alone system), but this will greatly limit the system's capabilities to access information and to interface with others, both as a receiver and transmitter of information. Even some large organizations will select option 1 for some of their internal subsystems to "ensure" security. As we shall see later, "stand-alone" systems certainly do not ensure system security. Most organizations that have large computer networks will deploy security mechanisms in their clients and at the perimeter of their information system, especially if they are using a public wide area network for their communications.

This book is for information system owners (or their named representatives) and security analysts who operate at an operational and analytical level (or even a managerial level) rather than at a detailed programming level. It is for people who wish to have a basic understanding of information system network security and the process for performing a security analysis, but do not need to understand the technical details of how the various security mechanisms operate nor how to write the programming code for providing security. Many of the references used for this book were obtained from Internet Web pages and the Uniform Resource Locators (URLs) for this information are identified in the text, such as the National Institute of Standards and Technology (NIST) Computer Security Resource Center (CSRC), which can be found at http://csrc.nist.gov/.

---

[2] The *perimeter* of a network is the entrance/exit point for the network, that is, it is the point where information enters or leaves the organizational system at a particular location.

At a very basic level, a security analysis consists of posing and answering this single, and seemingly simple, question:

> *"What countermeasures do I need for protecting my*
> *organization's information system network?"*

The contents of this book are devoted to the consequences, implications, and complexities of this one question. Unfortunately, it is not a simple question to answer. A security analysis should begin by addressing the following questions (either for formulating and resolving the issue or as an input):

- *Why* do I need to secure my information system? (This is the basis for the organization's Security Policy and its concomitant Security Rules Base.)
- *How* should the security activities be performed? (This is addressed by the Security Requirements for the information system.)
- *What* is required to perform the security activities? (This is a set of security mechanisms and methods that should be revealed in a Security Architecture.)
- *Who* should perform the security analysis? (This is a question that should be addressed by the owner or named representative.)
- *When* should the security analysis be performed? (This is another question that should be addressed by the owner or named representative.)
- *Where* should the security products be placed? (They need to be placed where they are most effective and the locations should be evident in the overall information system network architecture.)

The information and processes presented in this book will answer the above, and other, questions and should enable the owner or security analyst to minimize the various security risks to any extent desired and affordable. But there is no way to totally eradicate the risks completely. This is a characteristic of security risks to information systems. In addition, this book can be a supplemental guide for novices or even experienced individuals who need additional information and analytical processes in order to perform their required duties as security analysts. To achieve these objectives, a *systems engineering approach* has been taken so that the required procedures for conducting a security analysis are fully explained.

Rather than hiring security professionals or learning about security themselves, it is becoming common practice for many organizations to outsource[3] their security problems. For example, the National Science Foundation (NSF) has obtained Network Securities Technology to manage its security needs, primarily to comply with the Government Information Security Reform Act. This means that the organizations' owners (i.e., information system stakeholders) are totally dependent upon their selected security vendor[4] or vendors to ensure that the organization's system is secure.

---

[3] *Outsourcing* means hiring a vendor to perform some needed service.

[4] Security vendors (i.e., those companies who are the objects of security outsourcing) do not advertise their clients since advertising might very well cause a number of malicious hackers to attack the client's information system thereby defeating their efforts.

There are many organizations which provide managed security services, some of these are McAfee (www.mcafee.com), Sonicwall (www.sonicwall.com), Veritect (www.veritect.com), VeriSign (www.verisign.com), Meta Security Group (www.metases.com), JetNet (www.jetnet.com), Internet Security Systems (http://xforce.iss.net/), and Check Point (www.checkpoint.com). Many more can be found on the Internet using a search for "managed security services" from your home page or using the Yahoo or Google search engines.

However, before outsourcing any element of the organization, management should be aware that there are hidden costs in outsourcing of information technology efforts (e.g., writing the contract, transition costs, managing the vendor costs, and termination costs). The organization can reduce these hidden costs by not outsourcing critical or organization-unique activities, spend more time researching vendors, and hire managers with outsourcing experience (Betts 2001).

If a security breach should occur, then the security vendor is expected to respond to the problem, whatever this may involve. However, some owners are reluctant to respond to some malicious intrusions because this may require publicizing a particular vulnerability in their system. Unfortunately, if certain information is stolen and the vendor has promised that this specific event would not occur, then it may be the case that the owner may not hear about the loss of the information from the vendor. So the owner is somewhat at the vendor's mercy and may be sorry for this trust if the owner should somehow discover that this trust was inappropriate.

To preclude even the possibility of such an event, some security vendors allow for the owner's organization to co-manage the security system by being privy to the auditing or intrusion detection system results, that is, the auditing results are communicated and displayed at an owner's designated site. However, this privilege will only be as worthy as the owner's organization's abilities to both monitor security audit trails and understand these monitoring results. This places a significant onus upon the owner's organization since any intrusion should be identified by their organization whether or not the vendor detects and reports the incident.

Unfortunately, because malicious incursions may occur infrequently, owners and vendors may be lulled into a sense of complacency assuming that these (malicious) actions are not going to happen. It is very important for the owners to understand what threats affect their information system and what their system's vulnerabilities are. Some owners feel that a firewall will somehow magically block malicious activities launched against their system and that this firewall will offer all the protection that they need. This simply is not so, primarily because most of the malicious actions taken against an organization's information system emanate from within, that is, from internal users.

When security is considered, each owner's information system is different from every other information system so that there is no single silver bullet that will provide universal protection. Even a particular firewall may require a set of rules that will make it behave differently from all other firewalls. Each owner has an information system that may not consist of a unique set of devices, but its software and purposes will make it different in how these devices are used. And the larger the information system, the more complex it is and the higher the likelihood that this system is unique among all other existing information systems and hence most likely has unique security requirements. Although the systems may be different, it is possible that the security architectures are the same.

Because all systems (and organizations) are different, when considering the security requirements for a specific information system, a security analysis must be performed for this specific information system, and no existing security document or prior analysis will suffice for specifying its security requirements. Thus any security vendor that delivers a standard analysis to the customer is shortchanging the client and the client should be aware of this.

The initial task for any security analysis is to define the various security issues or problems affecting the information system. That is, a definition of the security problem for this specific system is required before attempting to discover a solution that will include a specific set of security devices, software, and methods.

All information systems require a periodic upgrading of the protection being provided because security requirements, threats, and technologies are constantly changing and today's security mechanisms that provide sufficient protection cannot be guaranteed to be sufficient for the life of the system nor to be cost-effective always.

This book contains an explanation of the various steps for performing a security analysis of an *information system* (i.e., a network of computing and associated devices). In its simplest terms, *security* for an information system consists of those mechanisms and methods that, when installed or implemented, offer improved protection of the system and its information from damage or theft. Thus an *information system security analysis* is the determination of those specific security mechanisms and methods (i.e., countermeasures[5]) required for cost-effectively improving the protection of a particular system and its information from damage or theft, and how these countermeasures should be implemented and tested. This analysis should also contain the rationale for why a particular set of mechanisms and methods was chosen from among the various alternatives available.

The analysis process is presented in Figure 1.1, *A Simple Overview of a Security Analysis Process*. The reader hopefully will find that the figures used in this book provide a simple and readily understandable graphical explanation of the various concepts presented here so that, for most ideas presented here, there is both a verbal and a pictorial explanation of the concepts involved.

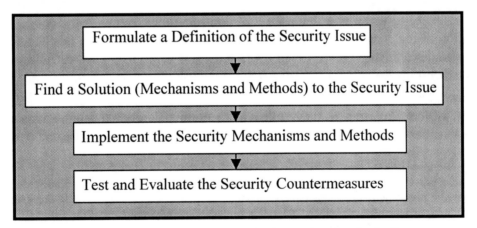

Figure 1.1 A Simple Overview of a Security Analysis Process

---

[5] A *countermeasure* is the specific safeguard required to minimize the potential negative effects from attacks based on a group of threats (one or more threat scenarios) and can be either a method (e.g., training) or a product (e.g., a firewall or anti-virus software).

Security mechanisms are the products or services that provide system protection, and *security methods* are those processes that ensure that system users or managers are knowledgeable of the behaviors required to ensure system protection. Together they represent the gamut of "countermeasures" for protecting the information system. An example of a security mechanism is a "firewall" and an example of a security method is a rule stating, "Never download an attachment in an e-mail message from someone whom you do not know" or "Never divulge any sensitive information to a caller, no matter who they say they are."

This method of acquiring internal information by a malicious hacker[6] about some organization is called *social engineering* and will be discussed later. Three types of architecture are discussed here:

1) *Reference architecture* - This is the architecture for the legacy system[7] into which the security architecture will be embedded.
2) *Security Architecture* - This is the architecture that will provide the protection needed by the reference architecture.
3) *Overall Architecture* - This is the complete architecture for the system that comprises both the reference and the security architectures.

To create a security architecture for a particular reference system, it is necessary for the analyst to understand the architecture of the reference system. This is so since the security architecture must be embedded in the reference architecture and the creation of the security architecture requires an understanding of the system into which it must be embedded.

The reference system plus the security system will be referred to as the "overall system" (or overall architecture as a model of the system), which can be described with an architecture model[8]. In most cases, the reference architecture (or legacy system model) will have some security mechanisms, but will require some improvements to meet the security requirements.

Multi-level security is not addressed in this book. Multi-level security is the attempt to provide all levels of security (Unclassified, Confidential, Secret, Top Secret, and Super Classified) on a single network going to single terminals. This objective has been either very difficult to achieve or impossible to achieve, and even when it is very difficult to achieve, the case has been that users may not trust it to do the job properly.

The need to operate safely is not new but as organizations increasingly rely on information system networks, as they endeavor to attain efficiency through shared resources, and as the intruders who perpetrate the threats to these systems proliferate and become more

---

[6] A *hacker* is anyone who uses the Internet or public wide area network to increase their knowledge or enter a local area network or computer in an unauthorized manner. *Malicious hackers* are those hackers who perform some sort of damage to the victim's system. These hackers are generally regarded as the external threat. A *cracker* is a special sort of hacker who enters a local area network and may or may not do some damage to the system.

[7] A legacy system is the existing system. It may have some security features but most often they are insufficient.

[8] The model of a network architecture will be referred to as an "architecture model" whereas the actual architecture will be referred to simply as the "architecture."

capable, the security posture of these systems is becoming more problematical. Organizational requirements can be characterized by the following (NSA 1998):

- The effectiveness of solutions must be judged based on their value to, and their effects on, the organization's mission (i.e., the countermeasures should assist the organization in meeting its mission objectives).
- The focus should be on providing adequate security solutions, balancing security risk with cost, and minimizing performance and operational effects (i.e., the countermeasures must be cost-effective).
- Affordability is a major consideration with all solutions (i.e., the countermeasures must be affordable to the organization).
- User friendliness is imperative, with features integrated into normal operations with minimal effects (i.e., the countermeasures must be easy to implement and use).
- Limitations of manpower availability drive all solutions to minimize personnel-intensive processes (i.e., the countermeasures must be user-friendly).
- Commercial off-the-shelf (COTS) products should be used to the fullest extent possible, augmented by Government-sponsored products or internally programmed products only when the needs dictate their use (i.e., the countermeasures must be COTS products, primarily, but may be internally programmed products if no appropriate COTS product exists). Because of the increasing usage by commercial organizations, the U.S. Government has instituted a requirement for COTS security products (Roback 2000).
- Security solutions must be based on those that are (or will be) available in a time frame that is consistent with the customers' needs (i.e., the countermeasures must be acquired in a timely manner).
- An upgrade path for implementation of all cost-effective security solutions must be provided so that customers can continue to take advantage of evolving information processing and network capabilities (both functional and security) (i.e., the countermeasures must be interoperable, portable, and scalable as provided by an "open architecture[9]").
- Security solutions should offer commonality, standardization, and interoperability, which can be achieved by working with various vendors to influence commercial offerings to this same end (i.e., the countermeasures must contribute to an open architecture).
- The judicious use of emerging technologies should be supported by offering customers the greatest operational capability with a minimum of programmatic risk

---

[9] Many use the term "open, standards-based architectures," but an architecture can be open only if it is based on standards, so the term "open, standards-based architectures" is redundant. *Interoperable* means the effective interconnection of two or more different computer subsystems (software or hardware) in order to support distributed computing and/or the exchange of data. *Portable* means a computer subsystem can be moved from one node in the system to another with little or no change required. *Scalable* means that a computer subsystem can be made to have more (or less) computational power by configuring it with a larger (or smaller): number of processors, amount of memory or storage, interconnection bandwidth, and input/output bandwidth, without interrupting service. Countermeasures that are interoperable, portable, and scalable with other subsystems (including countermeasures) are difficult to find.

(i.e., the countermeasures must be technologically up-to-date and supportable by the other technologies already embedded in the system).

To achieve these objectives requires the use of the latest information technologies, and a strong need for enterprise-wide interoperability. Much of the interoperability and cost-effectiveness is achieved using public wide area networks (WANs), such as the Internet, to reduce communications costs, however, the use of a public WAN also has its downside in that the threats are more numerous. Properly-selected network security solutions enable this connectivity and interoperability by allowing customers to maintain adequate protection of their information over these otherwise, unprotected communications systems (NSA 1998).

WAN topologies are network configurations that are designed to carry data over a great distance. *Point-to-point* means that the technology was developed to support only two nodes sending and receiving data. If multiple nodes need access to the WAN, then a Local Area Network (LAN) will be placed behind it to accommodate this functionality. A *T1 link* (1.544Mbps) is a full-duplex signal over two-pair wire cabling. T1s use time division multiplexing (TDM) to break two wire pairs into 24 separate channels. There are two common ways to deploy leased lines (or T1s):

1) The circuit constitutes the entire length of the connection between the two organizational facilities (this is the most secure link available but more costly than a public link), or
2) The leased line is used for the connection from each location to its local exchange carrier, such as frame relay.

While it is possible to sniff one of these circuits, an attacker would need to gain physical access to some point along its path. The attacker would also need to be able to identify the specific circuit to monitor.

The National Institute of Standards and Technology (NIST) concerning information technology product evaluation and validation states that "security is the protection of information from unauthorized disclosure, modification, or loss of use by countering threats to that information arising from human or systems-generated activities, malicious or otherwise." Countering threats to an information system network and mitigating risk helps to protect the confidentiality and integrity of information and ensure its availability.

Consumers of these products need to have confidence in their security features and to be able to compare various products to understand their attributes (such as capabilities and limitations). Assurance[10] for a particular information technology product can be based on:

1) The trusted reputation of the developer,
2) Past experience in dealing with the developer,
3) The demonstrated competence of the developer in building products through recognized assessments, or
4) Consumer testing of the product directly and obtaining the necessary results.

---

[10] *Assurance* is confidence that some product or service or system is of high quality.

The first two approaches lack measurable results but the third does provide a concrete performance capability on which the customer can rely. The fourth requires substantial and costly duplication of effort, if not managed properly.

A method that can be used to perform product test and evaluation is the Common Criteria (CC) Scheme (see the URL address http://csrc.nist.gov/cc/). The CC Scheme purports to overcome certain limitations and enable consumers to obtain an impartial assessment of an information technology product by an independent entity[11]. However, there are many schemes for test and evaluation that can be independent; that just means that the organization doing the evaluation of a product is not the organization that built the product.

The impartial assessment using the CC Scheme, or *security evaluation,* includes an analysis of the information technology product and the testing of the product for conformance to a set of security requirements. The specific information technology product being evaluated is referred to, in CC terms, as the *Target of Evaluation (TOE).* The security requirements for that product are described in its *security target* document. Information technology security evaluations are composed of analysis and testing, distinguishing these activities from the more traditional forms of conformance testing in other areas (NIAP 1997, NIAP 1999, NIST 1999).

Evaluations of information technology security products should be carried out in accordance with recognized standards and procedures. The use of standard information technology security evaluation criteria and an information technology security evaluation methodology contributes to the repeatability and objectivity of the results but is not by itself sufficient. Evaluation criteria usually require the application of expert judgment and background knowledge for which consistency is more difficult to achieve.

The final evaluation results can be reviewed by an additional independent party so as to increase the consumer's level of confidence in the evaluations. This review can provide an independent confirmation that the security evaluation was conducted in accordance with the provisions of the CC scheme and that the conclusions of the testing laboratory are consistent with the facts presented in the evaluation. The review, or *validation,* is intended to promote consistency of the security evaluations and comparability of results for all evaluations conducted within the scheme.

The impartial and independent validation of evaluation results, and the documentation resulting from these processes, provide evidential information for consumers about the security capability of the tested information technology products. However, consumers will still need to review this information carefully and assess its applicability to local needs, such as considerations of the situation and operating environment in which the product will actually be used.

Even after a complete security analysis has been performed and the appropriate countermeasures have been purchased and installed and tested and found to be operational, a security analysis must be performed again and a reaccreditation must be repeated for a system at least every three years or whenever the system has been significantly modified. This rule results from the rapidly advancing technologies of threats and security mechanisms.

It has been said that ignorance is bliss. Providing security for your system can be as simple as hiring a security vendor to do it for you and then completely trusting the vendor

---

[11] An *entity* is a person (i.e., user), application, service, or server.

that they will properly protect your system. The argument for the customer being "This is an approach, to obtaining security for my information system, in which I don't have to learn anything about security." This might be called the "lazy person's" approach.

For this reason, many security vendors hire salespersons who are adept at giving the customer the warm fuzzies. This trust will last until (which could be forever) the customer suddenly discovers that his or her system has been violated and that some important security incident has occurred that will cost them plenty. The more knowledgeable the customer is with respect to security, the higher the likelihood is that the customer will get proper security protection, even if it's from a security vendor. This is true for at least two reasons:

1) The customer will be less trusting of the vendor, and
2) The vendor must try harder to provide the appropriate countermeasures,

because the customer is knowledgeable about security issues.

To reap some of the benefits of understanding the basics of security a potential customer of information system security should (Scheier 2001):

- Understand the specific security functions (i.e., countermeasures) that you actually need because the service provider is not likely to tell you.
- Ensure that someone on your technical staff understands the basics of information system security.
- Don't give a sole source contract (usually called a service level agreement (SLA)) for your security to one provider because if their business should fail you are left defenseless. Be sure to include the names of the persons whom you should contact in case of an emergency.
- Develop a prior agreement on what functions the service provider will deliver and how its performance will be measured, in *specific, quantifiable terms* (e.g., number of malicious code messages identified versus number of malicious code messages not identified (i.e., those that got through and did some damage); or malicious code damage prevented (estimated) versus malicious code damage not prevented (actual); or costs of security services and missed malicious code message blockings versus savings from provider protection (estimated)).
- Consider hiring a qualified third-party[12] security consultant[13] to ensure that the service provider's technical expertise is appropriate and to monitor its performance.
- Plan what you will do when your security is breached, no matter where the attack originates.
- Don't expect your security service provider to offer perfect security against every possible threat, because they can't, even if they say they can, and do not try to word your contract so that you are asking for the impossible, even if the provider is willing to sign it.

---

[12] The first party is the client or customer, the second party is one or more vendors hired to provide system security, and the third party is a vendor hired to assure the client that he or she is doing the right thing.
[13] For example, TLA Associates., 6412 Beulah Street, Alexandria, VA 22310, csmith@tla.com.

Security vendors are probably capable of many aspects of security, but the question is "Are they capable of continued effectiveness?" This question can be answered by 1) the vendor's previous track record and, if hired, 2) your ability to realistically quantify the vendor's effectiveness during their contract with you. This requires collecting the appropriate data and then transforming this data into parameters (that are easy to understand) that represent the vendor's true performance.

Five easy steps to improved security are to take the following actions:

1) Turn off unneeded services in boxes attached to the Internet.
2) Never use a Web server for anything else.
3) Regularly apply security patches to critical machines.
4) Block all executable attachments at the gateway.
5) Use screen saver lockouts.

Hopefully, this book will provide security information that will serve to improve the knowledge that customers need to ensure that they get the appropriate amount of real information system network security that they actually need. A primary objective of this book is to provide the reader with a prescriptive process for finding the most economical and cost-effective set of countermeasures that will provide the maximum protection required for this organization's information system.

1.1.2 Overview

This first chapter, *An Introduction to Security Analysis*, provides an introduction to security analyses, the process for making a decision, the various aspects of security, and some countermeasures. Chapter Two, *An Overview of Security Analysis*, presents an overview of security analyses, provides some information on threats, vulnerabilities, and countermeasures. It also provides information on users, configuration management, security test and evaluation, and the certification and accreditation process. The next chapter, *Network Security Policy and Requirements*, covers both the development of security policy and security requirements. Chapter Four, *A Comprehensive Security Analysis Process*, presents a complete process for performing a security analysis, This complete process covers the development of security policy, security rules base, security requirements, risk assessment, migration plan, implementation, and test and evaluation. It also provides a description of alternative methods for choosing the most cost-effective countermeasures. Chapter Five, *Security Architectures*, is a presentation of the concepts that are components of security architectures, including the Web, communications networks, peer-to-peer concepts, and single sign-on.

Chapter Six, *Risk Assessment*, is a detailed examination of the risk assessment process. It presents detailed descriptions of the processes for identifying threats, vulnerabilities, and the performance of a comprehensive risk assessment. It also contains descriptions of the various potential countermeasures. The next chapter, *Countermeasures*, is a more detailed presentation of the various security mechanisms that might be purchased and installed to provide the desired security capabilities. Chapter Eight, *Migration Process*, is a description of the process for upgrading the current system to provide the capabilities of the security improvements that are cost-effective.

The next chapter, Chapter Nine, *Security Test and Evaluation*, contains a description of the process for performing a security test and evaluation of the upgraded system to ensure that the various installed countermeasures are operating properly. The last chapter, *Conclusions and Recommendations*, is a recapitulation of the book with recommendations for ensuring that a customer can acquire cost-effective security mechanisms for their information system

Appendix A, *Glossary*, contains definitions of terms relevant to information system security. Appendix B, *Acronyms and Abbreviations*, contains the acronyms and abbreviations use in the book or in the area of computer security. Appendix C, *Questionnaire: A Sample*, contains a sample of a questionnaire that might be needed to perform a security analysis. The next appendix is really two appendices, Appendix D.1, *Security Policy: An Example* and Appendix D.2, *Security Rules Base: An Example*. These appendices present examples of an organization's security policy and a security rules base. Appendix E, *Security Threats*, contains a table that includes most of the various threats to an information system and Appendix F, *Vulnerabilities*, contains a table of most of the potential vulnerabilities for an information system.

The next appendix, *Security Mechanisms*, contains a gamut of countermeasures that might be applied to protect an information system. Appendix H, *Mechanisms Versus Threats*, contains a table that contains the various mechanisms that might be deployed to protect against the indicated threats. Appendix I, *C2 Requirements*, contains a description of the requirements for controlled access protection from the DOD Orange Book. Appendix J, *Minimum Security Requirements*, contains a list of the fewest requirements a system might require for security protection. And the next to last appendix, *Rules for Architecture Development*, contains a set of rules that one can follow for developing a security architecture model from which the actual security architecture can be created. The last appendix contains a mathematical analysis of the question "Why is information system security needed?"

Carnegie Mellon University's (CMU's) Software Engineering Institute (SEI) (http://www.sei.cmu.edu/) has published the Operationally Critical Threat, Asset, and Vulnerability Evaluation[SM] (OCTAVE[SM]) Method Implementation Guide. A version written in HTML for browsing is also included. Information about the OCTAVE Method Implementation Guide is available for $400 at http://www.cert.org/octave/omig.html.

The guide is in 18 volumes, which is quite long and detailed. It includes a computer program for assisting users in performing a risk assessment. A brief table of contents for the OCTAVE guide is shown in Table 1.1, *OCTAVE Method Implementation Guide Table of Contents*.

Similarly, the process presented here is a prescriptive model (amenable to embedding in a computer to become a decision support system) for guiding and assisting knowledgeable personnel in performing a comprehensive security analysis (that includes a risk assessment) for identifying and selecting relevant cost-effective countermeasures for an organization's information system. The described procedure is a multidisciplinary decision process requiring inputs and decisions from managerial, financial, technical, and analytical personnel to arrive at an appropriate conclusion. It can best be implemented as a facilitator process requiring a knowledgeable person (i.e., the facilitator) to provide the appropriate overall guidance needed to ensure that the process is performed properly.

Table 1.1 OCTAVE Method Implementation Guide Table of Contents

Introductory Material	Method Material	Additional Materials
Preparation guidance Tailoring guidance Senior management briefing Participants briefing	For each phase and process - summary - detailed guidelines - worksheets - slides and notes	Asset profile workbook Catalog of practices OCTAVE data flow Complete example results

1.2   Logical Forms of Reasoning

Many organizations, vendors, and analysts seem to be more interested in creating a security solution to issues that they do not even understand. This seems to be a common malady among security enthusiasts who are bent on getting to a solution as quickly as possible. It is much wiser to proceed on a structured course that includes properly defining the issue prior to attempting to resolve it. This includes performing a formal Risk Assessment, which has been incorrectly identified as an unneeded step in the security analysis process (OMB Appendix III 1996).

Before charging out and defining some security architecture that ostensibly solves the problem (whatever it is assumed to be), it is very important to first define and understand the specific issues that need to be addressed (for the specific information system in question). To formulate the appropriate problem is to ask the right questions and to ask the right questions is to collect the right information that provides the data required to formulate these questions (Smith 1992).

Three forms of hypothesis (i.e., problem or issue) generation are available: 1) induction, 2) deduction, or 3) abduction. Following the collection of some data that indicates that there may be a security problem, the analyst can use one of the following methods for issue formulation:

- *Induction* - The collected data implies a hypothesis (particular to the general).
- *Deduction* - A postulated hypothesis based on initial collected data implies the existence of further data which, when collected, can be used to improve the hypothesis statement (general to the particular).
- *Abduction* - Knowledge of the pertinent environment plus initial collected data implies an innovative hypothesis (which was heretofore unimagined) and further data not yet collected that can be used to support and improve or modify the hypothesis.

In all the cases, alternative hypotheses can be developed and using further collected data, some of the hypotheses can be supported or unsupported (i.e., denied), until a single one emerges as the most promising candidate for an acceptable issue definition. The best method of formulating the most accurate hypothesis is *abduction*, although either of the other two methods will work. The appropriate knowledge that an analyst or group of analysts requires for a Sherlock Holmesian abduction process should consist of at least the following abilities, for both issue definition and issue resolution:

- A thorough understanding of all aspects of network security, such as identification and authentication, access control, authorization, auditing, intrusion detection, confidentiality, data integrity, availability, accountability, non-repudiation, and no object reuse,
- Knowledge concerning this specific system's various vulnerabilities,
- A thorough understanding of the owner's information system, including any security mechanisms and methods already implemented,
- Knowledge and understanding of the proper contents of a security policy,
- Knowledge and understanding of the security requirements for typical information systems,
- An understanding of the process of performing a risk assessment,
- Knowledge and understanding of the relevant threats,
- Knowledge and understanding of the methods available for discovering the vulnerabilities of an information system,
- Knowledge, or understanding of a knowledge elicitation process, for identifying the prioritized information system assets and their values (for this organization),
- Knowledge of the alternative security mechanisms that are available and the various risks (technical, schedule, and performance) associated with each,
- Knowledge of the security methods that are available for redressing various behavioral issues,
- Knowledge of a prioritized list of information system assets and their values according to the owner's organization,
- Knowledge of typical COTS security mechanisms and methods, and
- Knowledge of security mechanism vendors.

You will find that each of these will be discussed in enough detail to enable you to become sufficiently proficient to properly perform any of the security tasks required by the process shown in Figure 1.2, *A Security Analysis Process*.

The process and information presented here are sufficient to enable you to perform a computer-security analysis. Certainly the more systems one analyzes, the more knowledgeable that analyst will become and the more successful and capable the analyst will be for creating cost-effective security architectures (assuming of course that the analyst is using an appropriate process and is successful in his or her endeavors).

Issue formulation depends on both the collection of environmental data and the use of this data to ascertain the existence of security problems relevant to the organization. The issue statement is a hypothesis of the security problems for the given information system.

Albert Einstein, in a technical environment, said that one should explain things in as simple a manner as possible, ... but no simpler. My approach to any problem situation is to first explain the issue in very simple terms and then get progressively more detailed and complex, as the need for such requires. This allows me, and the reader, to better understand what is being described. Similarly, the approach to performing a security analysis, as explained here, is to first create a hypothesis that is a complete and understandable statement of the proper question (viz., a risk assessment) prior to formulating solutions to the issue. This issue formulation process will be explained in more detail later in Chapter Six, *Risk Assessment*.

In Figure 1.2, the term "solution" means countermeasures. The intention of the process is to identify the appropriate countermeasures and then implement them (acquire, install, test, evaluate, modify, and operate). The identification of the proper solution is quite dependent upon having created an appropriate and accurate question to be answered.

That is, the correct solution depends upon having posed and accurately answered the correct question.

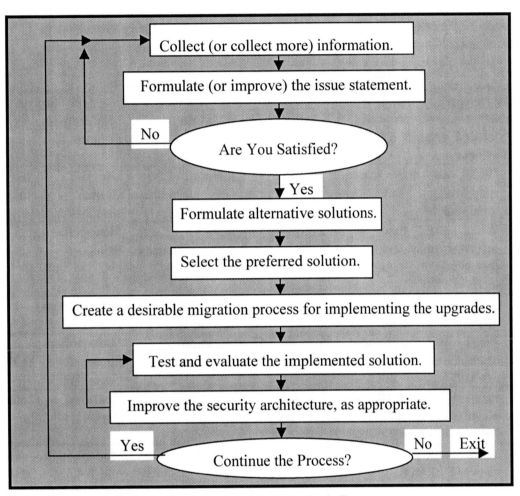

Figure 1.2 A Security Analysis Process

It appears that the secret to acquiring a cost-effective security architecture is the following two-step process:

1) Develop an excellent understanding of your system and its security requirements, and
2) Identify and hire top-quality security vendors for performing the basic tests of the system (namely, performing a scanning analysis to determine your vulnerabilities, performing a risk assessment, installing the technical security mechanisms, and performing a security test and evaluation of the upgraded system).

Any solution that is generated without having scanned the relevant environment for pertinent information to identify the appropriate question is almost assuredly not the correct solution.

1.2.1 Hypothesis Formulation

The formulation of the correct hypothesis depends upon addressing the pertinent questions. The pertinent questions for information system security are listed in Table 1.2, *Questions for a Security Hypothesis.*

Table 1.2 Questions for a Security Hypothesis

No.	Question	Who Answers	Data Sources
1	What is this organization's policy toward security of its information system?	Organization Owners and Named Reps	Management, Existing Policies
2	What are the relevant threats (physical, behavioral, and technical)?	Organization Users and Security Analyst	Users, Analysts, Books, Internet, Vendors
3	What are this organization's information system's vulnerabilities (physical, behavioral, and technical)?	Organization Users and Security Analyst	Management, Users, Books
4	What are the information system assets for this organization?	Organization Management	Management, Users, Books
5	What is the value (to the organization) of each of these assets?	Organization Management	Management, Financial Analysts, Users, Books
6	What is the annualized probability of attack for each relevant threat (i.e., each threat scenario)?	Security Analyst	Analysts, Internet, Books
7	What is the worst harm that each relevant threat might cost, in dollars?	Organization Users and Security Analyst	Management, Users, Internet, Books, Analysts
8	What is the risk associated with each vulnerability?	Security Analyst	Books, Internet, Analysts
9	For each vulnerability, what is (are) the appropriate countermeasure(s)?	Security Analyst	Books, Internet, Analysts, Vendors
10	What are this organization's preferred countermeasures?	Organization Users (and Security Analyst)	Management, Users (and Analysts)
11	Which of these preferred countermeasures are cost-effective?	Security Analyst	Analysts, Vendors

No.	Question	Who Answers	Data Sources
12	What are the remaining vulnerabilities? And how can they be managed?	Security Analyst	Analysts
13	What is the rationale for the selection of the countermeasures?	Security Analyst	Analysts

The procedure for formulating an issue definition requires a *risk assessment* that addresses each of the questions listed in the table. Following the risk analysis, a migration plan that can be presented in the form of a roadmap or Gantt chart can be used to guide the implementation effort to install, test, evaluate, and modify if appropriate, the various security mechanisms and methods recommended.

A formal risk assessment is the most difficult and time-consuming portion of a security analysis. This part of the analysis process will require the security analyst to develop an understanding of the threats to the system, the vulnerabilities of the system, the information system's assets and their values, alternative threat scenarios, the probability of attack for each threat, the potential losses for each attack, the alternative countermeasures for countering the threats, the preferred countermeasures, the remaining vulnerabilities, and, if desired, the development of the rationale for selecting the cost-effective countermeasures.

With each passing day the dangers to information systems, especially those connected to the Internet or other public WAN, are increasing. Referring to the Internet, Jeffrey Hunter, dean of the Heinz School of Public Policy and Management at Carnegie Mellon University (CMU), states "The environment is getting a lot more dangerous (McKenna 2001)." Sources of information system intrusions include not only the traditional malicious hackers but also sophisticated foreign intelligence and criminal intruders. The number of intruders and security incidents is growing exponentially (McKenna 2001).

With the increasing reliance on information systems, the potential cost of security incidents is also growing and the potential savings of implemented security countermeasures, to most organizations, is increasing. From 1999 to 2000, the number of security incident reports increased from 9,859 to 21,756 according to the Computer Emergency Response Team (CERT) Coordination Center at CMU. With the growth of the number of incidents, the severity of these incidents has also grown (McKenna 2001).

Many of the costly incidents have been electronic mail (e-mail) attachments, such as the "I Love You," "Anna Kournikova," and "SirCam" viruses and the "Code Red" worm. Each of the viruses and worm required that the e-mail recipient download an attachment, a no-no for e-mail receivers of messages from unknown sources. In some cases, the virus or worm uses the intermediate victim's e-mail address book to locate others to whom the malicious code is to be sent and thus the message may appear to be from someone whom the victim trusts.

A new buzzword for information systems security is the term "cybersecurity." This term follows from the term "cyberspace" which was coined to denote the public network domain for those inhabiting the Internet realm.

1.2.2 Solution Development

The next step in the decision process is to identify a solution to the issue that has now been formulated. Solving the problem requires the following:

- Understanding the issue (i.e., the initial step in the process),
- Formulating alternative solutions (i.e., countermeasures) and selecting the best ones, and then
- Implementing this preferred solution.

Fundamentally, this is the entire decision process and each step can be decomposed into many tasks, which need to be defined and performed. An attractiveness of the system engineering approach is that the processes required for performing the security analysis activities are defined and can be embedded in a computer system to facilitate the performance of the process steps. Unfortunately, many tools have been developed and sold without the vendors who sell these tools actually having a proper understanding of the entire analysis process.

This solution identification part of the decision process should result in the identification of the required countermeasures, a process for installing and testing them, and a process for operating the security system. The next phase consists of implementing the identified cost-effective countermeasures.

1.2.3 Solution Implementation

Once an appropriate set of security mechanisms and methods has been identified, in most cases it is necessary to install these mechanisms incrementally so that a migration process needs to be formulated. A *migration process* is a method of upgrading the current system to the target system in a manner that ensures minimal risks (viz., technical, schedule, and performance) for the complete upgrade as shown in Figure 1.3, *A Migration Process*. The process description is sometimes called a "roadmap" or "remediation plan[14]" or "Gantt chart."

In the case shown in the figure, there are anticipated to be **n + 1** upgrades to go from the legacy or current information system to the target system, based on having **n** interim systems, where *Interim System #1* is the first upgraded system. If the upgrade process (i.e., roadmap schedule) is spread over a sufficient length of time then further security analyses may need to be performed to ensure that the appropriate devices and methods are identified and implemented because the reference system may be upgraded and the threats, security technologies, and security requirements might be different.

This means that the devices and methods should not be procured all at once but over some period of time so that only those devices required for moving from the current system to the first interim system need be purchased and installed, initially. This also means that the security test and evaluation (ST&E) process is simpler and probably can be done more quickly, completely, and accurately, than if all of the security measures were undertaken all at once. However, it also means that the ST&E process must be performed more often.

---

[14] Remediation is an action to correct a deficiency.

To greatly diminish the likelihood of a successful attack on their information system, owners can be proactive by identifying and instituting safety measures that provide protection against potential malicious attacks on their system.

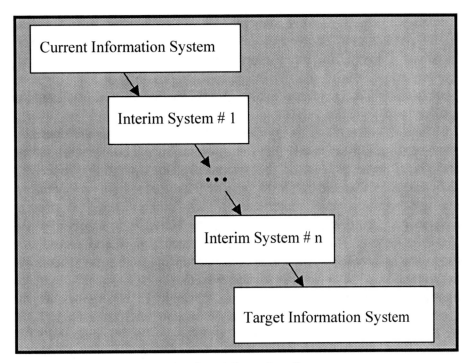

Figure 1.3 A Migration Process

Unfortunately, many organizations tend to feel that money spent on computer network security is money diverted from more productive avenues within the organization. However, when a security disaster occurs, such as a denial-of-service (DoS) attack that brings the system to its knees and causes a loss of availability for users or customers, the owners justifiably get upset. Some of these disasters can cost the organization a lot of money, such as the DoS incident[15] in February 2000 that essentially shut down Amazon.com, e-bay.com, and yahoo.com and cost them many millions of dollars in lost customer orders. Today, there are software algorithms (i.e., rules) that can be placed in a firewall, which can greatly diminish the possibility of a successful DoS attack.

1.3 Security Test and Evaluation

There are two fundamental types of ST&E processes that the analyst can implement, 1) an end-to-end evaluation and 2) a subsystem evaluation. An *end-to-end evaluation* is an ST&E for the complete system, from the clients throughout the entire enterprise computer network system to other clients or applications that reside on local networked servers anywhere in the total system. Some communications may be among intranets (LANs for internal users of this organization) and extranets (LANs for external users of

---

[15] An *incident* is any event that affects the security level of an information system, such as denial-of-service attacks.

another organization), and include the Internet or some other public WAN that provides the linkage for all of the intranets.

On the other hand, a *subsystem evaluation* is the testing of a single LAN or an intranet. Of course, one user's intranet is another organization's extranet, and vice versa. So an end-to-end evaluation will usually involve sites that are not collocated, in fact, the sites may be spread over the entire nation or even around the world. Such systems are referred to as *distributed systems*. Clearly, it is much more difficult to perform an end-to-end evaluation than a subsystem evaluation. It is difficult (perhaps impossible, at least for some cases) to extrapolate subsystem ST&E results to end-to-end ST&E results and I will not attempt to do so in this book.

The ST&E report, when all of the system's security capabilities are operating properly, provides evidence (in a logical sense) that the information system network is operating in a secure mode. Or, if some aspects of the system are not operating properly, it provides recommended upgrades that are required in order for the system to operate properly. If modifications are required and implemented, then this evidence is presented to the Designated Approving Authority (DAA) who has the authority to essentially place a stamp of approval on the system. This approval, namely a certification and accreditation, implies that the information system has sufficient security according to the stated security policy and requirements so that it can be declared a trusted system.

For either distributed or local systems (i.e., a single LAN), the owner may wish to perform ST&E on a subset of the entire enterprise system, such as a single LAN or even a subset of the LAN, say a particular application package, and in this case the ST&E is called a *subsystem evaluation*. These types of ST&E are simpler to perform but it is important for the tester to stipulate the assumptions about the tests so that all the conclusions drawn from the test are logical and follow properly from the evidence presented.

A more detailed and comprehensive illustration of a security analysis process is presented in Figure 1.4, *An Overview of a Comprehensive Security Analysis Process.* At every step in this process, the analysts can return to an earlier step and repeat an earlier step based on knowledge gained from completing, or performing in, a later step. The more detailed process will be explained later in Chapter Four, *A Comprehensive Security Analysis Process.*

1.4 Connections to Other Networks

Another concern of establishing the security capability of your particular network is the fact that your network may be linked to another network (that is trusted) without a firewall in between. This can cause a diminution of the security quality of your network since any intruder into the subnetwork can then access your network. Thus only by testing all of the linked subnetworks, or by placing firewalls between your network and all of the subnetworks (which then implies that they are not trusted, or at least not trusted enough) can you be comfortable with your system's security assessment and that it represents the actual security of your system.

If you are convinced that a trusted subnetwork has the same or higher security protection than your system has, then it can truly be trusted and there should be no concerns about whether your system might have vulnerabilities through the trusted subnetwork. Other aspects of a trusted network are discussed below.

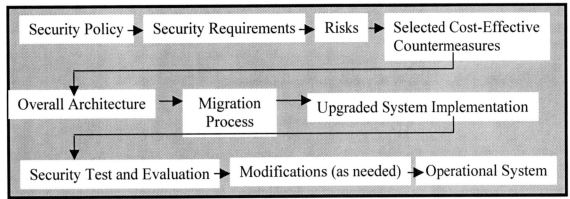

Figure 1.4 An Overview of a Comprehensive Security Analysis Process

1.5 Operating Systems

There are two fundamental types of operating systems (OSs):

1) Client OSs and
2) Network OSs.

A *client operating system* is the software that enables all other software packages residing in the client computer (i.e., workstations) to operate. It controls the access to the system, the execution of all software instructions, the use of peripheral devices (such as printers), and communications among the network elements. Examples of client operating systems are the proprietary systems such as the Mac OS 10.0 from Apple, the Windows 2000 OS from Microsoft, and the Linux OS from Red Hat, and the open system, UNIX OS.

A *network operating system (NOS)* provides the basis for which all communications among the various hosts on a network can depend. Examples of network OSs are the Netware 4.0 NOS from Novell, Windows NT 4.0 NOS from Microsoft, and the Cisco IOS from Cisco.

The Cisco IOS network operating system is a widely deployed software package that can deliver intelligent network services using a flexible networking infrastructure and can enable rapid deployment of Internet applications. Its capabilities are being constantly expanded. The Cisco IOS is embedded in most Cisco routers and switches. These network devices carry most of the Internet traffic today. NOSs can recognize, classify, and prioritize network traffic, optimize routing, support voice and video applications, and more.

NOSs provide a wide range of capabilities including basic connectivity, security, and network management to technically advanced services that enable organizations to deploy their applications (e.g., real-time trading, interactive support, on-demand media, and unified messaging).

The OS can provide many aspects of the gamut of security capabilities, namely identification and authentication, authorization, availability, and auditing (or accountability). However, vendors, applications, and middleware can provide some security capabilities.

A *security kernel* is the central element of an operating system and consists of the hardware, firmware, and software elements of a Trusted Computing Base (TCB) that

25

implement the reference monitor concept. This kernel must mediate all access, be protected from modification, and be verifiable as correct. A *reference monitor* enforces the authorized access relationships between subjects and objects of the system. A reference monitor has three design requirements:

1) *isolation* (from access by other softwre elements),
2) *completeness* (impossible to bypass), and
3) *verifiability* (can be assessed as a secure package).

Another OS is the FreeBSD, an open advanced operating system that is Intel ia32 compatible as well as with DEC Alpha and PC-98 architectures. It is derived from BSD UNIX, the version of UNIX developed at the University of California, Berkeley, and it is free and comes with full source code. This OS offers advanced networking, performance, security, and compatibility features today, which are still missing in other operating systems, even some of the best commercial ones (Free BSD Organization 2001). Unfortunately, FreeBSD has lost its financial support but now is supported by Wind River.

FreeBSD can be used as a network operating system for either Internet or intranet servers. It provides robust network services, even under the heaviest of loads, and uses memory efficiently to maintain good response times for many simultaneous user processes (even thousands). This OS can be installed from a variety of media including CD-ROM, floppy disk, magnetic tape, an MS-DOS partition, or if you have a network connection, you can install it *directly* over anonymous File Transfer Protocol (FTP) or Network File System (NFS). All you need is a pair of blank, 1.44MB floppies and direction available on the Internet (http://www.freebsd.org/) or for more information you can contact bod@FreeBSDFoundation.org.

Another version of this OS is the BSD/OS Internet Server Edition (v4.2), which is an effective and reliable alternative to proprietary systems for Internet services, such as Web, FTP, email, dialup, and others. BSD/OS maintains good reliability with extended uptimes for use in mission-critical applications. Organizations worldwide have implemented BSD/OS for quality performance and reliability for Internet services (Wind River 2001). The most versatile version is the NetBSD, which is a free, secure, and highly portable UNIX-like operating system available for many platforms, from 64-bit Alpha servers and desktop systems to handheld and embedded devices. It has a clean design and has advanced features that make it a top performer in both production and research environments, and it is user-supported with complete source code (Net BSD 2001). The OpenBSD OS (v3.0) has a spotless security record. The OpenBSD OS default security has never been breached by blocking remote privileged access by an unauthorized user to an OpenBSD server (Yager 2001).

Security issues in NetBSD (v1.5.2) are handled by the OS's security officer and the security alert team. As well as investigating, documenting, and updating code in response to newly reported security issues, the OS's security alert team also performs periodic code audits to search for, and remove, potential security problems. NetBSD has integrated Kerberos and Secure Shell (SSH) into its code. In addition, all services default to their most secure settings, and insecure services are disabled by default for new installations. This OS also contains full support for IPSEC for both IPv4 and IPv6 (Net BSD 2001).

1.6   Information System Security

Information system network security consists of a) physical controls, b) behavioral controls, viz., training and awareness, and c) technical controls.

1.6.1   Physical Controls

Physical controls are those devices and actions that ensure that only authorized persons have physical access to the information system facility and information subsystems. These controls include guards, patrols, badges, fences, gates, cyber locks, and facilities with limited access.

1.6.2   Behavioral Controls

Behavioral controls are methods used to modify the behavior of the various personnel associated with the information systems, namely, management, users, support personnel, and others who must interact with the information system and its network.

Generally, behavioral controls comprise three areas, 1) awareness (i.e., knowledge improvement), 2) training, and 3) background checks.

1.6.2.1 Security Awareness

Security awareness is the process of ensuring that all employees are knowledgeable of the security policies and rules of the organization. This information can be provided by computer-based training (CBT) or by ensuring that the rules are stored in databases that are easily accessible by all employees. Also, the relevant material can be learned from a Security Policy and Procedures Manual that is either in hard copy form or placed on the network where all employees can access it at any time from their client computer.

1.6.2.2 Training

Training is the process of ensuring that all employees are knowledgeable of the security policies and rules of the organization. This information can be learned in a classroom during a training session or it can be learned from a Security Policy and Procedures Manual that is either in hard copy form or placed on the network where all employees can access it at any time from their client computer.

1.6.2.3 Background Checks

Background checks are methods used to determine if a person can be trusted to occupy a particular position. For example, for system administrators, a very stringent background check is required, whereas for most users, a more modest background check is usually all that is needed. However, in environments where the system and its information are very sensitive, all users, technicians, and administrators should be subject to a stringent background check.

1.6.3　Technical Controls

Technical controls are the controls that are provided by the computer system infrastructure. Technical controls include the following capabilities:

1) Identification and authentication,
2) Authorization,
3) Access control,
4) Accountability,
5) Availability,
6) Data Integrity,
7) Confidentiality,
8) Non-Repudiation,
9) Least privilege, and
10) Contingency planning, and
11) Non-catastrophic failures.

1.6.3.1 Identification and Authentication

The *identification and authentication (I&A)* capability of the OS is provided by software that can recognize a user's identity (User ID) and then determine if the user has provided the correct password (or other form of authentication) or not. Some I&A systems are more complicated (such as biometrics) and can use other methods for authentication that can further ensure that the entity[16] logging on can truthfully answer the question "Are you who or what you say you are?"

I&A is a set of security services that is used with other services (such as access control) as a first step to determine the identities of the entities who are participating in some activities on the information system. Some aspects of I&A depend on the network operating system and middleware servers such as for single sign-on. Single sign-on is discussed later.

1.6.3.2 Authorization

Authorization is the process of ensuring that an entity has access to only those applications, databases, parts of databases, and other files for which they are entitled.

An entity's authorization is sometimes based on the distinguished name (DN) in the entity's certificate. On the basis of the DN (or other identification), the operating system will provide certain access privileges to the entity. Depending on the entity's authorization, the entity can access a variety of applications and databases. Authorization addresses the question of "Do you have the need to access this file?"

One of the capabilities of authorization is *data separation*. This is an attempt to ensure that only authorized entities have access to certain types of information. For example, a typical user cannot have access to payroll information and a manager cannot create travel checks.

---

[16] An *entity* can be any one of the following: person, application, service, or server.

1.6.3.3 Access Control

Access control is provided by a software subsystem that can grant access to applications, databases, parts of databases, or other files depending on the specific privileges that an entity has. The most appropriate form of access control for commercial organizations (and many government organizations) is Role-Based Access Control (RBAC), which is discussed later.

1.6.3.4 Accountability

Accountability is the process of holding each entity that performs any act on the network to be responsible for their actions. This requires the each entity is identified and authenticated, and that each action is recorded (auditing).

In security terms, *auditing* is the process of performing the accountability security function, which is ensuring that any entity will be held responsible for any action that they take on the information system network. Auditing is a software process whereby system activities are recorded so that the entity performing the activity and characteristics of the activity are recorded and audit trails are generated so that the entity can be identified and the activity can be reproduced if required, for example if the activity turns out to be malicious.

Another process for performing accountability is the Intrusion Detection Systems (IDS), which is an attempt to determine automatically, that some entity is attempting to perform some malicious action against the information system. The IDS can be an offline system that reviews the audit trail information or an online system that attempts to identify malicious users based on some trait of their messaging to the network to determine whether they are malicious or not.

Many COTS packages provide for auditing the actions of the information system, such as operating systems, database management systems, and other middleware.

1.6.3.5 Availability

All information and subsystems should be capable of quickly responding to any entity's requests for service. This capability is called *availability*. Availability is an important aspect of security, but there are characteristics of availability that are not considered to be a part of security concerns, but in this book, all aspects of availability are considered to be a part of the security area.

1.6.3.6 Data Integrity

Data integrity is the process of ensuring that all information (stored or transmitted) is not altered by an unauthorized entity. This is usually provided by including a checksum in the message but also can be provided using encryption.

1.6.3.7 Confidentiality

Confidentiality is the process of providing assurance that stored or transmitted information is kept private, that is, that the information is not viewed (in a meaningful way[17]) by unauthorized entities.

Confidentiality is provided by the encryption of information. *Encryption* is the process of transforming plaintext information into a form that is unreadable by an unauthorized entity.

Many organizations feel that there are only three aspects to security, namely, availability, data integrity, and confidentiality. They also believe that these three areas can be rated according to the manner in which they are used to handle sensitive but unclassified information as defined below:

High   Extremely grave injury accrues to organizational or U.S. interests if the information is compromised; could cause loss of life and/or major financial loss; or require legal action for correction.

Medium   Serious injury accrues to organizational or U.S. interests if the information is compromised; could cause significant financial loss; or require legal action for correction.

Low   Minor injury accrues to organizational or U.S. interests if the information is compromised; could cause only minor financial loss; or require only administrative action for correction.

For example, a system and its information may require High degree for both data integrity and availability, yet have a Low degree need for confidentiality. An example of this is an organization that has a great need for accurate information that is always available but does not care whether the information is seen or acquired by unauthorized persons.

Encryption is a countermeasure that can be used for either good or evil. It can be used to protect information from being used by malicious people or it can be used by malicious people to hide their communications from those who could stop them from carrying out some malicious action if discovered. However, even without encryption, malicious users can camouflage their communications so that commercial or Government organizational eavesdropping will not gather any useable indications. However, there are other means for determining intent that the black world can use. However, these methods are beyond the scope of this book.

1.6.3.8 Non-Repudiation

Non-repudiation is the process of ensuring that no entity that transmits a message can later deny having transmitted the information, nor that any entity that receives a message can later deny having received the message. Non-repudiation is provided by a digital signature,

---

[17] The information is unreadable to anyone who cannot decrypt it.

which requires that an entity possess a digital certificate, or by a "digitized signature," which is a bit-map of an actual signature.

1.6.3.9 Least Privilege

A system that provides *least privilege* is one that ensures that no entity has access to any information beyond that which is required to perform their duties.

Similar to the data separation privilege, the least privilege limitation on the access for authorized users can be provided by the authorization and access control software capabilities.

1.6.3.10 Data Separation

Data separation is the system's capability to ensure that no user has access to information that can be conflicting. Access to conflicting information can result in the user having the capability to perform actions on both sides of an issue. For example, a person cannot have the privilege of stipulating an amount of money needed for a business trip and the authority to write the check also.

1.6.3.11 Contingency Planning

Contingency planning includes the process of planning for a contingency[18], creating the plan document, providing storage capability for replication of critical information at a remote site, providing a new temporary location, and providing for recovery, in case of a contingency, at the new location.

1.6.4 Non-Catastrophic Event Protection

Non-catastrophic failures are events such as minor water damage or loss of electrical power from the local electrical company. Protection from these occurrences includes: uninterruptible power supply systems; backup heating, ventilation, and air conditioning systems; and clustered servers.

1.7 Security Compendium

Users of network-based distributed applications require assurance that their interactions and data are reliable, private, accessible, and readable only by authorized entities. Such assurance is provided by mechanisms, which make use of the following security services:

1) Authentication,
2) Access control,
3) Data integrity, and
4) Confidentiality.

---

[18] A *contingency* is an event that causes the loss of an information system facility site.

Effective security depends on the use of the four services listed above. The four security services are interdependent, that is, all of them must be applied in order for each of them to be effective (Schuermann 1998).

Security in information systems is often dependent on cryptography. Information system security depends on the cryptographic algorithms used to implement these services. Since the effectiveness of cryptographic algorithms depends on computing speed, encryption algorithms will lose their effectiveness over time as computing speed increases. Thus, encryption countermeasures can be circumvented by decrypting information used to support these mechanisms. But if the encryption methodology is made complex enough, then the time required to decrypt the encrypted information is far too long to be of any use to the interceptor.

Faster computers, such as distributed parallel computers, can significantly improve the speed at which encrypted information can be decrypted. For example, the 56-bit RC5 encryption that would require a single desktop computer thousands of years to decrypt a message was broken (in 1997) after 250 days of a massively parallel computational effort by tens of thousands of ordinary desktop computers. Computers are much faster today and algorithms that use the parallel computational power of many PCs connected to the Internet have been developed.

Therefore, a basic principle in the design of a security architecture is that the security interface be separated from the implementation of the security algorithms, allowing the implementation to be upgraded to keep pace with advances in decryption methodologies and ensuring that system security can be maintained at a capability level sufficiently high to satisfy the organizational requirements for authentication, access control, integrity assurance, and confidentiality in the future.

By making countermeasures modular, and assuming that the countermeasures are standardized, then the customer can change out an installed countermeasure from one vendor and replace it with a countermeasure from another vendor with little or no down time. For example, if an anti-virus software package from one vendor (that is compatible with a particular operating system, say Mac OS 10.0) is deemed to be inferior to another package, then the superior package can be installed as a replacement (assuming of course that it will also run under the Mac OS 10.0). Similarly, if the architecture ensures that a firewall is installed as a modular device, then the installed firewall can be replaced by a different firewall, assuming that the two firewalls operate according to the same standards. If the standards are different, then some alterations of the system software may be required, thus making the replacement upgrade somewhat more complex and time consuming. However, the desired improvement may be well worth the effort.

As technologies and threats change, the need to improve the installed countermeasures through their replacement will continue. Thus the portability, interoperability, and scalability of hardware devices and software packages through the use of standards-based architectures (called "open architectures") with modularized countermeasures serves to ensure that the most cost-effective products can be installed in an efficient manner.

A commonly employed metaphor for managing groups of people according to a classification or a relationship to a group or organization is reflected in the Role-Based Access Control (RBAC) model (Ferraiolo and Kuhn 1992), which was developed by the National Institute for Standards and Technology (NIST). Administrators already think of individuals in terms of their job position or their role within an organization (Schuermann 1998).

Use of the Public Key Infrastructure (PKI) for authentication, authorization, confidentiality, data integrity, and non-repudiation requires the necessary support for certificate and key management. PKI systems also provide handshake capabilities that ensure that both the sender and receiver are both authenticated to one another through the use of public keys.

Public keys play an important role in computer communications because they permit two entities to trust each other's identity and then engage in private communications without either entity ever having communicated previously. Prior to public key technology, exchanging secret keys was the only means of ensuring secure communications. With PKI, communications between two entities may be encrypted and decrypted with a private key and the other party's public key. Each party has access to the other's public key, but neither can access the other's private key. Because the overhead associated with the use of asymmetric encryption (using public and private keys) can be very time consuming, the current methodology is to use PKI for transmitting a session key (a nonce) and then encrypting the traffic using symmetric encryption with the single session key.

A public key establishes no identity, unless at least one other trusted entity verifies that identity. Verifying identities is the responsibility of Certificate Authorities, whose authority is trusted by virtue of their established presence and widely-accepted trust, and whose signature on someone's public key (i.e., their digital certificate) guarantees the identity of the owner of that key.

Digital signatures provide a means of authenticating information, and when based on public keys allow entities to verify the authenticity of the sender and the message. This is also closely related to ensuring the data integrity of the message.

Considering that RBAC is a natural representation of existing administrative policies and procedures, the adoption of RBAC lowers the cost of administering a system's access control policy and reduces the chance of errors when compared to other access control mechanisms, in particular when such mechanisms directly associate users with specific permissions. The RBAC model also supports non-trivial access control policies without compromising efficient implementation, which is an important consideration when large-scale environments need security policies that are enforced through such an RBAC model. An object could make authenticated requests to other objects to obtain additional authorization beyond what the RBAC model offers, and thus extend these basic services in ways that are dictated by the unique needs of a particular object.

Authentication and confidentiality services may be provided through the use of security systems that support public key algorithms. Such systems include the Secure Socket Layers (SSL) protocol, secure shell (SSH), CORBA, COM, and Kerberos.

Although public key certificates are in widespread use, there are serious organizational and legal policy questions that must be decided by an organization before using the technology. A user's public key certificate may be provided by the user directly, or by a known public key server, provided that the user has registered his/her public key with such a server. The public key supplied by a third party in this manner may require an additional verification step by which the public key server's identity is authenticated. The public key server would need to present an authenticating certificate that has been signed by a known and trusted certificate authority.

In addition to its own services, the RBAC model is flexible enough to allow the actual implementation to choose from, or implement custom, authorization systems that match the installation's need, size, and other aspects of its environment. In this way, RBAC not only adapts itself to existing policy models, but may also be extended as the need arises (Barkley 1995).

In order for objects to be securable (i.e., able to take advantage of authentication, access control, integrity services, and confidentiality assurance) these objects must implement a so-called security interface. Through this interface, arbitrary objects can be queried for certain security-related properties, thus making these objects securable.

Methods (or functions) that implement the security interface are:

1) *Authorized Roles* - Obtain information on the authorized roles that may access the object.
2) *Digital Certificate* - Obtain the object's digital certificate.
3) *Invoke Secure* - Invoke the object's purpose/function using security credentials.
4) *Scenario Walk-through* - Perform an analysis of the implementation of the security mechanisms for a particular sequence of events.

Using the following scenario, the security services can be demonstrated.

John dials up and establishes a connection (i.e., handshake). John's computer and the server present to each other their respective public keys. The server's certificate has been signed by a certificate authority that is recognized and trusted by John's browser software. Similarly, John's certificate was signed by an entity that the server recognizes and trusts.

John and the server can both be certain of the identity of the other, and both can use their respective public keys to encrypt any and all data that they exchange. Their communications are confidential.

John now decides to resume previous work with Widget Kit. Calling up the Widget Kit software causes the Widget Kit application to request from the server authorization to offer its services to John. Through John's session RBAC information, the server knows that John is a member of "The A-Team." Members of this group are allowed to use the Widget Kit, according to the RBAC services used by the server, so the Widget Kit application receives permission for John to access the Widget Kit application.

After John completes his work, the Widget Kit object creates a performance record that becomes part of John's permanent profile. This performance record is digitally signed by the Widget Kit software (for authentication and non-repudiation), so that its authenticity can be verified at any time in the future. Anyone recognizing the identity of the signing entity (the Widget Kit software, in this case) will recognize that John has obtained the performance record through legitimate use of the Widget Kit software.

1.7.1   Sensitive Information

Sensitive information is any information that is considered to be critical to the organization. It is often referred to as "sensitive but unclassified." The mission of the information system will determine the information that is processed. The mission and associated information will influence the environment and information sensitivity requirements applicable to the processed information.

Information categories are defined by their relationships with common management principles and sensitivity requirements promulgated by the organization's security policy for each information category. Processing, transmission, and storage of data of more than one category of information does not create a new category but instead inherits and must satisfy all the sensitivity requirements of the assigned categories. The two information categories, 1) non-sensitive unclassified and 2) sensitive information, are defined below.

Non-sensitive unclassified is the category of information that includes all information that is not sensitive as defined below.

Sensitive information is the category of information that includes information for which the loss, misuse, or unauthorized access to, or modification of, could adversely affect the organizational or national interests or the conduct of federal programs, or the privacy to which individuals are entitled under the Privacy Act of 1974 (Public Law 93-579), but that has not been specifically authorized under criteria established by an Executive Order or an Act of Congress to be kept secret in the interest of national defense or foreign policy. Systems that are not national security systems, but contain sensitive information are to be protected in accordance with the requirements of the Computer Security Act of 1987 (Public Law 100-235).

In many cases, it may be useful to further characterize the sensitive information by determining the subcategory. This may indicate additional national, DOD, Service, or Agency requirements, which are imposed by processing this type of information. Descriptions for the eight different categories for sensitive information are shown in Table 1.3, *Category Descriptions for Sensitive Information*.

"Organizational Sensitive" can be interpreted to mean "Trusted Information," namely, Category 7. The various categories for any organization may be different from those listed in the table above and the analyst must decide that the information is either sensitive for some reason, or it is not sensitive, just unclassified. Classified information, Category 8 in the Table 1.3, is not addressed in this book.

Table 1.3 Category Descriptions for Sensitive Information

CAT.	TITLE	DESCRIPTION
1	Proprietary	Information provided by non-government sources on the condition that it not be released to other non-government sources. Examples include proprietary data, contract bids, and pre-award survey information.
2	Privacy	Personal and private information as defined in the Act of 1974 or the Freedom of Information Act (FOIA).
3	Personal gain	Information that could be changed to provide personal benefits, such as performance ratings or education levels.
4	Accounting	Quantitative data that provides official accountability records, such as balances on hand, asset amounts, credits, and debits. Does not include similar data when used only for reference or research purposes.
5	Asset loss	Data that contributes to the automated decision to transfer or pay out a tangible asset, including asset routing information (e.g., data involved in creating payments via check or electronic fund transfer).
6	Security control	Data associated with the security mechanisms (e.g., passwords) that control access to the system, contain audit records, and assure the integrity of the Trusted Computing Base and its extensions.
7	Trusted information	Information that when received is accepted as authentic (e.g., electronically prepared Internet messages such as e-mail).
8	Classified	Information protected under the DOD Information Security Program.

1.7.2 Middleware

Middleware is the software that resides between the user (client) and the applications or other software such as database management systems that reside on the servers or mainframes. It is the technology that facilitates the interface between a user or client device and a "back office" server infrastructure[19] (Vulpe 2001). It can facilitate interfacing two apparently incompatible applications or back office servers. Middleware can make otherwise incompatible information available throughout an organization and in fact, throughout the world to many organizations, if it is made available on the Internet. A powerful new standard for transporting information is the Extensible Markup Language (XML), an Open Information standard (Castro 2001). Middleware enhances a system's open architecture capabilities to provide scalable, interoperable, and portable hardware devices and software packages by enabling these systems to be viewed as compatible subsystems by virtue of the transparent operations of the middleware.

---

[19] An *infrastructure* of a system consists of the elements that are required for the system to operate. It comprises the software, hardware, procedures, management, and network elements.

Middleware provides the connections between the users and the various software packages that provide the applications and services required to perform the information system's expected duties. Usually middleware does not play a role in the ST&E process but is an important ingredient in the proper operation of the system. Middleware extends existing services instead of replacing systems or adding new systems.

Middleware provides the capability to facilitate fully secured data manipulation and content management services and is characterized by universal access, guaranteed scalability, transparent interoperability, and application independence, in any combination (Vulpe 2001). It is the kind of software that resides between the user's application and operating systems. It can conceal the complexities of networking protocols[20], database dialects, and alternative operating systems by interpreting data requests for routing through an organization's information system network.

Middleware can provide a cost-effective infrastructure that can result in transparency, seamless integration, and an ease-of-access or user-friendliness demanded by modern software packages and information technology environments.

An example of middleware is the Object Request Broker (ORB). ORB is a component in the Common Object Request Broker Architecture (CORBA) programming model that acts as the middleware between clients and servers. In the CORBA object-based computing model, a client can request a service without knowing anything about what servers are attached to the network. The various ORBs receive the requests, forward them to the appropriate servers, and then hand the results back to the client. Another example is the Network Basic Input/Output System (NetBIOS) that provides for the application-programming interface (API[21]) for Microsoft products. NetBIOS is a protocol that provides the underlying communications mechanism for some basic Windows operating system functions, such as browsing and communications between network servers (Pachomski 2000).

There is a Java Message Queue that can be categorized as Message Oriented Middleware (MOM). It will handle transactions between applications. It is in wide use across major corporations and has proven to be a practical, cost-effective, and architecturally sound subsystem.

In some cases, middleware can make new technologies affordable when they otherwise might be too expensive (Vulpe 2001).

1.7.3 Database Management Systems

Database management systems (DBMSs) are the software, usually COTS, that provides for the management and control of information that is stored in databases. The user requests for database or information control is usually based on the Structured Query Language (SQL) or a similar language. The DBMS shields the user from hardware details (Date 1988).

The DBMS handles all access to the database files by the following activities:

1) User issues an SQL request,
2) The DBMS intercepts the request and analyzes it,

---

[20] A *protocol* is a set of standards and rules that need to be followed exactly.
[21] "API" is a message and language format that allows programmers to use functions within another program.

3) The DBMS inspects the external schema for that user, identifies the appropriate external mapping, provides the conceptual schema, identifies the appropriate conceptual mapping, and defines the proper storage structures, and then
4) Executes the necessary operations on the stored database file (Date 1988).

The information can be relational (i.e., tabular) or object-oriented. DBMSs enable customers to access a wide variety of enterprise and network data, with access from remote locations and with high availability for replication of data to storage areas where data may be centralized for an entire organization. They offer robust, scalable solutions across a variety of platforms controllable by a variety of operating systems such as Windows, Unix, Mac OS, and Linux. The capabilities and graphical application development tools of the DBMS can deliver increased performance, yielding the desired speed needed to integrate diverse data sources and applications without sacrificing information integrity.

Many organizations are adopting the object-oriented (OO) technology but others have not, some because they say that there is no mathematical foundation for OO as there is for the relational approach, namely E. F. Codd's seminal research (Booch 2001). Because of the increasing popularity of OO and the fact that the proliferating Java languages are OO, many vendors are adopting OO languages for their security software.

1.7.4 Database Information

Database information can be stored in relational form (i.e., tabular) or as objects. There are many issues concerning the security of database information (Date 1986), in particular, when database information is accumulated and stored in a single location, such as is possible in the case of a Storage-Area Network (SAN). A SAN is a high-speed, special-purpose network that connects servers to storage devices by enabling storage devices and servers from different vendors to work together. They improve network efficiency by allowing network administrators to easily add, reconfigure, and reallocate shared storage resources in the network in response to changing storage requirements.

SANs can contribute to disaster-recovery solutions, but these networks can be tricky to set up and manage (Zyskowski 2000). The principal advantage of a SAN is that information can be offloaded so that capacity requirements now are the responsibility of a separate network that is optimized for storage. This means that applications on the main network will run faster and more efficiently. The SAN is accessible from any host in the organization's network.

Many of the security issues for any information system apply to database information, such as:

- Not allowing access to only authorized users,
- Not controlling access to databases and not enabling access to only parts of databases,
- Not ensuring the accuracy of the data (e.g., no data integrity capability),
- Not ensuring that users are accountable for any damage they do to the database, and
- Not ensuring that database information is always available to any user.

Database information is supplied to computers using a DBMS, which can be used to store, retrieve, create, modify, or delete information. The information may be stored in the memories of servers; on internal hard disks, external hard disks, or redundant arrays of inexpensive disks (RAIDs); on transportable media (floppy disks, zip disks, super disks, or compact disks); or on magnetic tape. The costs, response times, and physical requirements are different for each storage mode. The analyst should decide what types of storage media are appropriate for this organization based on the desirable attributes of the storage devices, e.g., all costs (device acquisition, installation, storage media (i.e., disks), etc.), response time, store time, retrieve time, ease-of-use, volume of devices, reliability, availability, speed, bit error rate, etc.

It is convenient for all information to be stored in a central location, such as the case for some SANs; however, in some cases the actual data is not at the central location, only a pointer to where the data is actually placed. To improve availability and reliability, many SANs use RAIDs or a simpler version called "just a bunch of disks" (JBOD) subsystems. A capability similar to the SAN is the network attached storage (NAS) system. NAS products can implement one or more distributed file system protocols to allow both clients and servers to access files that are stored (usually in RAID products) in a common shared storage pool. By storing files in this common area, the following advantages are realized (LSI 1999):

1) Files can be easily shared among users,
2) Files are easily accessible by the same user from different locations,
3) Demand for local storage at the PC is reduced,
4) Storage can be added more economically and partitioned among users,
5) Data can be backed up from the common repository more efficiently than from PCs,
6) Multiple file servers can be consolidated into a single managed storage pool, and
7) High availability storage can be cost-effective when provided to PCs.

It is important to ensure that all information (databases, applications, operating systems, and middleware) is properly replicated for backup and recovery and that all sensitive information is protected during storage and during transmission. Database information can be stored in at least two different forms, relational (or tabular) and object-oriented.

Two other, but rarely used, format forms for database information are hierarchic and network. Hierarchic information is stored in the form of trees usually displayed with the "root" at the top and the rest of the tree shown at different levels, down to the "leaves" of the tree. The network form can be considered an extension of the hierarchic form (Date 1986). The network database consists of two sets, a set of *records* (i.e., the information) and a set of *links* (i.e., a description of how the information is related).

1.7.4.1 Relational Data

For relational information, using the Structured Query Language, there are eight operations that can be performed on two or more databases (A, B, and C). These operations are (see Figure 1.5, *Relational Algebra*) (Date 1986):

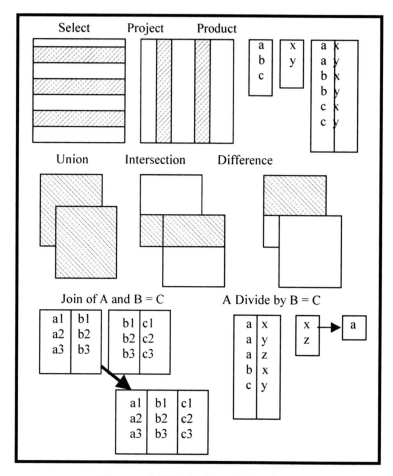

Figure 1.5 Relational Algebra

- SELECT - Selects specified tuples (rows) from a specified table.
- PROJECT - Selects specified attributes (columns) from a specified table.
- PRODUCT - Builds a table from two specified tables by concatenating tuples, i.e., C = A TIMES B.
- UNION - A new table consisting of all tuples appearing in either of two tables, C = A UNION B.
- INTERSECT - A new table consisting of all tuples appearing in both tables, C = A INTERSECT B.
- DIFFERENCE - A new table consisting of all tuples in the first table but not the second table, C = A - B.
- JOIN - A new relation consisting of special relations of tuples from two tables specifying a special relation, C = A JOIN B.
- DIVIDE - A new table that takes two tables, one binary and one unary, and form a new table, C = A DIVIDEBY B.

1.7.4.2 Object-Oriented Data

Much of the information available in today's computers is *object-oriented data*. Information such as images, video, and data for object-oriented applications are stored in object form. In object-oriented jargon, variables are often called *objects* or *object instances*. The data types are called *classes*. The term "object" is more often used when referring to a variable that has been declared using an object-oriented programming language class as a data type.

Most DBMS products available today have both relational and object-oriented capabilities.

Other security concerns involve sniffers, scanners, stimulators, and scenarios. These will be discussed next.

1.7.5 Sniffers, Scanners, Stimulators, and Scenarios

Sniffers are electronic devices that can monitor a network to gather information concerning the traffic traveling along the network. Ordinarily, it is not possible to determine if a sniffer (sometimes called a packet sniffer) is attached to your network except by physical inspection, but sensing devices that can detect the potential presence of a sniffer based on an inappropriate resistance on a link are possible. A malicious hacker can utilize the information collected by a sniffer to determine how your network might be attacked, such as by identifying passwords (if transmitted in plaintext (or plaintext, i.e., not encrypted)) or by packet sniffing to gather information that is transmitted in plaintext or an easily decrypted encoding. One way to defeat sniffing is to encrypt (using an appropriate encrypting method) all sensitive traffic, such as passwords or critical information, so that the sniffed information is unintelligible.

Sniffers can be used for either malicious or non-malicious purposes. For performing ST&E, they can record relevant information about encryption and other security aspects of system traffic. A sniffer or network analyzer is effectively a computer operating in a promiscuous mode[22].

Sniffers can be placed anywhere inside the LAN or outside the LAN depending on what the owner wants to find out. For example, a sniffer can be used to determine how voice quality can be managed and how a Voice-over-Network (VoN) can be installed to improve the voice quality or make the system more reliable. A new product for sniffing Voice-over-IP messages is the "Sniffer Voice" sniffer from Sniffer Technologies, Inc. (www.sniffer.com/dm/voice.asp); see Figure 1.6, *Sniffer Voice Technology*.

A new and controversial sniffer created by the FBI is the "Carnivore[23]" which has been renamed the "DCS1000." The Carnivore system consists of an ordinary personal computer running Microsoft Windows 2000 or Windows/NT and some proprietary software. Its primary purpose is to intercept large volumes of electronic mail and other forms of electronic communication passing through a network. This listening process remains passive

---

[22] A computer on a network, operating in this mode, is constantly monitoring the transmissions of all other stations on the network.

[23] *Carnivore* is the e-mail sniffing software, similar to a wiretap that the FBI uses to monitor alleged criminals.

at all times, and like all other sniffers it alters no data and prevents no messages from continuing on to their intended destinations (Mitchell 2000).

For Carnivore to gain access to this much data, its hardware must be plugged directly into the network at a central location. Because most Internet-based communications in the United States flow through large Internet Service Providers (ISPs), the Carnivore computer typically would be installed inside an ISP data center. Controlled physical and network access improves the system's overall security (Mitchell 2000). Unfortunately, Carnivore was a quick and dirty hack and suffers from many of the maladies that quickly developed software has (Hayes 2000). One way to defeat or diminish the capabilities of the Carnivore sniffer is to encrypt the e-mail messages (Mitchell 2000).

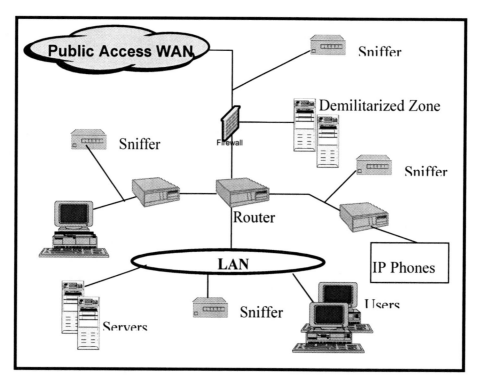

Figure 1.6 Sniffer Voice Technology

Scanners are electronic devices that are useful tools for determining the vulnerabilities of your information system (Canavan 2001). These devices can be used to determine what the vulnerabilities are and also some of them will provide a fix for your system provided it is something that the scanner has done before, that is, the problem is familiar to the scanner system. Some scanning tools that are available are the following:

- *Security Administrator's Tool for Analyzing Networks (SATAN):* This tool can assist system administrators in identifying vulnerabilities in the system. It identifies the security issues and generates a report on the problems along with information that explains each issue, the possible results of the issue, and how to fix the vulnerability. It is available at the following Web site: http://ciac.llnl.gov/ciac/SecurityTools.html.
- *Security Administrator's Integrated Network Tool (SAINT):* This is an updated and improved version of SATAN. It can be downloaded from: www.wwdsi.com/saint.

- *Computer Oracle and Password System (COPS):* This tool is actually a collection of tools that can be used to check for common configuration problems on UNIX systems. It checks for security issues such as weak passwords, anonymous file transfer protocol (FTP), and inappropriate permissions. It generates reports with detailed findings that a security administrator can use to strengthen a system's security. It is available at the following Web site: http://ciac.llnl.gov/ciac/SecurityTools.html.
- *TITAN:* This tool is similar to COPS except it makes an attempt to correct issues that it discovers. Like COPS and SATAN, TITAN checks for different system vulnerabilities.

It is recommended that security analysts use two or more of these scanners since they are independently developed and perform their scanning in different ways so that the vulnerabilities that one scanner does not discover may be discovered by another vendor's scanner. There are other scanners, such as TIGER, TCPWrapper, NetSonar 2.0 from Cisco (Cisco NetSonar 2001), and Tripwire.

Another scanner is the CyberCop Scanner from PGP Security, a part of Network Associates. This scanner can identify security holes to prevent intruders from accessing mission-critical data. It identifies weaknesses, validates policies, and enforces corporate security strategies. It tests NT and UNIX workstations, servers, hubs and switches, and performs thorough perimeter audits of firewalls and routers. This scanner combines architecture design and security data to make system security more certain by (NAI 2001):

1) Validating the effectiveness of security systems and policies.
2) Ensuring that the information system has been tested with the most complete list of security checks available.
3) Providing in-depth assessment details to help strengthen network security.

Stimulators are computer programs used to provide inputs (such as an operator's keyboard entries) to an information system so that the system can be operated as a black box. Stimulators are useful tools for providing inputs to an information system that should cause the system to react in known ways and if it doesn't, the analyst can conclude that something is not working properly. By collecting the relevant data, the analyst can evaluate the data to define and understand the problem, and its likely cause.

Simulators are usually embedded software that will produce the same results as some physical mechanism or phenomenon, such as the behavior of a particular type of market or competitor. An advantage of a stimulator is that it is repeatable and cannot input misinformation due to a keyboard typing error.

Simulations are computer programs, and perhaps a platform also, for computing realistic inputs for driving the system in a mode required for ST&E. Simulated inputs for driving the upgraded system include the following two types of stimuli:

- *User activity inputs* - (These simulate the actions of the various users who have access to the system's applications, and ideally it can simulate the exact manner in which the users provide their inputs.)

- *Data inputs* - (These present the information required by the subsystem in order to perform its duties. This information may be either push or pull[24].)

Pushed information is data sent to the information system for stimulating some process. *Pulled information* is specific data that is requested by a service or application in order to operate.

Scenarios are sets of sequential events that should occur in the case of a particular dynamic situation. Scenarios are useful in developing an understanding of a particular threat conjuring up how the threat might be employed. Scenarios can be used to provide a stimulator or simulator with the input data it needs to provide the inputs to drive an information system with accurate data that occurs in a realistic manner as defined by the scenario timeline. Other devices pertinent to security include routers and firewalls.

1.8   Client Systems

Client systems can consist of desktop personal computers (PCs), laptop PCs, and Portable systems (i.e., either Personal Digital Assistants (PDAs) or cellular phones) that are connected to one another and to the various servers via Internet Service Providers (ISPs), modems, and wireless networks.

1.8.1   Desktop Personal Computers (Thin and Fat Clients)

Desktop computer systems are personal computers (PCs) that reside on a user's desktop and can be classified as either thin or fat clients. These desktop computers include both Macintosh personal computers and IBM-compatible computers, both of which are referred to here as PCs. *Thin clients* are PCs that have a minimal amount of internal software and a small operating system that offers connections to servers where applications and database software resides. *Fat clients* are the currently available general personal computers that have large memories for storing many applications and databases and are controlled by a large complex operating system.

Some attributes of both thin and fat client systems are shown in Table 1.4, *Security Attributes of Thin and Fat Client Systems*. Fat clients are often used as servers, in some cases, because of their large capacity and high speed. In many client/server networks, PCs can become servers and vice versa, although a server that becomes a client most often was originally a client.

Table 1.4 Security Attributes of Thin and Fat Client Systems

Thin Clients	Fat Clients
Less expensive.	More expensive.
Cannot accommodate anti-malicious code software.	Can accommodate anti-malicious code software.
Less likely to be attacked by a malicious hacker or virus.	More likely to be attacked by a malicious hacker or virus.
Easier to protect.	More difficult to protect.

---

[24] *Push* systems put information out to specified files whereas *pull* information must be accessed.

Thin Clients	Fat Clients
Less susceptible to hacker or virus attacks since it has a minimal amount of software and a minimal operating system.	More susceptible to hacker or virus attacks since it has a large amount of software and a large operating system.
If a server is successfully attacked, then its effects will be felt at many clients.	Many attacks are launched against clients in general and, if successful, will infect many Fat clients.
Database servers are located in Thin Client Systems and may be the object of a malicious hacker attack.	Many database servers are located in Fat Client Systems so Fat Client Systems are not immune to server attacks.
System owner can concentrate on countermeasures for protecting the servers.	System owner must concentrate on protecting both clients and servers.
Clients can only be a client and hence are vulnerable only to attacks against clients.	Clients can become either a client or a server and hence are vulnerable to attacks against both server systems and client systems.

Thin clients are easier to protect since most of the software they use is located on the network servers. The security focus thus is on the server devices rather than the clients. Additionally, since the operating system for a thin client is relatively small, it can be reviewed more easily by the programmers that developed it for any potential holes that can be readily fixed prior to releasing it as a shrink-wrapped package. Fat clients offer a much more desirable target since many of the applications packages and data are on the client and hence are subject to being attacked by malicious hackers and viruses.

1.8.2 Laptops

Laptop computer systems are PCs that can be placed in a user's lap and are considered to be portable computers yet they can also be desktop PCs when a docking device is used (Compaq 2001). Laptops operate on a battery (except when used with the docking device in which case they can be plugged into an electrical wall socket) and have essentially the same performance characteristics as desktop PCs.

1.8.3 Portable systems

Portable systems are personal computers that can be easily carried by a person, such as Personal Digital Assistants (PDAs) or cellular phones. One of the best selling *PDAs* is the Palm M105, which has 8 Megabytes of random access memory (RAM), a black and white display, a battery life of two months, a Palm OS 3.5 operating system, and weighs less than 4 ½ ounces (Dell Net 2001).

Cellular phones are portable or mobile telephones that can be used most anywhere (i.e., anyplace where the phone can connect to a cellular network through a wireless mechanism) including ambulatory locations, homes, offices, aircraft, trains, buses, and automobiles. A popular cellular phone is the Nokia 8890 that has the following characteristics: voice dialing built-in modem for digital data, and wireless communications that can operate in over 120 countries (Nokia 2001).

The security software for portable systems has the capability to encrypt both stored and transmitted information (Pointsec 2001). The security requirements for these systems are contained in Federal Information Processing Standard (FIPS) 140-1. FIPS 140-1 specifies the security requirements that are to be satisfied by a cryptographic module utilized within a security system product that protects sensitive but unclassified information within any computer or telecommunications system (including voice systems).

The security for cellular phones, PDAs, and laptops with wireless antennas can be assured using the same precautions and countermeasures that are used for desktop computers, namely, identification and authentication, encryption, non-repudiation, but adding the safety of the device itself, for example, not losing it or having it stolen (Peterson 2001).

1.8.4 Wireless Local Area Networks

Wireless local area networks (WLANs) are networks that are connected to other computer networks via any method that can communicate without requiring a hard connection, such as microwave antennas or satellites. Millions of people exchange information every day using pagers, cellular telephones, and other wireless communication products such as personal digital assistants. With the success of wireless telephony and messaging services, wireless communication is beginning to be applied to the areas of personal and business computing. The WLAN standard is IEEE 802.11, which is still in review. WLANs are primarily for communicating within a building or single structure.

The average distance for WLAN communications is 1000 feet. This is a sufficient distance to enable malicious persons to intercept WLAN traffic, even if they are not in the building. To improve the security capabilities of wireless networks, it is recommended that users implement the following five capabilities, as they apply (Meserve 2001):

1) Use VPN software on any computer connecting to a corporate LAN over a wireless network.
2) Implement the Wired Equivalent Privacy standard that protects WLAN traffic that leaks outside a building.
3) For WLANs in a public area of a building, keep traffic separate from that on the wired network.
4) Minimize the amount of transmission power used in a WLAN so as to decrease the communications distance, but ensuring that it is sufficient for the desired building communications.
5) Customize WLAN settings since malicious hackers can use the default settings to invalidly enter the WLAN.

The major benefit from wireless LANs is increased mobility for the users. Many network users, such as mobile users in businesses, hospitals, manufacturing plants, and university settings, can benefit from the added capabilities of WLANs (Lough 1997). Being free from conventional network connections, WLAN users can move about almost without restriction while accessing wired LANs from nearly anywhere.

Medical professionals can obtain patient records, real-time vital signs, and other reference data from the patient's bedside without relying on paper charts and physical paper handling. WLANs with real-time sensing can allow a remote engineer to diagnose and maintain the health and status of manufacturing equipment, even on an environmentally hostile factory floor. Wireless smart price tags, complete with liquid crystal display readouts, allow storeowners to eliminate discrepancies between stock-point pricing and scanned prices at the checkout lane (Lough 1997).

The range of applications for wireless network access is limited only by the imaginations of application designers.

1.9   Server Systems

Server systems are those computers that are dedicated to performing some service for the networked client systems. The servers include Web servers, file transport protocol (FTP) servers, domain name service (DNS) servers, and other types of servers.

1.9.1   Web Servers

Web servers are those computers that provide the connection to the World Wide Web (Web), or the Internet. The Web or Internet is a network of computer systems and subnetworks that enable anyone in the world to connect to anyone else (provided they have the appropriate hardware and software) who is also connected to the network. Originally, the Internet was used by various commercial organizations as an extension to their corporate network with applications that could (Minoli and Schmidt 1999):

1) Provide electronic mail for outside parties, vendors, and colleagues,
2) Provide access to research, vendor, or advertiser information,
3) Provide connections for electronic commerce,
4) Provide supporting capabilities such as help desk, call center, or data entry functions from remote users, and
5) Provide access to remote network devices for downloading statistics and for performing other network management tasks.

More recently, the Internet has come to be used for other functions, namely (Minoli and Schmidt 1999):

1) Decreasing costs by sharing the Internet for a large number of users,
2) Decreasing connection costs by managed Internet Service Providers,
3) Decreasing communication costs by using the Internet rather than purchasing dedicated lines,
4) Decreasing costs by using a variety of communications speeds and modes,
5) Decreasing costs and communications complexities by the standardization brought about by the Internet (e.g., the TCP/IP protocols),
6) Decreasing costs due to the vendor competition, and
7) Improving communications services by using Quality of Service support.

1.9.2   FTP Servers

The File Transfer Protocol (FTP) servers provide for the transmission and reception of messages and files that are encoded using the file transmission protocol.

1.9.3   DNS Servers

The *Domain Name Service (DNS) servers* map the local names of host systems to their actual network addresses. This facilitates the operation of the local area network.

1.9.4   Other Servers

Other servers are those systems that provide single sign-on, cryptographic encoding and decoding, remote authentication, portal, or e-mail (viz., SMTP) and other such services.

Other computers systems that are situated on the network include routers and firewalls.

1.10   Routers

Routers are computer systems that can mask off the header of a message to examine the Internet Protocol (IP) addresses of a packet's source and destination locations and forwards the messages (i.e., packets) to the proper destinations via the next most appropriate router. It has a very minimal operating system that enables the router to operate at high speeds. Routers are often referred to as *gateways* since they are used to connect logical networks. In some cases, a router on the customer's premises is called a *premise router*, or customer premise equipment (CPE) router.

Routers can use either static or dynamic routing tables. *Static routing* is based on predefined routing tables or a predefined mapping of the network and is a more secure method for building the network's routing tables. *Dynamic routing* is based on routing tables that are updated dynamically and this can enable an attacker to feed false routing information to the router (Brenton 1999). This false information is then propagated to other routers and can be used to provide spoofing (i.e., providing a false address of the attacker).

Some routers will provide for blocking of traffic based on the packet's source or destination addresses. Companies using a firewall to isolate corporate IP networks from the Internet may have a false sense of security. Although using a firewall to isolate corporate IP networks from the Internet is a good security practice, many organizations believe that their firewall is their first and only line of defense. A firewall system is definitely recommended for any company connected to the Internet, but some organizations fail to include the router used to connect to their Internet Service Provider (ISP) in their repertoire of countermeasures.

A router not used in conjunction with a firewall can result in the demilitarized zone (DMZ) being penetrated, possibly leading to a compromise in front of the firewall that may go undetected. A remote attacker can cause a number of malicious actions, including monitoring Internet traffic by using a sniffer program or gaining access past the firewall.

Even if the DMZ isn't vulnerable, the Internet boundary router itself can be a target for an attacker. Published flaws in outdated firmware may allow tailored attacks to reboot the router or cause problems with operation, resulting in a Denial of Service (DoS) or intermittent and confusing behavior of the router.

While assisting in a security audit, some Internet routers have no administrative access controls or traffic filters at all. This lack of administrative access controls invites hackers with the possibility of using telnet to gain access to the command prompt or use Simple Network Management Protocol (SNMP) to monitor and control a router without being authenticated to the router. Without filters, there is a possibility that an attacker could spoof the router into passing traffic that looks like it originated from the router or the router can be flooded with garbage IP traffic.

It is critical for organizations to consider their perimeter router when implementing Internet security. The analyst can begin by ensuring that the router's firmware is current. For example, some routers may be using a router software release that is several years old. Router vendors constantly update and improve the router operations with bug fixes, and security.

A perimeter router should always be configured to process IP traffic for IP networks assigned to the organization's connection. This will ensure that unwanted traffic is blocked, and can greatly reduce the work of the firewall. Some older routers do not directly support traffic filtering. It is a good policy to invest in one that does.

A router without administrative access controls, simple administrative user IDs and passwords, or default administrative logon information can be an invitation to malicious hackers. Administrative access control should always be used to secure an Internet router. If feasible, the router can be programmed to filter IP traffic even further by adding rules that will:

- Limit protocols (e.g., by dropping ICMP packets from the Internet),
- Disable all protocols that are not needed, and
- Disable other routing protocols.

These actions will screen your router and firewall from unwanted traffic and improve the performance of both.

1.11   Firewalls

Firewalls provide protection for LAN users from malicious users on public networks, such as the Internet. Firewalls are subsystems or groups of subsystems that assist in enforcing an access control policy. They operate on the basis of one of two principals:

1) block everything except what is specifically allowed through, and
2) allow everything through except what is specifically to be blocked.

Firewalls are computers that have a simple and fast operating system and a set of rules for blocking or passing traffic. *Software firewalls* are programs that reside on an existing platform to provide system protection whereas *hardware firewalls* are simply software firewalls with their own computer platform.

The utility of a firewall is only as good as the policies of the security administrator. If the administrator is willing to bypass the firewall for some remote user who calls in and wants to be patched to the internal network, then the firewall is as secure as a house with a locked steel front door that has many of its windows open.

Generally, firewalls operate on the basis of packet inspection and are in essence packet switchers. This fact makes routers and firewalls very similar. The *blocking/passing countermeasures* are usually firewalls, but may be routers. However, these two devices are becoming more alike and may eventually merge into a single platform. However, it may be the case that the resulting response delays caused by combining these may force owners into using separate platforms to ensure a desirable throughput rate.

Firewall loading can be addressed using clustered servers (as the platform) with load balancing and automatic failover software. The most common types of firewall are:

1) Static packet filtering,
2) Dynamic packet filtering, and
3) Proxy.

Static packet filtering firewalls use header information inside the packet. The header information is compared against the stored predefined or static access control policy rules *Dynamic packet filtering firewalls* are the same as stateful inspection firewalls and are more difficult to fool than the static filter (Brenton 1999).

Proxy firewalls represent an intermediate point that provides protection for IP addresses of internal users and are sometimes referred to as application gateways. To ensure that the firewall can only be programmed by authorized users, in some cases it is recommended to use an out-of-band[25] system for programming the firewall.

Firewalls, if properly configured, are intended to protect network resources from the following kinds of attacks (Cisco Firewall Web Page 2001):

- *Passive Eavesdropping/Packet Sniffing*—The attacker uses a packet sniffer to glean sensitive information from traffic between two sites or to steal user ID and password combinations, either on a private carrier or a public network. Even if applications were to encrypt traffic within their own streams, a sniffer could still detect sites using a form of traffic analysis. The attacker could then concentrate on transmissions involving the application being used for the transmissions.
- *IP Address Spoofing*—An attacker pretends to be a trusted computer by using an IP address that is within the accepted range of IP addresses for an internal network.
- *Port Scans*—This is an active method of determining which ports on a network device a firewall is listening. After attackers discover the "holes" in a firewall, they can concentrate on finding an attack that exploits the applications that use these identified ports.
- *Denial-of-Service Attacks*—This method differs from other types of attack because, instead of seeking access, the attacker attempts to block valid users from accessing a resource or gateway. This blockage can be achieved through SYN flooding a

---

[25] An out-of-band subsystem is a device (usually a personal computer) that is connected to the primary network only through a single subsystem (such as a firewall). Therefore, no network information can travel through it; hence, it is an out-of-band subsystem. It can also mean the use of other means of transporting information, such as writing it on a floppy disk and carrying it to another location, sometimes called "sneaker communications" because a person wearing sneaker shoes carries the information.

network resource to exhaustion through using half-open sessions (sending TCP packets with the SYN bit set from a false address) or by creating packets that cause a resource to perform incorrectly or even to crash.

- *Application-Layer Attacks*—These attacks take many forms, exploiting weaknesses in server software to access hosts by obtaining the permission of the account that runs an application. For example, an attacker might use the Simple Mail Transfer Protocol (SMTP) to compromise hosts that run older versions of *sendmail* using undocumented commands in the *sendmail* application.

Another method of attack is to use the Trojan horse, where the victim runs malicious code by being misled into believing the program (into which the Trojan horse is embedded) is something other than what it really is. Also, some advanced application-layer attacks exploit the complexity of new technologies such as the Hypertext Markup Language (HTML), Web browser functionality, and the Hypertext Transfer Protocol (HTTP). These attacks include object-oriented Java applets and ActiveX controls to pass malicious code across a network and transfer and download them with Web browsers, such as Internet Explorer or Netscape Communicator.

The newer firewalls can offer all of the benefits of today's firewalls with Virtual Private Network and Intrusion Detection System capabilities, including the capability to form up to as many as 500 virtual firewalls (Greene 2001).

Other attacks, such as a *man-in-the-middle attack* can be launched with a ping-of-death command (an ICMP command containing an oversized datagram), which brings down a system so that the malicious hacker in the middle can take over.

The operating system and the firewall's security rules base are intended to counter many of these attacks, but it is virtually impossible to guarantee that all types of attacks will always be countered by any single firewall. Capabilities usually accompanying firewalls include virtual private network encryptions, stateful inspection, and demilitarized zones.

1.11.1   Virtual Private Networks

Virtual Private Networks (VPNs) provide security at the IP level by encrypting and decrypting messages using the Internet Protocol Security (IPSEC) protocols, namely the Authentication Header (AH) and Encapsulating Security Payload (ESP). VPNs are a relatively low-cost alternative to link encryption (R. Smith 1997) that use widely available access to public networks (e.g., The Internet) to connect remote sites together using a tunneling method that allows users to transmit sensitive information over public links. VPN servers operate in a server-to-server mode and will only work when there is a complementary server to decrypt a tunneled message encrypted by another server.

Many VPN users are concerned about the security situation that exists at the other end of their VPN connections. The VPN termination location is as important as the overall connection itself (Gittlen 2001). If the security of the end-point is poor, then it is entirely possible that there is no protection for the encrypted information from the source (since the information can be unprotected after it has been decrypted by the VPN server or firewall).

Sometimes, a client with a highly secure VPN network will require that their customers adhere to a particular security policy in order to accept their VPN traffic. In some cases, a user with a highly secure network will actually pay for upgrades to end-point networks to ensure that the destinations of their VPN connection have sufficient security, and the end-point organizations may even be asked to sign a liability document. One of the methods for ensuring end-point VPN security is to configure the system so that the VPN information is routed to a DMZ with a separate network card (in the firewall or server) for the VPN traffic (Gittlen 2001).

Some firewall vendors (e.g., Check Point) offer VPN capabilities with their firewall software.

1.11.2 Packet Switching

Packet switching (Cisco Web Site 2001) is the detection of each packet in the link traffic and using the source and destination addresses (which are masked off the packet) to decide where the message should be forwarded. In the case of a firewall, the decision may also include blocking the message should it meet the criteria for disallowance.

1.11.3 Stateful Inspection

Stateful inspection (Check Point Web Site 2001) is the process of using the *contents* of the message (not the source or destination addresses) to decide if the message should be blocked or allowed to pass. Stateful inspection is a patent of Check Point Software, Inc. Their stateful inspection engine enforces the security policy on the gateway on which it resides and looks at all communication layers and extracts only the relevant data, enabling support for a large number of protocols and applications, and easy extensibility to new applications and services.

Stateful inspection has become a *de facto* standard and is embedded in firewalls made by other vendors, such as the Raptor firewall by Cisco. The Check Point Web site is www.checkpoint.com/ and the Cisco Web site is www.cisco.com/.

1.11.4 Demilitarized Zones

Demilitarized zones (DMZs), a term taken from the military, means an area of the information system network where unauthorized external users have access to organizational applications and information but not the organization's internal capabilities. A DMZ concept is contained in Figure 1.5, *An Illustration of a DMZ.*

The DMZ is a location for organizational Web, File Transport Protocol (FTP), portal, caching, or e-mail servers where public users (untrusted) have access but are separated and blocked off from the organization's internal users and servers. If a malicious user should do any damage to the information (e.g., mar your Web page or destroy a database) then the damage can be readily detected and the appropriate information can be uploaded from an internal server to the relevant server to replace the damaged information.

The figure shows a dual routing system with intrusion detection systems attached to both the DMZ and the LAN with a Web server or Mail server on the DMZ. An external user (who is unauthorized), for example, can access the organization's Web page or portal and get any information that is allowed for the public. Should the external unauthorized user, for whatever reason, decide to mar the Web page or to clobber database information that is available, then when this malicious behavior has been detected, the system administrator can simply upload the Web page (say in Java, HTML, or whatever language it's in) and replace any database information that has been destroyed, or whatever else needs to be fixed in the Web server. This is the purpose of the DMZ, namely to allow external access to only data or applications that can be replaced easily but does not have access to users, applications, or information that are internal to the organization.

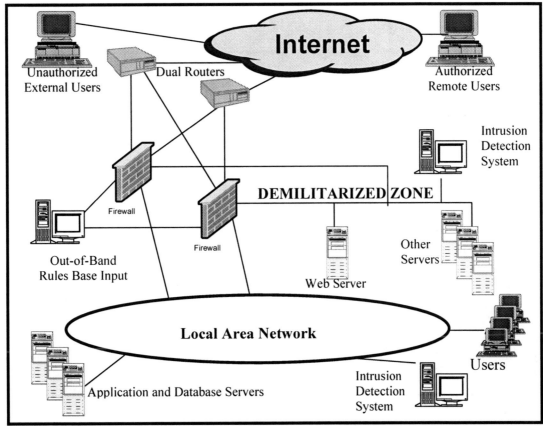

Figure 1.5 An Illustration of a DMZ

A type of server that one might wish to place in a DMZ is a *caching server*. In Web jargon, a *cache* is a place where temporary copies of objects are kept. Essentially, once the object pointed to by a URL has been cached, subsequent requests for the URL will result in the cached copy being returned, and little or no extra network traffic (Janet Web 2002).

Most modern computer systems use this principle in a number of places, to improve the performance of the main processor(s), speed up disk accesses, and so on. A server where Web pages or other projects are temporarily stored is called a *caching server*.

By placing firewalls at internal perimeters, even internal communications between any two users could be subjected to having to pass through a set of firewall rules (C. Smith 1999) as shown in Figure 1.6, *A Perimeter Architecture*.

Authorized external users have access to the internal LAN or the DMZ, whichever they choose. This means that the router and the firewall must be capable of recognizing authorized users. This can be done with User IDs and Passwords, but more confidently, with a single time password or biometric identifier. Smart cards offer another alternative for improved authentication, but they can be lost or stolen, and with social engineering practices, malicious users who happen to obtain a smart card can gather relevant information (such as the personal identification number (PIN)) and subsequently gain access to the information system and certain information.

In most architectures, internal users are not confronted by a firewall, but they could be if internal users were not trusted. Based on statistical data, it is more likely that a malicious action will come from an internal user than an external user. Thus the system administrator should be concerned about detecting and blocking malicious actions by internal users as well as unauthorized external users.

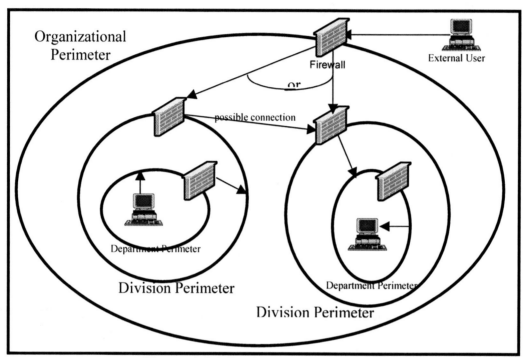

Figure 1.6 A Perimeter Architecture

1.12   Tradeoffs

Whatever the analyst concludes about the organizational needs for security, the final solution of which countermeasures will be required will be subjected to a tradeoff analysis that will regard various considerations. For example, certain social issues such as personal privacy may preclude the implementation of certain countermeasures. Also, the state of the environment must be considered because subsequent to a national tragedy such as the terrorist events on September 11, 2001, the nation's attitude toward increased security will be greatly modified.

For example, authentication measures (such as using facial characteristics), which might have been considered to be invasive to personal privacy before September 11, 2001, may be considered to be perfectly okay after September 11, 2001.

Considerations of personal rights or other issues will cause the diminution of the level of security, and correspondingly, strictness towards personal liberties in an austere environment will serve to increase security levels.

Countermeasures should be selected and implemented with a recognition of the rights and legitimate interests of others as well as the environmental situation existing at the time (Swanson 1996). The tradeoffs between the organizational desired security level of an information system and societal norms is not necessarily antagonistic because considerations of the existing national attitudes toward security may actually serve to increase the level of security.

REFERENCES AND BIBLIOGRAPHY

1.1.1 Bernstein, Terry, et al. (1996). "Internet Security for Business," Wiley Computer Publishing, 1996.

1.2 Canavan, John, (2001)."Fundamentals of Network Security," Artech House, Inc., 2001.

1.3 Date, C. J., "An Introduction to Database Systems," Addison-Wesley Publishing Company, Inc., 1988.

1.4 Smith, Richard E., (1997). "Internet Cryptography," Addison-Wesley Publishing Company, Inc., 1997.

1.5 Smith, Sr., Charles L. (1992). "A Theory of Situation Assessment," PhD Dissertation, George Mason University, Fairfax, VA, May 1992.

1.6 Smith, Sr., Charles L. (1998). "Computer-Supported Decision Making: Meeting the Demands of Modern Organizations," Ablex Publishing Corporation, Greenwich, Connecticut and London, England, 1998.

1.7 Roback, Edward A. (2000). "Guidelines to Federal Organizations on Security Assurance and Acquisition/Use of Tested/Evaluated Products," Computer Security Division, National Institute of Standards and Technology, NIST Special Publication 800-23, August 2000.

1.8 Keller, Sharon S. (1999). "Modes of Operation Validation System for the Triple Data Encryption Algorithm (TMOVS): Requirements and Procedures," Computer Security Division, National Institute of Standards and Technology, NIST Special Publication 800-20, October 1999.

1.9 Cisco Web Page (2001), www.cisco.com/, 2001.

1.10 Check Point Web Page (2001), www.checkpoint.com/, 2001.

1.11 OMB (1996). "Memorandum For Heads Of Executive Departments And Establishments, Subject: Management Of Federal Information Resources," Circular No. A-130, February 8, 1996.

1.12 OMB Appendix III (1996). "Appendix III to OMB Circular A-130, Security of Federal Automated Information Resources," Appendix to Circular No. A-130, February 8, 1996.

1.13 NSA (1998). "Network Security Framework," Security Solutions Framework, Network Security Group, National Security Agency, December 1998.

1.14 Reason, James (1987). "Generic Error Modeling System (GEMS): A Cognitive Framework for Locating Common Human Error Forms," in J. Rasmussen, K. Duncan, and J. Leplat (Eds.), *New Technology and Human Error*, (pp. 63-83), Chichester, UK, John Wiley & Sons.

1.15 Vulpe, Michel (2001), "Achieving Data Integration: The S4 DataPipes Solution," Infrastructures for Information, Inc, 5 February 2001.

1.16 Brenton, Chris (1999). "Mastering Network Security," SYBEX Network Press, 1999.

1.17 Smith, Sr., Charles L. (1999). "A Process for the Development of a Security Architecture for an Enterprise Information Technology System," Command and Control Research and Technology Symposium, U.S. Naval War College, Newport, RI, June 1999.

1.18 Smith, Sr., Charles L. (1997). "A Survey to Determine Federal Agency Needs for a Role-Based Access Control Security Product," International Symposium on Software Engineering Standards '97, IEEE Computer Society, Walnut Creek, CA, June 1997.

1.19 Compaq (2001). "Compaq Docking Solutions," http://www5.compaq.com/products/notebooks/docking/index.shtml, 2001.

1.20 Dell Net (2001). "Tope 10 Sellers, PDAs," http://eshop.msn.com/softcontent/softcontent.asp?scmId=603, 2001.

1.21 Nokia (2001). http://www.nokiausa.com/, 2001.

1.22 Pachomski, Jason (2000). "TCP/IP Primer," Tech Republic, www.techrepuboic.com, 2000.

1.23 Zyskowski, John (2000). "Good Things Are In Store," Federal Computer Week, www.fcw.com/, 11 September 2000.

1.24 Cisco NetSonar (2001). "NetSonar 2.0," http://www.cisco.com/warp/public/cc/pd/sqsw/nesn/, 2001.

1.25 Iomega Zip Drive (2001). "Zip Drives," Iomega, http://www.iomega.com/, 2001.

1.26 NIAP (1997). "Common Evaluation Methodology for Information Technology Security, Part 1: Introduction and General Model," Version 0.6, National Information Assurance Partnership (NIAP), CEM-97/017, January 1997.

1.27 NIAP (1999). "Common Methodology for Information Technology Security Evaluation, Part 2: Evaluation Methodology," Version 1.0, National Information Assurance Partnership (NIAP), CEM-99/045, August 1999.

1.28 NIST (1999). "Information Technology Security Testing—Common Criteria," Version 1.1, National Voluntary Laboratory Accreditation Program (NVLAP), U.S. Department Of Commerce, National Institute of Standards and Technology (NIST), Draft Handbook 150-20, April 1999.

1.29 Pointsec (2001). "Pointsec Solutions," http://www.pointsec.com/solutions/solutions.asp, 2001.

1.30 FIPS 140-1 (1994). "Security Requirements For Cryptographic Modules," Federal Information Processing Standards (FIPS) Publication 140-1, NIST, http://csrc.nist.gov/publications/fips/fips140-1/fips1401.htm, January 1994. (Now superseded by FIPS 140-2, 25 May 2001.)

1.31 Castro, Elizabeth (2001). "XML for the World Wide Web," Peachpit Press, 2001.

1.32 Scheier, Robert L. (2001). "Watching the Watcher," COMPUTERWORLD, pp. 36-37, August 20, 2001.

1.33 McKenna, Ed (2001). "Help in a Dangerous World," Federal Computer Week, pp. 38-40, September 3, 2001.

1.34 NAI (2001). "CyberCop Scanner," Network Associates, Inc., http://www.nai.com/, 2001.

1.35 Greene, Tim (2001). "Vendors Pitch All-In-One Security Boxes," Network World, pp. 45-46, 10 September 2001.

1.36 Mitchell, Bradley (2000). "Carnivore, Sniffers, And You," Computer Networking, http://compnetworking.about.com/library/weekly/aa071900a.htm?once=true&iam=d pile&terms=carnivore, 19 July 2000.

1.37 Hayes, Frank (2000). "Quick and Dirty," Computer World, 11 December 2000.

1.38   Swanson, Marianne and Barbara Guttman (1996). "Generally Accepted Principles and Practices for Securing Information Technology Systems," SP 800-14, NIST, September 1996.

1.39   Peterson, Shane (2001). "Securing Thin Air," Mobile Government, pp. 10-13, September 2001.

1.40   Schuermann, Udo and John Barkley, "Security," NIST Paper, June 1998.

1.41   Barkley, John (1995). "Implementing Role Based Access Control using Object Technology," National Institute of Standards and Technology, 1995.

1.42   Gittlen, Sandra (2001). "VPN Security Requirements Debated," Network World, p. 14, September 17, 2001.

1.43   Meserve, Jason (2001). "The Scoop On Wireless LAN Snoops," Network World, p. 17, September 17, 2001.

1.44   Lough, Daniel et al. (1997). "A Short Tutorial on Wireless LANs and IEEE 802.11," Virginia Polytechnic Institute and State University, http://www.computer.org/students/looking/summer97/ieee802.htm, 1997.

1.45   Betts, Mitch (2001). "Hidden Costs of IT Outsourcing," COMPUTERWORLD ROI, p. 6, http://www.computerworld.com/roi, September/October 2001.

1.46   LSI (1999). "The Intelligent Storage Hub," LSI Logic Corp., http://www.boostsystems.com/files/NAS%20Backgrounderb.pdf, June 1999.

1.47   Booch, Grady (2001). "Objectifying Information Technology," Rational Corporation, http://www.rational.com/products/whitepapers/394.jsp, 2001.

1.48   GAO (2001). "Critical Infrastructure Protection: Significant Challenges in Developing National Capabilities," General Accounting Office (GAO), GAO-01-323, page 39, April 2001.

1.49   Free BSD Organization (2001). "FreeBSD," http://www.freebsd.org/, 2001.

1.50   Wind River (2001). "BSD/OS," http://www.wrs.com/products/html/bsd_os.html, 2001.

1.51   NetBSD (2001). "Net BSD," http://www.netbsd.org/, 2001.

1.52   Yager, Tom (2001). "BSD's Strength Lies in Devilish Details," Info World, p. 58, November 5, 2001.

1.53   Ferraiolo, David and Richard Kuhn (1992). "Role-Based Access Controls," Proceedings of the 15th National Computer Security Conference, Volume II, pp. 554-563, 1992.

1.54   Minoli, Daniel and Andrew Schmidt (1999). "Internet Architecture," Wiley Computer Publishing, 1999.

1.55   Janet Web (2002). "The National Janet Web Cache Service," http://wwwcache.ja.net/intro.html, 2002.

CHAPTER TWO

AN OVERVIEW OF SECURITY ANALYSIS

2.1   An Overview of Information System Network Security

An *information system network* is the collection of computers, network devices, and software that provide an integrated capability to process, transmit, receive, display, and store information. Because of the myriad threats that exist for information systems that are connected to other network systems, some sort of protection is required. *Network security* is the concept of ensuring that an information system is maximally protected against certain threats.

The security capabilities for an information system network should protect the system and its information. This protection is provided by a set of countermeasures that include physical, behavioral, and technical devices (e.g., routers and firewalls) and methods (e.g., training and education).

A formal security analysis is required to identify the appropriate set of countermeasures required to provide cost-effective protection for your information system. This analysis should also present a comprehensive rationale that supports the purchase and installation of these methods and mechanisms. A countermeasure (sometimes called a safeguard or security mechanism) can be either a process (e.g., training) or a product (e.g., a firewall or anti-virus software).

A security analysis is not unlike a certification and accreditation process. The Department Of Defense Information Technology Security Certification And Accreditation Process (DITSCAP) (DOD 1999) was intended to develop a standard certification and accreditation process for the DOD that meets the security policies mentioned in DOD Directive 5200.28, Public Law 100-235, OMB Circular A-130, and other DOD Directives. However, the purposes of the DITSCAP and this book are quite different for at least two reasons:

1) This book is for organizations that are not required to protect classified information, and
2) This book recommends a formal risk assessment as part of a comprehensive security analysis.

There is a divergence between the recommendations and processes described in this book and the DITSCAP, for example, since Public Law 100-235 (The Computer Security Act of 1987 (amended in 1997) states that a formal risk assessment is not required. This book stands in direct contradiction to that since a formal risk assessment *is required* if your organization ever wishes to understand the rationale for why some vendor is recommending that you purchase and install an expensive set of countermeasures.

A process document, very similar to the DITSCAP, which was written for civilian organizations, is the "National Information Assurance Certification and Accreditation Process (NIACAP)," (NSTISSC 2000).

Most organizations seem to be content to not worry about adding security until there is a security incident. A *security incident* is any breach of security that is recognized by the organization and is significant enough that the organization wishes to protect itself against any other such incidents. Unfortunately, many such incidents result in great costs to the organization, but they highlight the need for information system network security.

2.2   Threats

Threats to your information system consist of all entities that can do some harm to the system. These threats can be physical (unauthorized person enters your information system facility and knocks a server out of order with a hammer), behavioral (a manager faces a novel situation (to her) and does not know what to do), or technical (a malicious hacker manages to gain electronic access to your information system and deletes a database from hard disk memory).

Top five threats or security issues, in the author's opinion, are:

1) System administrators creating access holes for important users.
2) System administrators not removing former user names (IDs) from system access control lists.
3) System administrators not installing software patches as soon as they are available.
4) System software not enforcing User ID and Password (or other authentication mechanism) for granting user, server, or program application and database information access.
5) System administrators not enforcing organizational security policy due to a lack of knowledge of, understanding of, or access to, the policy.

The top five threats (or security issues) are quite subjective and will vary as much as there are those who attempt to identify them. It will be based on the answerer's experience and an analyst's experience with different information systems will vary considerably from analyst to analyst. My experience has been that most of the problems with system security emanate from a laxness toward security on the part of information system administrators and my top five threats reflect this. For a sample of the answers to this question, see http://www.computerworld.com/community/security (Computerworld 2001).

2.3   Vulnerabilities

A *vulnerability* is any situation that exists in your information system facility that makes the system susceptible to some threat. All information systems are vulnerable to some threats (although the owners may not be aware of it) and one of the tasks for a security analyst is to identify these vulnerabilities and determine the seriousness of these vulnerabilities. Vulnerabilities can be categorized as physical, behavioral, or technical.

The *physical and behavioral* vulnerabilities can be identified through interviews with owners and users, sometimes using questionnaires[26], see Appendix C, *Questionnaire*, for a sample questionnaire, and through observations of the information system facilities. *Technical* vulnerabilities can be identified by:

---

[26] A *questionnaire* is a method of gathering information from an identified set of persons that consists of a set of questions (concerning some particular area of interest) that should be asked in the same order each time they are presented to an interviewee.

1) scanning the network,
2) performing an analysis of the system using representations and descriptions of the system, or
3) comparing this system's architecture with other similar systems that have known vulnerabilities.

The extent of the identified vulnerabilities depends on:

1) The countermeasures currently implemented,
2) The size of the system,
3) The volume of business done by the system,
4) Value of information stored in the system,
5) Notoriety of the system,
6) Perceived susceptibility of the system to non-catastrophic threats, namely, power outages, minor fires, minor water damage, random server failures, and other such events, and
7) Other factors such as system familiarity among potential malicious hackers.

The potential damage that can be done to a system with a particular set of vulnerabilities depends on the assets and the values of these assets to the organization. Thus, the identification of the information system assets and the determination of the values of these assets is an important task for any security analyst.

Vulnerabilities are often divided into three categories:

1) *Critical* - functions or services that, if lost, would prevent the organization from exercising safe control over its assets.
2) *Essential* - functions or services that, if lost, would reduce the capability of the organization to exercise safe control over its assets.
3) *Routine* - functions or services that, if lost, would not significantly degrade the capability of the organization to exercise safe control over its assets.

These categories can be of use for deciding which vulnerabilities are the most important and for performing a risk analysis of the vulnerabilities that are not countered because it was not cost-effective to do so.

A list of standardized names for vulnerabilities and other information security exposures can be found in the Common Vulnerabilities and Exposures (CVE). The CVE is a collaborative effort hosted by The MITRE Corporation to standardize a set of names for all publicly known vulnerabilities and security exposures (MITRE CVE 2001). The MITRE Web page contains links to a dictionary of relevant terms and the CVE list.

The CVE list contains the name and description of all publicly known facts about computer systems that could allow somebody to violate a reasonable security policy for that system. These potential violations are referred to as "vulnerabilities."

The broad use of the word vulnerability usually refers to any fact about a computer system that is a legitimate security concern, but only within some contexts. For example, since the **finger** service reveals user information, there are reasonable security policies that disallow **finger** from being run on some systems. Thus **finger** may be regarded as a vulnerability according to this usage of the word.

However, a more focused perspective is that some security-related facts fall short of being actual vulnerabilities. With respect to the presence of the **finger** service, one can argue that since **finger** behaves as it was designed to behave, this application should not be considered to be a vulnerability.

Some examples of vulnerabilities given on the CVE Web page are:

1) **phf** (remote command execution as user "nobody"),
2) **rpc.ttdbserverd** (remote command execution as root),
3) world-writeable password file (modification of system-critical data),
4) default password (remote command execution or other access),
5) denial of service problems that allow an attacker to cause a Blue Screen of Death, and
6) **smurf** (denial of service by flooding a network).

Some CVE examples of exposures include the following:

1) running services such as **finger** (useful for information gathering, though it works as advertised),
2) inappropriate settings for Windows NT auditing policies (where "inappropriate" is enterprise-specific),
3) running services that are common attack points (e.g., HTTP, FTP, or SMTP), and
4) use of applications or services that can be successfully attacked by brute force methods (e.g., use of trivially broken encryption, or a small key space).

2.4   Assets and Values

Assets are the elements of an information system that are of value to an organization, such as the different types of information and the information system itself. These elements may be of value to the organization's users or to customers of the organization, depending on the type of business the organization does. The assets can have values to the organization and this value is dependent on the type of business the organization does, the amount of business done, and other relevant business parameters.

For example, a research and development organization will most often value the information that it generates as having much value whereas an organization that sells products may not value its internal information but will value the information it has on customers and what they purchase.

2.5   Threat scenarios

Threat scenarios are situations developed for some threat and usually some system vulnerability where the system can suffer some sort of damage. The amount of damage and the probability that the threat is successful are used to compute the security risk.

2.6   Countermeasures

A *countermeasure* is any method or product that can be implemented to minimize the potential damage that exists from a particular set of threats.

The commercial market provides a great many products for countering the threats that are currently known. One of the objectives of this book is to provide the reader with the following capabilities:

1) An understanding of the gamut of countermeasures,
2) Their attributes,
3) Methods for comparing or rank ordering them, and
4) How to determine which ones are cost-effective.

Only those countermeasures that are cost-effective should be purchased and installed. The determination of the cost-effectiveness of any countermeasure is always relative to the specific information assets and their values that it protects.

For security one can do any (or all) of the following:

1) Identify the users on the phone and then manually plug them in only after identity confirmation and then unplug them when they are through (this process requires a fool-proof method of human identification and authentication as well as being aware of when the user is through with the line and capable of unplugging the user),
2) Disconnect the information system from all public wide area networks, such as the Internet, and use it as an internal system,
3) Institute private communications links that are accessible only by approved individuals, and
4) Purchase and install security methods and mechanisms to protect the system.

Some actions that one can take to limit information system vulnerabilities are:

1) Require all employees to sign a document to confirm their understanding of security policies.
2) Develop a more specific acceptable-use policy that clearly lays out the boundaries for computer usage.
3) Require that employees lock down their system (log off the network or institute a screen saver that requires a password when not in use.)
4) Institute a formal security-awareness program to keep these issues in the forefront of employees' minds.
5) Get rid of some minor, non-critical services that are running on the servers.
6) Ensure that all employees have passwords of sufficient complexity.
7) Make sure that users are not employing the same password for the network and their e-mail services.

8) Institute a badging system to differentiate employees from visitors.

9) Strengthen security at the server rooms and take down the "Computer Room" sign.

10) Train employees to temper their friendliness with a healthy wariness of possible security exposures.

11) Install pick plates on all external and critical internal doors to hamper break-ins and provide evidence of security breaches.

12) Require that employees remove business cards from their desks and keep them in a more secure area.

2.6.1 Physical Countermeasures

Physical countermeasures are the devices and methods designed to protect the information system against intruders who wish to do physical damage to the system. These types of countermeasures include badges, fences, gates, cyber locks for secure server rooms, guards, and guard patrols.

A physical security office may be formed to be responsible for developing and enforcing appropriate physical security controls, in consultation with computer security management, program and functional managers, and others, as appropriate. Physical security should address not only information system installations, but also backup facilities and office environments. In government and industry, this office is often responsible for the processing of personnel background checks and security clearances. Even those who deal with sensitive information must have background checks.

2.6.2 Behavioral Countermeasures

Behavioral countermeasures are those activities designed to protect the system by affecting the behavior of the relevant personnel in a positive manner. These types of countermeasures include training and education for users, administrators, and other relevant personnel.

Most organizations will form a training group that has the responsibility of developing course material and presenting training and educating material to personnel. This material can be developed and presented by employees or by special consultants.

2.6.3 Technical Countermeasures

Technical countermeasures are those devices designed to provide protection of the information system against technical threats. These countermeasures include both hardware and software subsystems. Some are really both, such as firewalls or routers. These devices can include the following capabilities: identification and authentication systems, anti-virus software, anti-denial-of-service software, advanced identification and authentication (such as using biometrics or tokens), digital signatures, encryption methods (symmetric and asymmetric), hashing functions, cyclic redundancy checksums, virtual private networks (VPNs), authorization systems, intrusion detection systems, cabling, and auditing systems.

The various capabilities for providing technical countermeasures for an information system network are shown in Figure 2.1, *An Illustration of Technical Countermeasures*, below.

Both internal and external users may attempt to gain access (properly or improperly) to the information system network. The large majority of these users are non-malicious; that is, they have a legitimate purpose for using the information system; however some users may have malicious goals.

Some of the countermeasures will attempt to block malicious external users (the solid arrow going away from the blocking/passing countermeasure), but may inappropriately pass these in some cases (the dashed arrow going toward the information system network). Similarly, for non-malicious users, either external or internal, and for internal malicious users, some of these users may be inappropriately blocked (the three dashed arrows going way from the blocking/passing countermeasures, but most will be passed (the three solid arrows going toward the information system network).

Of course, there are a number of other countermeasures embedded in the information system as shown in Figure 2.1. All of the countermeasures are illustrated in Figure 2.2, *Security Countermeasures*.

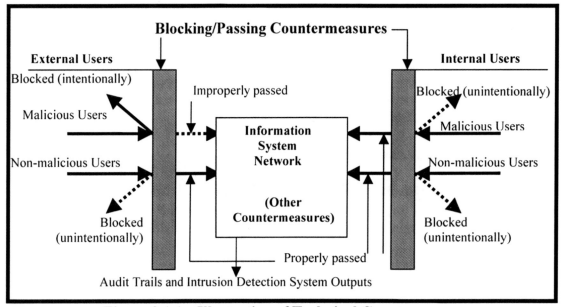

Figure 2.1 An Illustration of Technical Countermeasures

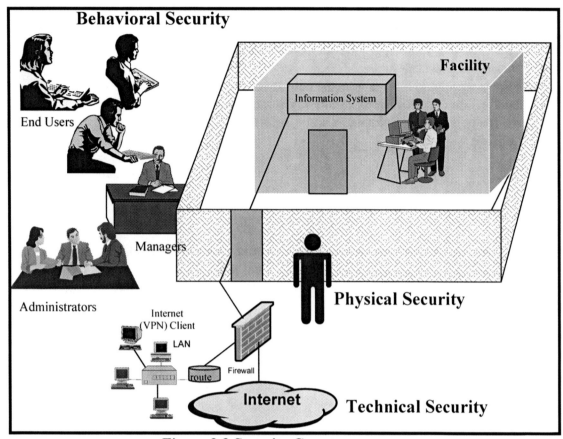

Figure 2.2 Security Countermeasures

Information system users are generally grouped into two categories: internal and external. An entire list of users is contained in an Access Control List (ACL), which identifies all of the users, whether internal or external. Ideally, the ACL is kept at one site and is modified whenever a new entity is added or an entity no longer needs access and is immediately deleted from the list. Unfortunately, some organizations have several sites where an ACL is required and these are manually updated and delivered to each site as required.

2.7   Organizational Structures

Most modern organizations have an information system for performing some facet of the organization's business, such as administration, payroll, research and development, information access and retrieval, employee communications, document creation and transmission, information storage, and interfacing with the public for its business. The importance of this information system varies but in some cases, it is the primary or single point-of-contact with the buying public and hence is essential to the organization's business capabilities (e.g., amazon.com).

Some organizations believe that by limiting its information system users to employees only and ensuring that the system is not networked with any outside system, that its security needs are diminished to nearly nil. However, these organizations fail to realize that statistics show that over half of all malicious activities regarding information systems are committed by internal users (i.e., employees). Employees can remove sensitive information from the organization's premises using floppy disks, compact disks, super disks, or zip disks hidden in their pockets or brief cases.

For some organizations, it is a legal requirement that certain organizational information is retained for some period (e.g., automotive manufactures must keep records of vehicle crash tests for at least three years). This is information that can be stored in information system archives.

2.8 Users

2.8.1 Internal Users

Internal users include those entities (users, administrators, applications, or servers) who use the information system from an internal local area network (LAN, often called an intranet) and those who use the system from a remote site (often called remote users and may be extranets).

Most malicious actions against information systems are the result of internal users. In fact, "Study after study has shown that the biggest security threat is from people inside the organization, not outside," (William 2001), such as users who have logged on to the system and then left their computers unattended.

2.8.2 Actual Internal Users

The *actual internal users* are those users who are located within one of the organization's LANs. Because internal users are authorized users who have proper access to the information system, they cannot be identified based on the implemented Identification and Authentication procedures (such as User ID and Password because they possess such), but their actions can be identified as malicious if the auditing and intrusion detection system (IDS) processes are properly performed. Some malicious internal users may be blocked, but this is unintentional and is usually caused by some misunderstanding by the security administrator and/or the user, that is, either a mistake or a slip[27].

Thus, only a good auditing and IDS capability can protect the information system from internal malicious users. This includes internal users who are malicious through accidents, lack of knowledge, or incompetence. All of these malicious actions and users can be identified and an appropriate action can ensue, if the owner should choose. In some cases, the organization's security policy may state that non-malicious authorized users who frequently (but inadvertently) damage the system may be disallowed from using the system, especially if their actions result in frequent and/or costly mitigations. Some users believe

---

[27] A "mistake" is an error that happens due to a misunderstanding of the proper principles involved (e.g., a programming error either due to an erroneous requirement or poor programming) whereas a "slip" is an error that occurs due to an entity (user, administrator, application, or service) inappropriately performing some action (Reason 1987), e.g., hitting the wrong key or misspelling a word. Normally, errors performed by an application should be found and mitigated during security testing of the application. However, this may not always be the case.

that information system owners cannot be protected from user stupidity, but this is just not true.

Based on statistics from organizations that have responded, the proportion of malicious attacks on their information system networks (that have been identified) is anywhere from 50 percent to 80 percent, but certainly over 50 percent. This may astound some because most news reports concern stories about external malicious hackers who gain access to an organization's information system network from an external source, such as the Internet or a public wide area network (WAN).

2.8.3   Remote Internal Users

Remote users, who are also internal users, are those users who are at remote sites and enter the organization's network though their modem, which enables them to use the Internet communications via an Internet Service Provider (ISP). In some cases, the remote users are connected via a leased private line and although private lines offer a certain amount of security advantages, they are much more expensive than the Internet.

2.8.4   External Users

External users are those entities that enter the information system from a source outside of the organization's enterprise system, such as through an Internet Service Provider. Note that remote users *are not* external users. External users are entities that do not have a user ID or password (otherwise they would be an internal user or they have access to internal user information such as that provided through a social engineering method, to be discussed later). Malicious users who are not blocked (for whatever reason), but somehow gain access to the information system, will have their actions identified by the auditing and IDS capabilities of the system. In some cases, these actions can be identified in near real time by the IDS and these users may have their actions immediately discontinued as well as being identified and thereby suffering some sort of negative response.

Within the information system network are other countermeasures that are designed to deny malicious users, such as anti-virus software, auditing capabilities, encryption of information, and other nefarious capabilities. When malicious code enters the information system, these countermeasures are designed to block the negative effects or record their activities and associate the identity of the malicious source with their actions.

Firewalls are a security device for protecting the internal system and users from the external world. If you can afford the cost and response delay, it is wise to have more than one type (i.e., vendor) of firewall in tandem because the methods used to block and pass messages by the different vendors are complementary and serve to enhance one another. Thus, it is more likely that a malicious user who attempts to enter the system will be blocked if two or more firewalls (from different vendors and in sequence) are used.

2.9   Configuration Management

A *system configuration* is a description of how the hardware elements of a system are arranged. For example, in a LAN, certain printers, routers, firewalls, or computer systems may be disconnected or disabled. Similarly, a *software configuration* is a description how particular elements of the software system are arranged. For example, some applications or services may be disabled. *Configuration management* is the process of ensuring that a system is appropriately defined. A process is shown in Figure 2.3, *Configuration Management,* below.

Configuration management is a collection of processes and tools that promotes network consistency, tracks network changes, and provides up-to-date network documentation and visibility. By building and maintaining configuration management best-practices, one can expect several benefits such as improved network availability and lower costs. These advantages include the capabilities to:

1) *Lower support costs* due to a decrease in reactive support issues.
2) *Lower network costs* due to device, circuit, and user tracking tools and processes that identify unused network components.
3) *Improve network availability* due to a decrease in reactive support costs and improved time to resolve problems.

Most organizations have a *Configuration Control Board (CCB)* that is responsible for ensuring that all changes to the system are reviewed and approved prior to their implementation. All proposed changes to the network should be presented and approved by the CCB prior to their implementation.

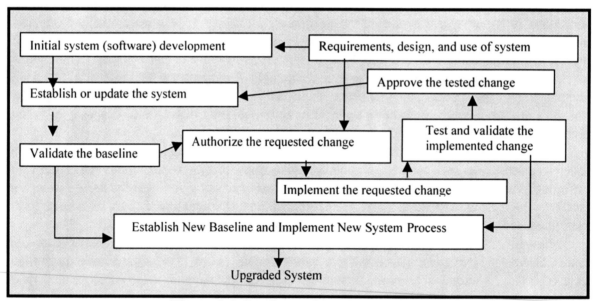

Figure 2.3 Configuration Management

2.10   Knowledge Management

When information is stored in a knowledge base, it represents knowledge (i.e., rules) about some topic of interest to the organization, such as how to perform a security analysis. In most cases, this knowledge is quite sensitive and needs to be protected. It can be stored in encrypted form so that only privileged users can view the information. By making the knowledge base available for others on the Web, then the knowledge can be easily accessed by anyone on the Web.

The placing of knowledge on the Web enables organizational intranets to put information in the base and to read the rules that reside there (Throne 2000). By ensuring that the information transferred to the knowledge base is transparently encrypted by the system, or decrypted for those extracting information, for certain privileged users, the data is protected from malicious hackers.

Many organizations either have or wish to institute a knowledge base for storing the technical and managerial knowledge that has been accrued by the organization. This knowledge can be used for the following:

1) Referencing in cases where a decision maker can use some assistance,
2) Provide training for novices in areas where the organization has knowledge,
3) Provide a repository for organizational knowledge, and
4) Provide inputs to decision support systems (DSSs) or expert systems, such as a Security Analysis DSS.

2.11   Security Test and Evaluation

A *security test and evaluation (ST&E)* process is essential for providing evidence and rationale for supporting the assurance that the implemented security mechanisms and methods are operating properly and providing the necessary protection for the information system network.

Often, the ST&E input process can be performed by a security analyst at the keyboard of a client computer. In cases where the information system applications are Web-based, the inputs can be provided from a client computer located anywhere as a host on the Internet. However, in some cases, additional devices are required to stimulate the system and to collect the relevant data.

Devices used for performing ST&E inputs and for collecting the system data needed for evidence of proper operation are:

1) Simulations (to provide inputs),
2) Sniffers[28] (to collect relevant data), and
3) Auditing mechanisms (to collect relevant evidence data).

Each of these was explained and discussed in Chapter One.

After collecting the appropriate data and then analyzing and evaluating it, the analyst then writes a Systems Security Authorization Agreement (SSAA) (or Security Certification and Authorization Package (SCAP)) to inform the Designated Approving Authority (DAA) of the potential security capabilities of the information system.

The SSAA (or SCAP) report is a living document and should be updated whenever any major[29] modifications are made to the information system.

---

[28] Many organizations will not allow the security analyst to place a sniffer on the network, even for test purposes. However, some organizations will allow the sniffer to be attached but only after the attachment has been approved by the appropriate individuals.

2.12   Certification and Accreditation

After the ST&E process has been completed and the analyst has issued the ST&E report, one of the addressees of the report should be the DAA who will read the report and make a decision regarding the information system under consideration. If the ST&E report is favorable, then the DAA will grant the information system a Certification and Accreditation (C&A) to operate. Even after the system has been given a C&A to operate, this is C&A good for either three years or until an upgrade is implemented that requires that another ST&E process be performed.

Organizations allowed to provide Certification and Accreditation as well as information on all aspects of network security include: The MITRE Corporation (**www.mitre.org/**), VeriSign (**www.verisign.com/**), and the SANS Institute (**http://www.sans.org/newlook/home.htm**).

---

[29] A system modification is *major* if it changes the security capabilities or alters the needs of the information system.

REFERENCES AND BIBLIOGRAPHY

2.1   DOD (1999). "Department Of Defense Information Technology Security Certification And Accreditation Process (DITSCAP)," Application Document, Department of Defense Manual, ASD (C3I), Number 5200.40-M, 21 April 1999.

2.2   Bernstein, Terry, et al. (1996). "Internet Security for Business," Wiley Computer Publishing, (1996).

2.3   Canavan, John, (2001)."Fundamentals of Network Security," Artech House, Inc., (2001).

2.4   Date, C. J., "An Introduction to Database Systems," Addison-Wesley Publishing Company, Inc., (1988). Smith, Richard E., (1997). "Internet Cryptography," Addison-Wesley Publishing Company, Inc., (1997).

2.5   Smith, Sr., Charles L. (1998). "Computer-Supported Decision Making: Meeting the Demands of Modern Organizations," Ablex Publishing Corporation, Greenwich, Connecticut and London, England.

2.6   Roback, Edward A. (2000). "Guidelines to Federal Organizations on Security Assurance and Acquisition/Use of Tested/Evaluated Products," Computer Security Division, National Institute of Standards and Technology, NIST Special Publication 800-23, August 2000.

2.7   Keller, Sharon S. (1999). "Modes of Operation Validation System for the Triple Data Encryption Algorithm (TMOVS): Requirements and Procedures," Computer Security Division, National Institute of Standards and Technology, NIST Special Publication 800-20, October 1999.

2.8   Cisco Web Page (2001).

2.9   Check Point Web Page (2001).

2.10   OMB (1996). "Memorandum For Heads Of Executive Departments And Establishments, Subject: Management Of Federal Information Resources," Office of Management and Budget, Circular No. A-130, February 8, 1996.

2.11   OMB Appendix III (2000). "Appendix III to OMB Circular A-130, Security of Federal Automated Information Resources," Appendix to Circular No. A-130, February 8, 1996.

2.12   NSA (1998). "Network Security Framework," Security Solutions Framework, Network Security Group, National Security Agency, December 1998.

2.13   Reason, James (1987). "Generic Error Modeling System (GEMS): A Cognitive Framework for Locating Common Human Error Forms," in J. Rasmussen, K. Duncan, and J. Leplat (Eds.), *New Technology and Human Error*, (pp. 63-83), Chichester, UK, John Wiley and Sons, 1987.

2.14   Throne, Adam (2000). "The Evolution of Knowledge Management," Call Center Magazine, www.callcentermagazine.com/, April 2000.

2.15   William, Jon (2001). "Mobilizing Security," Federal Computer Week, pp. 30-31, 13 August 2001.

2.16   Computerworld (2001). "COMPUTERWORLD Security Web Page," http://www.computerworld.com/community/security, 2001.

2.17   NSTISSC (2000). "National Information Assurance Certification and Accreditation Process (NIACAP)," National Security Telecommunications and Information System Security Committee (NSTISSC) Instruction No. 1000, April 2000.

2.18   MITRE CVE (2001). "Common Vulnerabilities and Exposures," The MITRE Corporation, http://www.cve.mitre.org/, 2001.

CHAPTER THREE

NETWORK SECURITY POLICY AND REQUIREMENTS

3.1   Information System Security Policy

Information System Security Policy refers to the statements made by owners and high-level managers of organizations (commercial, civilian Government, or some aspects of the Defense community) to establish procedures, rules, repercussions, and responsibilities on information system access and safeguards. The Defense community says Information System Security Policy refers to the rules relating to the clearances of users and the access to information that must be marked with the appropriate classification. This latter definition is not an objective in this book although knowledgeable readers will find that the material covered in this book is generally applicable to the Defense community. A security policy should state the rules that are enforced by a system's security countermeasures (Russell 1991).

Briefly, a security policy is essentially a high-level set of requirements for the information system. These statements should be vendor independent, that is, one should not stipulate a particular vendor's product as a requirement in a policy statement, nor even in a requirements statement. Policy should be a statement by the organization's management of the desired capabilities of the security mechanisms and methods (i.e., countermeasures), the responsibilities of the employees, and the potential responses for any security violations.

Commercial and civilian Government organizations have information that is called *sensitive but unclassified*. Commercial organizations define this information as data that could be detrimental to the organization if it is either destroyed or divulged to an unauthorized person or organization. The Computer Security Act provides a definition of the term "sensitive information":

> "Any information, the loss, misuse, or unauthorized access to or modification of which could adversely affect the national interest or the conduct of federal programs, or the privacy to which individuals are entitled under section 552a of title 5, United States Code (the Privacy Act), but which has not been specifically authorized under criteria established by an Executive Order or an Act of Congress to be kept secret in the interest of national defense or foreign policy."

The above definition can be contrasted with the long-standing defense-based classification system for national security information (viz., UNCLASSIFIED, CONFIDENTIAL, SECRET, TOP SECRET, or higher security classifications). This system is based only upon the need to protect classified information from unauthorized disclosure. However, the Defense Department is also concerned with protecting its sensitive but unclassified information.

A definition for sensitive but unclassified information that applies to commercial users is the following:

> Sensitive information is any information, the loss, misuse, or unauthorized access to, or modification of, which could adversely affect the organizational interest or the conduct of organizational programs, or the privacy to which individuals are entitled under section 552a of title 5, United States Code (the Privacy Act).

Thus, contract information, employee information, customer information, and private technical information satisfy this definition.

When constructing an organization's security policy, the author should avoid being vendor-specific or even method-specific. For example, if confidentiality is treated in the organization's security policy statement, it may be more cost effective to have a private link than to use a public link with encryption. Thus, if confidentiality is a policy of the organization, the security policy author should be wary of offering any potential solution to the issue in the policy statement.

The information system network is often referred to as a Trusted Computing Base (TCB) if it provides a certain level of security, and has been tested to ensure this. The TCB is defined as follows (Russell 1991):

> The TCB is the totality of protection mechanisms within a computer system—including hardware, firmware, and software—the combination of which is responsible for enforcing a security policy. It creates a basic protection environment and provides additional user services required for a trusted computer system. The ability of a TCB to correctly enforce a security policy depends solely on the mechanisms within the TCB and on the correct input by system administrative personnel of parameters (e.g., a user's clearance level) related to the security policy.

For our purposes, namely ensuring the protection of sensitive but unclassified information, user clearances are not an issue[30]. The terms used above can be interpreted in this book as follows:

1) The "Trusted Computing Base" is our "security infrastructure,"
2) "Protection mechanisms" are our "countermeasures," and
3) The "Trusted computer system" is our "overall information system."

The OMB Circular A-130 and its Appendix III state that a formal risk analysis is not required for Government information systems. This raises the issue of, "What is meant by 'formal' in this context?" It is not important to address this question since it is important to conduct a Risk Assessment (or analysis) since without one, it is not possible to ever rationally select the appropriate cost-effective countermeasures.

Having a Government agency state that a formal risk assessment is not required is probably a relief to many Government agencies, however, it should not be a relief since this statement (if followed) will likely result in the purchase of an improper set of countermeasures for any information system for each Government or commercial organization that has decided to abide by this OMB directive. For commercial organizations that must be profitable to remain in existence and who must have real protection against the

---

[30] In many cases, even access to sensitive but unclassified information requires a background check that is organization-specific, such as the Department of Transportation. Thus, an employee who wishes to have access to the information system must have the proper authorization.

various threats to their information systems, cost-effective countermeasures might be the critical difference between:

1) Being a profitable organization that can continue to operate even in the face of various types of attacks against its information system, and
2) Possible failure of the organization.

Thus, the organization's security policy should unequivocally state that there must be a formal Risk Assessment performed to ensure that the organization identifies the most cost-effective countermeasures available for their information system. The adjective "formal" means that the assessment must be procedurally, mathematically, and logically rigorous. A summary of the OMB Circular A-130 security policy review is provided at the end of this chapter in section 3.9, *A Summary of OMB Circular A-130 Security Policy.*

Even if the organization has a security policy and it has been around for a long time, the policy can change, for a variety of reasons, among them being:

1) The environment changes, such as society's attitude toward personal privacy because of some event,
2) The organization's mission is modified by management (top down policy), or
3) The organization's security requirements are altered and the policy must be changed to reflect this (bottom up policy).

So an existing security policy is not a set of rigid rules that will last for the life of the organization.

A program (PoliVec Builder 1.0) from e-Business Technology, Inc. (www.ebiz-tech.com) can be used to generate a security policy for your organization (Andress 2001).

3.1.1 Rationale for Computer Security

Computer security is an integral part of any organization's capabilities. The purpose of computer security is to protect an organization's valuable information system resources, such as hardware (e.g., computers and network devices) and software, (viz., firmware, applications, and information, especially sensitive information). Through the selection and application of appropriate safeguards (i.e., countermeasures), security helps the organization achieve its mission by protecting its physical and financial resources, reputation, legal position, employees, and other tangible and intangible assets.

Security is sometimes viewed by the users as upsetting the organization's mission by imposing rules and procedures (that may be considered as bothersome or poorly defined) on users, managers, and systems. Security policies, procedures, rules, and requirements that are well defined, thoughtful, and rigorous do not exist for the sole purpose of inflicting difficulties on the users; instead they are put in place to protect important system assets and thus support the organization's overall mission. In many cases, it is only after the occurrence of an unfortunate security incident that an organization will begin a serious effort to identify and install countermeasures.

In a commercial business, having good security is often secondary to the need to make a profit. But security ought to increase the firm's ability to make a profit and security ought to assist in improving the organization's services provided to their customers. Thus there ought to be a quantitative (or at least a qualitative) value for a return on investment (ROI) for resources spent on network security. However, computing an accurate value for the ROI on security investments is somewhat problematical. The ROI computation possibilities will be discussed in Chapter Six, *Risk Assessment*.

Computer security policy is often partially based on mandates. For example, the NIST Computer Security Policy is based primarily of the following three Government mandates (NIST 1995):

1) The *Computer Security Act of 1987*, which requires agencies to identify sensitive systems, conduct computer security training, and develop computer security plans.

2) The *Federal Information Resources Management Regulation* (FIRMR), which is the primary regulation for the use, management, and acquisition of computer resources in the federal government.

3) *OMB Circular A-130* (specifically Appendix III), which requires that federal agencies establish security programs containing specified elements.

Computer security policy should formally state the organizational dictates and objectives for the information system network. A computer security policy, which in essence is a high-level requirements document, should:

1) Be compatible with the organization's overall security policy and follow from it.

2) Explicitly state the various comprehensive needs (i.e., all of them) for organizational security for its information system.

3) Explicitly state the responsibilities for each job function within the organization.

4) Explicitly state the potential responses and punishments for violations of the organization's security policy.

5) Support the mission of the organization.

6) Be periodically evaluated and modified, if required.

7) Be a living document by being responsive to changes in: a) the organization's mission and direction, b) information technology, c) threats, and d) relevant societal factors.

Although policy should be independent of technology, often it isn't so that is why technology is included in the list above (item 7.6). Security Policy can be divided into three categories: 1) physical, 2) behavioral, and 3) technical. *Physical policy* addresses the various facilities and processes required to ensure that no person (either an employee or an intruder) can have improper physical access to sensitive systems or sensitive information. *Behavioral policy* addresses the training and awareness facts that all employees (technical, system users, managers, and other personnel) should appropriately know. *Technical policies* addresses the technical aspects of the information system, namely, the computer hardware, the computer software, network systems, intrusion detection systems, firewalls, as well as water damage control systems, fire damage control systems, power failure control systems, heating and ventilation and air conditioning systems, and other such systems.

3.1.2   Mandated Security Policies

There are several mandates that are relevant to the security policies of many organizations. Some of these are the following:

OMB Circular Number A-130 (Management of Federal Information Resources, 1996) provides uniform government-wide information resources management policies as required by the *Paperwork Reduction Act of 1980*, as amended by the *Paperwork Reduction Act of 1995*, 44 U.S.C. Chapter 35. This Transmittal Memorandum contains updated guidance on the "Security of Federal Automated Information Systems," and makes minor technical revisions to the Circular to reflect the Paperwork Reduction Act of 1995 (Public Law 104-13).

Appendix III to OMB A-130 establishes a minimum set of controls to be included in Federal automated information security programs; assigns Federal agency responsibilities for the security of automated information; and links agency automated information security programs and agency management control systems established in accordance with OMB Circular No. A-130. The Appendix revises procedures formerly contained in Appendix III to OMB Circular No. A-130 (50 FR 52730; December 24, 1985), and incorporates requirements of the Computer Security Act of 1987 (Public Law 100-235) and responsibilities assigned in applicable national security directives.

The Privacy Act of 1974 (Public Law 93-579) is a law to protect personal privacy from invasions by Federal agencies such as actions by employees who have no need-to-know who work for the Internal Revenue Service or the Social Security Administration.

The Foreign Intelligence Surveillance Act of 1978 (Public Law 95-511) is used to obtain electronic surveillance and physical searches without warrant, but under court order in cases of foreign intelligence, international terrorism, or sabotage activities.

The Electronic Communications Privacy Act of 1986 (Public Law 99-508) provides for personal privacy of digitized voice, data, or video that is transmitted over any public network.

The Computer Security Act of 1987 (Public Law 100-235) states that government and industry should provide for ensuring the privacy of individuals when information about them is placed on information systems.

The Paperwork Reduction Act of 1980 and 1995 (Public Law 104-13), 44 U.S.C. Chapter 35 declares that government shall provide electronic information sources to replace those provided by paper today.

The Telecommunications Act of 1996 (Public Law 104-104) provides for a pro-competitive, de-regulatory national policy framework.

The Information Technology Management Reform Act of 1996 (Public Law 104-106) (later renamed the Clinger-Cohen Act of 1996) states that OMB should provide guidance, policy, and control for government information technology procurement. It also requires government agencies to have a CIO. It also emphasizes OMB, NIST, and agency responsibilities regarding information security. This act can be found at URL address: http://policyworks.gov/policydocs/2.pdf.

The National Defense Act of 1997 (Public Law 104-201) directs the President to submit a report to Congress that sets forth national policy on protecting the national information infrastructure against strategic attack.

The Economic Espionage Act and National Infrastructure Protection Act of 1996 (Public Law 104-294) provide for prosecuting various espionage actions and addresses protecting the confidentiality, integrity, and availability of data and systems.

The Health Insurance Portability and Accountability Act of 1996 (HIPAA) provides for health insurance coverage for workers and their families when they change or lose their jobs. The Health Care Financing Administration (HCFA) is responsible for implementing various unrelated provisions of HIPAA; therefore HIPAA may mean different things to different people. It provides for the privacy of personal health care information and gives patients greater access to their own medical records and more control over how their personal health information is used (HIPAA 2001).

The Financial Services Modernization Act (1999) or the Gramm-Leach-Bliley Act does the following: facilitates affiliation among banks, securities firms, and insurance companies, provides for functional regulation, provides for insurance, provides for unitary savings and loan holding companies, requires privacy, provides for federal home loan bank system modernization, and contains other provisions (FSMA 2001).

PDD[31] 39, "U.S. Policy on Counterterrorism," June 21, 1995, directs measures to combat terrorism.

---

[31] Presidential Decision Directive

PDD 62, "Combating Terrorism," May 1998, addresses the national problem of countering terrorism in all of its varied forms.

PDD 63, "Protecting America's Critical Infrastructures," May 22, 1998, focuses specifically on protecting the Nation's critical infrastructures from both physical and cyber attack. This directive is the culmination of an intense, interagency effort to evaluate those recommendations and produce a workable and innovative framework for critical infrastructure protection. The President's policy sets a goal of a reliable, interconnected, and secure information system infrastructure by the year 2003, and significantly increases security for government systems by the year 2000, by:

a) Immediately establishing a national center to warn of, and respond to, attacks.
b) Building the capability to protect critical infrastructures from intentional acts by 2003.

The President's policy:

a) Addresses the cyber and physical infrastructure vulnerabilities of the Federal government by requiring each department and agency to work to reduce its exposure to new threats;
b) Requires the Federal government to serve as a model to the rest of the country for how infrastructure protection is to be attained;
c) Seeks the voluntary participation of private industry to meet common goals for protecting our critical systems through public-private partnerships; and
d) Protects privacy rights and seeks to utilize market forces.

It is meant to strengthen and protect the nation's economic power, not to stifle it. The policy seeks full participation and input from the Congress.
PDD-63 sets up a new structure to deal with this important challenge:

a) A *National Coordinator* whose scope will include not only critical infrastructure but also foreign terrorism and threats of domestic mass destruction (including biological weapons) because attacks on the U.S. may not come labeled in neat jurisdictional boxes;
b) *The National Infrastructure Protection Center (NIPC)* at the FBI which will fuse representatives from FBI, DOD, U.S. Secret Service, Energy, Transportation, the Intelligence Community, and the private sector in an unprecedented attempt at information sharing among agencies in collaboration with the private sector. The NIPC will also provide the principal means of facilitating and coordinating the Federal Government's response to an incident, mitigating attacks, investigating threats, and monitoring reconstitution efforts;
c) An *Information Sharing and Analysis Center (ISAC)* is encouraged to be set up by the private sector, in cooperation with the federal government;

d) A *National Infrastructure Assurance Council* drawn from private sector leaders and state/local officials to provide guidance to the policy formulation of a National Plan;

e) The *Critical Infrastructure Assurance Office* will provide support to the National Coordinator's work with government agencies and the private sector in developing a national plan. This office will also help coordinate a national education and awareness program, and legislative and public affairs.

One approach to catching malicious persons who would perpetrate a cyberattack on America's critical infrastructure is to assign top-notch analysts with massively parallel computers to cracking codes used by persons suspected of being terrorists. In this manner, the encrypted messages sent by these suspects could be read and acted upon.

PDD-67, "Enduring Constitutional Government and Continuity of Government Operations," 21 October 1998. Among other things, PDD 67 requires Federal agencies to develop Continuity of Operations Plans for essential operations.

House Report 106-945, Public Law 106-398, "Government Information Security Reform Act (GISRA)," 2000. This law requires agencies to submit vulnerability assessments and security improvement plans by September 2001. It describes the security requirements for government managers and agencies.

"Uniform Computer Information Transactions Act (UCITA)," National Conference of Commissioners on Uniform State Laws, July 2001. This act has been adopted by Virginia and Maryland but has not been adopted in most states as yet. This act is an effort to provide uniform state laws for information products and services. It was drafted in response to the fundamental changes in the U.S. economic system due to the increased reliance on information technologies and need for clarity in the law. This law says little about information system security except in terms of the testing of implemented security products and services.

Any organization's security policy must also consider issues specific to the organization such as:

a) The organization's mission,

b) The products or services provided by the organization (i.e., the nature of the organization),

c) The attitude of the owners of the organization towards security and breaches in security,

d) The importance of internally generated information to the organization's success or failure, and

e) The size and structure of the organization.

Referring to the critical infrastructure sectors of Government information systems, Senator Joseph Lieberman (D-Connecticut) stated, "While it has never been easy to protect our critical infrastructure from conventional attacks...It is even more difficult to protect against cyberattacks (Dorobek 2001)." The task of protecting America's cybersystems from attack is becoming ever more critical and important. The attack on one front (the September 11, 2001, terrorist attacks on the World Trade Center and the Pentagon) is pushing government to make sure that all fronts are secured (Dorobek 2001).

3.1.3 Government Cyber-Protection Programs

In response to PDD 63 and other government mandates, and the various cyber threats, many groups have been formed to counter the potential threats, see Table 3.1, *Government Cyber-Protection Programs* (Barrett 2001a).

3.1.4 Organizational Security Policies

Organizational security policies are those policies that are specific to one's organization and need to be compatible with the organization's policy and procedures statements. Thus, development of a security policy is dependent upon having an overall policy statement, sometimes referred to as the Standard Operating Policies and Procedures (SOPs) document, that stipulates the policies and procedures for the organization and should follow from the organization's *mission statement*.

If the organization does not have an SOPs document, then the process for creating the security policy, as explained here, will consider that possibility.

Examples of a security policy for a fictitious organization are presented in Appendix D.1, *A Security Policy: An Example* and Appendix D.2, *Security Rules Base: An Example*. If an organization does not have a security policy and does not know how to create one, or does not wish to create one for fear of doing it incorrectly, then a security analyst can create one for you. The security analyst can create a security policy for your organization with knowledge of other policies and through interviews with top-level management of your organization. Security knowledge and knowledge of your organization are the required inputs to a process for creating an organizational security policy.

A security rules base can be derived from the organization's Security Policy and/or an understanding of the various security needs for this particular system.

3.2 Security Rules

A set of *security rules* is needed to ensure that the information system is operating properly and that the firewalls are implemented with the proper rules for passing and blocking traffic. The set of security rules should imply an appropriate set of countermeasures. An example of a set of security rules for an extreme case for security is presented in Appendix D.

Table 3.1 Government Cyber-Protection Programs

Name	Description
Critical Infrastructure Assurance Office (Dept. of Commerce)	This group is charged with assisting with the integration of federal information infrastructure protection initiatives.
Federal Computer Incident Response Center (General Services Administration)	Provides a central coordination analysis facility that deals with computer security issues affecting the civilian agencies and departments of the federal government.
National Coordinator for Security Protection and Counter-terrorism (National Security Administration)	This group is responsible for the government's critical infrastructure protection efforts, as well as foreign terror threats of mass destruction.
Computer Emergency Response Team (Department of Defense)	This group tracks and responds to computer attacks within the armed services.
Computer Emergency Response Team (U.S. Air Force)	This group tracks and responds to computer attacks within the U.S. Air Force.
Computer Emergency Response Team Coordination Center	This is a federally funded threat tracking and warning organization operated by Carnegie-Mellon University. It provides information to all of the other computer analysis groups.
National Infrastructure Protection Center (Federal Bureau of Investigation)	This group is the government's focal point for threat assessment, warning, investigation, and response for threats or attacks against critical infrastructures, such as energy, transportation, and justice.
Joint Task Force Computer Network Operations (Defense Information Systems Agency)	This is an integrated defense center design to protect DOD computer systems from attack. It also coordinates attacks against enemy systems.
National Infrastructure Assurance Council	This group is charged with improving the partnership of the public and private sectors to address threats to the nation's critical infrastructure.
Information Sharing and Analysis Centers (ISAC)	These are Presidentially mandated hubs for the sharing of threat information and solutions by industry sector. Four ISACs are operational, financial services, energy, telecommunications, and transportation.
Computer Hijacking and Intellectual Property Units (Department of Justice)	These are new teams dedicated to prosecuting computer crimes.

The uses of the rules base include the following three objectives:

1) Inputs to the creation of the Security Requirements,
2) Inputs to the identification of the various countermeasures for this system, and
3) Inputs to the creation of a Security Architecture.

The security rules base also can be used in the creation of the rules embedded in firewall and router software for implementing the organization's access policy relevant to external or remote users. However, there are some differences since the passing or blocking of packets is a somewhat different problem than stating organizational policy with regard to the various security issues. However, these devices should reflect the organization's security policy.

3.3 Legal Issues

Legal issues can be violations of either criminal or civil laws. *Criminal Laws* require the perpetrator to go to jail or be fined if they are violated. Criminal laws may be federal or state or local. Federal laws include:

1) The *Computer Security Act of 1987* (privacy act, PL 100-235);
2) The *Presidential Decision Directive 63* concerning information system security concerns;
3) The *Paper Reduction Act* concerning the actions needed to reduce the use of paper forms for federal activities; and
4) The *OMB Circular -130 and its Appendix III*, to ensure the proper protection of information systems.

Civil Laws require that the perpetrator pay a plaintiff some amount to offset losses incurred due to actions by the plaintiff (i.e., the perpetrator). Laws may be federal or state. Civil law includes tort, contract, or intellectual property. An example of a tort law break is to commit denial-of-service on another person's network. An example of a contract law break is the breaking of a promise to do something. Intellectual law covers patent and trademark rights (Bernstein 1996).

Legal issues include at least the following actions:

1) Defamation of another person's or organization's reputation that can be interpreted as Libel or Slander,
2) Sexual Harassment,
3) Privacy, and
4) U.S. Export laws on export of cryptography methods.

Other issues that can be considered legal issues or at least issues that can get an organization or its employees into problems, including getting fired are:

1) Viewing pornographic literature on the Internet,
2) Sending spam e-mail,
3) Sending e-mail that is intended to make the receiver angry (called flaming),
4) Viewing or acquiring private information that is of a personal nature,
5) Not observing common courtesy when on the Internet, or
6) Failing to provide sufficient security that causes a loss to someone else.

Legal issues also appear when attempting to create nondisclosure agreements (NDAs) and service level agreements (SLAs). Writing these agreements so that they are acceptable to both the vendor and the client is a difficult task. However, it is important to the client to ensure that these agreements contain statements that include proper protection for the client so that should some unfortunate incident occur, the client will have legal grounds on which to sue the vendor. Similarly, the vendor will wish to exclude such statements from the contracts and will provide rebuttal arguments against contracts making these statements.

Companies have information that is very sensitive, about network intrusions, computer viruses, and other problems that (if shared) could help others harden their systems and improve their information security preparedness. However, without adequate protections from inappropriate disclosure of information shared with the federal government, commercial organizations are concerned that such information sharing could lead to potentially damaging public release of confidential data (ITAA 2001).

3.4   Security Requirements

The *security requirements* for the upgraded system should follow from the security policy and security rules base for the target security system. However, this usually turns out to be an iterative process in that the analyst will likely return to the policy and the requirements statements after having completed the requirements analysis. Then after restating the requirements, the analyst may wish to return to amend the policy or rules base, and so forth, until either the time for these activities has been expended or the analyst is satisfied with the three statements (policy, rules, and requirements). This iterative process also will likely include the creation of the security architecture (at some point in time). A useful tool for constructing and storing the security requirements is the Dynamic Object-Oriented Requirements System (DOORS) by Telelogic (Telelogic 2001).

The security requirements should be formulated at three levels,

> 1) *User security* requirements (least detail),
> 2) *Architecture security* requirements (more detail), and
> 3) *Software and hardware security* requirements (most detail).

The *user security requirements* are usually formulated based on the organization's security policy and the expressed desires of the organizational owners and users. These high-level security requirements form the basis of the more detailed security requirements that will follow later. From this initial set of user security requirements, the analyst can create a concept design for the security architecture. But to create the security architecture, the security analyst will likely require a more detailed set of requirements, called the *architectural security requirements*.

To define the security architecture, the architectural requirements must be further detailed to achieve the software and hardware security requirements, as described in the lifecycle process presented in Section 4.2, *The Security Analysis Process*.

The *software and hardware security requirements*, sometimes referred to as the software and hardware *security specifications*, should be detailed enough to ensure that the software and hardware subsystems are well defined. That is, the specifications should be sufficiently detailed that the needed products can be defined and obtained either by purchasing appropriate hardware or COTS shrink-wrapped packages (e.g., secure operating systems and middleware) and/or by building the software subsystems using the stated specifications. This usually means that *technicians, engineers, programmers, and programmer/analysts* must be involved in the development of the security specifications, as derived from the security policy and rules base.

Similarly, specific security hardware subsystems (e.g., routers and firewalls) can be defined from the hardware specifications and hence *hardware engineers* are usually included in the writing of these specifications.

Ideally, all lower level security requirements should be traceable to higher level requirements, see Figure 3.1, *An Example of a Requirements Lattice* (this lattice is also a tree), that is:

1) The *User Security Requirements* should follow from the Security Policy and Procedures statements but they are also the statements from the users (and owners) of needed security capabilities,

2) The *Architecture Security Requirements* (i.e., medium level security requirements), should follow from the User Security Requirements and

3) The *Software and Hardware Security Requirements* (i.e., detailed level security requirements or specifications) should follow from the Architecture Security Requirements.

Thus any specific lower level requirement (viz., a "child" statement) should be such that it can be traced back up thorough the requirements base to a "parent" statement in the security policy. In object-oriented terms, these elements should produce a class. In fact, it is a good idea for the database format used for storing and retrieving requirements to be an object-oriented database.

By traceable, I mean that if one chooses any element in the requirements lattice[32], say the black one in Figure 3.1, *An Example of a Requirements Lattice*, then there must be a direct link back to a previous element, and this linkage should continue all the way back to some element in the security policy.

---

[32] A *lattice* is a mathematical term meaning a set of elements that have a "child-parent" relationship. Except for the root element (or elements), each element is a child and connects with a parent, and each parent connects with at least one child. A *tree*, which is a special form of lattice, has only one parent for any child.

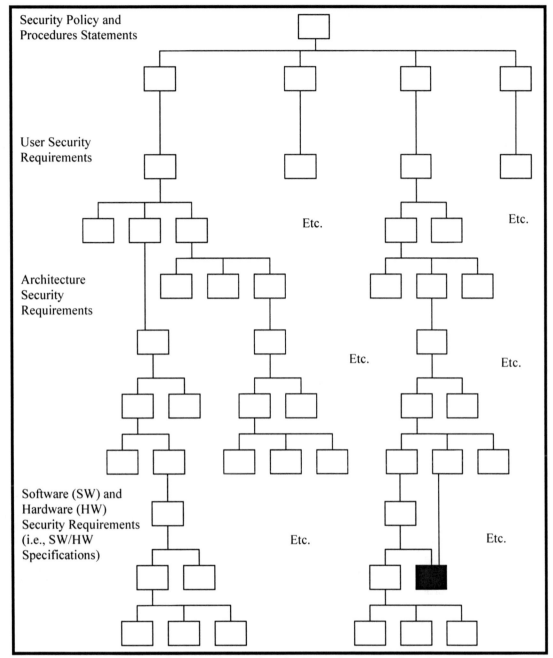

Figure 3.1 An Example of a Requirements Lattice

If there is no linkage back to a previous element, then one of two things is incorrect,

1) This selected element is not really a requirement (because it does not follow from any previously-expressed need), or

2) There is a missing element in the lattice above the selected element that is a higher order requirement (because the lower level element is part of a decomposition of the higher order element).

It is not possible, nor is it feasible, and it certainly isn't desirable, to attempt to create an information system security architecture without first having a set of security requirements, although some owners may wish for this to be done and are even willing to pay someone to do this and there are many who will do such a thing. The user security requirements can be decomposed into a more detailed set of requirements that is suitable for a definition of the security architecture and these are called the *architectural security requirements*. This means that the routers, servers, firewalls, intrusion detection systems, and other hardware security products can be identified (what they are and where they are) using the architectural security requirements, but the specific specifications for these is dependent upon having a formal risk assessment.

A requirements analysis is one of the most important steps in hardware and software development. Some organizational characteristics that should be considered in the development of security requirements are presented in Table 3.2, *Organizational Traits* (Smith 1998).

Table 3.2 Organizational Traits

Organizational Traits	Potential Effects
Type of culture	If the culture is oriented toward consensus decision-making, then a group decision process is appropriate.
Capability Maturity Model Level (CMM) level	An "ad hoc," or level 1 CMM, implies that a sophisticated computer support system may not be appropriate.
Type of mission	User security requirements should relate to the organization's mission.
Users are non-technical and time-constrained	Explanation of a sophisticated tool can be a waste of time.
Inappropriate users may defer to the appropriate users	Relevant users are knowledgeable and other users may not wish to provide the needed information since they are not experts.
Number of users is too large	If too many users are involved, the meeting progress may be hampered.
No existing requirements set	User themselves are the repository of organizational functional knowledge and must be intimately involved in the development of the user security requirements.

There are two extremes in the continuum of assumptions about security requirements for information systems,

1) *Perfection*, which states that security requirements must be perfect and the analyst must develop and iterate a set of security requirements until they are assessed as being perfect.

2) *Inexactness*, which states that the analyst must accept that the security requirements will never be perfect and hence it really doesn't matter how poor they are because they will have to be changed anyway.

Both of these assumptions lead to an untenable situation (Smith 1998). The first, perfection, because one will never achieve the condition of perfectly stated security requirements and the second, inexactness, because if the security requirements are too poorly written, then the later lifecycle activities will be muddled and inaccurate.

A useful process for identifying security requirements is the following process:

1) Research through relevant documents (e.g., Security Policy, Security Rules Base, and a concept of operations, if available),
2) Conduct interviews with management and users, and
3) Perform analyses of collected material.

Constraints on the security requirement should be considered, namely, government agency mandates, available funds, and definitions of data standards. It is important to spend some extra time ensuring that the security requirements are top quality since errors or omissions will propagate through the entire lifecycle process and it will be much more difficult and costly to correct any mistakes later.

By a *formal risk assessment*, I mean a well-defined set of procedures to follow in the performance of an analytical assessment process and that this process is repeatable within the reasonable bounds of subjectivity required for some parts of the process.

3.5   Risk Assessment Considerations

One of the principal purposes of a formal risk assessment is to improve the organization's knowledge of the security requirements so that they better represent the true needs of owners and users of the system.

A formal risk assessment is dependent on having the following well-written documents:

1) Security policy,
2) Security rules base, and
3) Security requirements.

If any or all of these do not exist then the analyst must generate something similar in order to be able to create a formal risk assessment.

3.6   Security Requirements Considerations

After the risk assessment has been completed, the analyst should revisit the Security Requirements to see if any knowledge gained from the risk assessment can be used to improve the requirements statement. Additionally, the Security Policy and Security Rules Base should be revisited in case these statements also need to be improved.

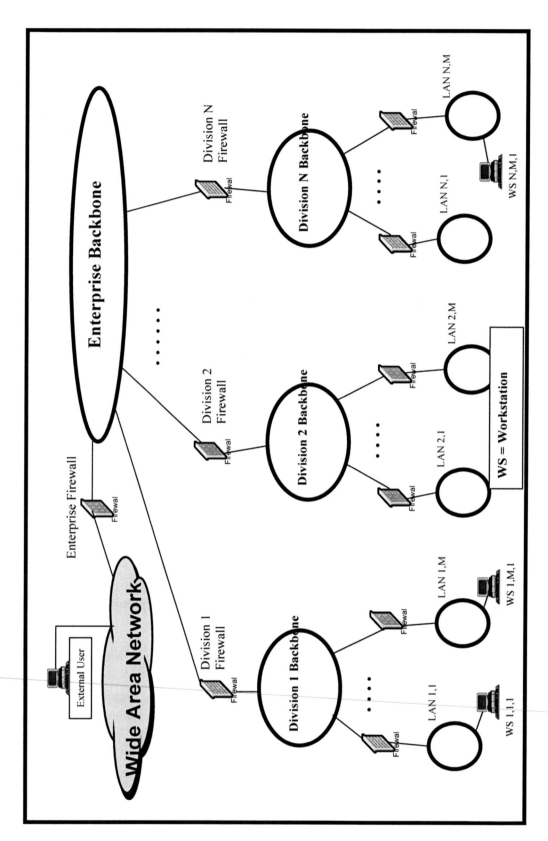

Figure 3.2 A Security Architecture for a Three-Tiered Network

92

In some cases, the analyst may wish to create a prototype system to test the selected countermeasures. If this is done, then the lessons learned from the prototype can be used to amend the security requirements.

3.7 Security Architecture Considerations

When a risk assessment has been completed and the security requirements have been adjusted, as required, to accommodate the findings in the risk assessment, the analyst is then prepared to create a security architecture. A generic security architecture for a three-tiered network is shown in Figure 3.2, *A Security Architecture for a Three-Tiered Network.*

According to the statements in the Security Requirements document, the analyst should create the Security Architecture to reflect the needs of the owners and users. If the organization has produced a Concept of Operations (CONOPS) document, this also should be used in the generation of the Security Architecture.

3.8 Security Infrastructure

Once a security architecture has been developed, the security analyst can recommend a set of countermeasures (physical, behavioral, and technical) that will provide the desired protection. The physical and behavioral countermeasures should be clearly defined and their implementation should be described.

The security infrastructure consists of the various cost-effective security mechanisms required to provide the technical security for the information system. The technical countermeasures and their precise location within the reference system produce the security infrastructure and can be shown as in Figure 3.2.

Of course, the final decision for funding and implementation of the recommended security countermeasures lies with the information system owners.

3.9 A Summary of OMB Circular A-130 Security Policy

The Office of Management and Budget (OMB) issued a revised comprehensive policy on computer security that provided a structure that was useful for both the commercial and government sectors. The policy was contained in the revised OMB Circular A-130, Appendix III, *Security of Federal Automated Information*, which is mandatory for executive branch agencies, but many other commercial and government sector organizations found its accepted business practices useful in developing information security practices in current and emerging information technology environments.

The parenthetical expressions noted below are mine. According to the OMB, IT policy can be divided into the following ten categories:

1. **Individual Responsibility** - The main thrust of security is that the responsibilities of security ultimately depend upon the users and managers of computer systems and information. Since computers and electronic access are available to almost everyone, this approach is necessary to address security in current information system environments. Previous computer security policies and programs have focused on securing data processing centers and large custom applications.

2. **General Support Systems and Major Applications** - Users and managers need a framework that can embrace many technological possibilities, so a suggested structure encompasses two categories: 1) general support systems and 2) major applications. *General support systems* include local area networks, wide area networks, personal computers, workstations, servers, networks and all manner of information technology including data processing centers. General support systems are normally a collection of computers, networks, and other IT components. They can run a huge variety of commercial off-the-shelf applications such as word processing, email, productivity tools, databases, and custom applications, although any one general support system may run only a few applications or many. These routine applications are part of the general support system. The lines that separate general support systems from each other are often managerial rather than physical or electronic.

A *major application* is a critical business or mission resource. Although major applications are, like routine applications, resident on general support systems, they need to be given special management attention because of the organization's reliance on them. In government organizations, a typical major application is providing citizen benefits. Agencies are most able to identify their major applications.

OMB Circular A-130 does not distinguish between sensitive and non-sensitive systems. Rather, consistent with the Computer Security Act of 1987, the Circular recognizes that systems are procured and operated to serve particular agency needs of varying sensitivity and criticality. All general support systems contain some sensitive information and, therefore, require protection, including a security plan. This should help prevent arguments about what is sensitive and allow that energy to be spent securing systems. (This statement essentially says that all commercial and government information should be considered sensitive).

3. **Responsibility, Plans, Review, and Authorization** - In most organizations, it is appropriate to delegate decisions about computer security to line managers and to retain agency-level control of major applications. The methodology for managing computer security is based on four inter-related management controls:

    1) assigning responsibility for security,
    2) security planning,
    3) periodic review of security controls, and
    4) management authorization.

A goal of this process is to create management accountability for security decisions and implementation. To be accountable for a decision, a manager needs authority. Although management controls are required for both general support systems and major applications, significant differences determine how they are implemented. These are described below.

3.a Assigning Responsibility - Circular A-130 requires that a single individual should be assigned operational responsibility for security. This individual must be knowledgeable about the information resources used and how to secure them. For major applications, the assigned individual must be able to give special management attention to the security of the application. By assigning a knowledgeable security officer, management should receive better security information thereby causing them to desire someone who is knowledgeable and skillful.

3.b Security Planning - Good security planning is essential, but it must be more than simply the generation and review of paper. Circular A-130 prescribes a series of specific planning activities rather than a theoretical framework. The activities include the development of rules, security training, and the implementation of other operational, management, and technical controls. Plans for major applications should be reviewed by the manager of the primary support system that the application uses.

3.c Review of Controls - The security of a system or application degrades over time, as technology and threats evolve and as staffing and procedures change. Organizations should use security reviews to assure that management, operational, and technical controls are appropriate and functioning effectively. These review requirements are much broader than the certification review required under previous policies. The security plan should be the basis for the review thus enhancing the usefulness of the security planning process. For major applications, reviews must include an independent review or audit. Independent audits can be internal or external but should be performed by someone free from personal and external constraints that could negatively affect their independence and they should be independent of any other department in the organization.

3.d Authorization - The authorization of a system to process information, granted by a management official, provides an important quality control. By authorizing processing of a system or application, a manager accepts the associated risk. The authorization, which some agencies refer to as an accreditation, should be based on the review of controls. The authorization of major applications will generally occur at a very high managerial level, either by a political appointee or a senior career employee.

Under the revised policy, agencies are required to develop security rules. They are the decisions made about security-related options and require tradeoffs, since all desired security objectives will probably not be achievable. The system-specific policy, stated as operational rules, will have technical and operational implications. The requirement for rules is designed to force people to address and document security-related decisions. Some of the types of issues for which rules are needed are:

1) Are employees allowed to put work data on their home PCs?
2) How often do passwords change?
3) Who is allowed to have accounts on what computers?

Charles L. Smith, Sr.

Rules should be developed using a risk-based approach. However, a formal risk assessment is not required. (This is not a good suggestion for reasons stated earlier.) Organizations require the flexibility to select decision-making processes that fit their environments.

The practical effect of this policy is that agencies no longer need periodic risk assessments of their computer systems. Agencies may still choose to perform a traditional risk assessment, which remains a valuable tool. Risk assessments are most effective in areas where risks and safeguards can be quantified or otherwise discretely measured or described.

Many security personnel point out other benefits of traditional risk analysis, especially the visibility to upper management through system review and authorization. Circular A-130 attempts to keep these important benefits as a part of the authorization of systems.

4. **Personnel Controls** - Since the greatest threat to most computer systems comes from authorized users, agencies should institute personnel controls such as the following: 1) least privilege, 2) separation of duties, and 3) entity accountability (i.e., auditing and intrusion detection systems to ensure accountability of all entities using the system).

   This is a much broader view of personnel security than in the old version, which only addressed personnel screening. Screening[33] is required for personnel, such as system or security administrators, emergency personnel, and others who can bypass technical and operational controls and therefore may not always be subject to other security controls such as least privilege, separation of duties, or entity accountability. This is a much smaller set of people and should result in significant cost savings to agencies.

5. **Incident Handling** - Organizations need an incident handling capability, which is the ability to detect and react quickly and efficiently to disruptions in normal processing caused by malicious threats. Since information technology is so complex and widely distributed and users are often unfamiliar with the technology, an incident handling capability is imperative to provide security support. Many organizations do not currently have a capability to handle, or even to recognize, computer security incidents. The development of an incident handling capability does not have to involve a separate staff; it could be a service of a Help Desk, with appropriate training. Agencies are directed to share information about common vulnerabilities so that the Federal government can improve its overall ability to respond to security threats.

6. **Training** - Like planning, training is an area of computer security that receives more praise than action. Users should be trained about the specific general support systems or applications they use, based on the system rules, specifically including how to handle incidents. This requirement cannot be met solely with organization-wide

---

[33] *Screening* is the investigation of the background of potential employees.

training programs that address basic computer security. The training should use a media appropriate for the audience and the risks (such as those due to social engineering attacks). Training need not be formal classroom instruction; it could use interactive computer sessions or well-written and understandable brochures or even computer-based instructions available to anyone at any time. Specialized training of users is required for major applications.

7. **Network Interconnectivity** - Very few general support systems will exist as closed systems. Nearly all are networked to other organization systems and to external public and private networks (such as the Internet). The gateways, where networks meet, serve an important security role. System rules in the external network may be very different or enforced differently. These interconnections should be explicitly approved by organization managers.

    One important type of interconnection gateway is a firewall or secure gateway. Firewalls block or filter access between two networks, often between a private network and a public network.

8. **Contingency Planning** - Contingency planning is a critical element in an information system security program. Not only should contingency plans be developed, but they should be tested periodically to ensure that relevant personnel can respond properly to catastrophic events in a timely manner. Federal agencies and private sector organizations have been expanding the scope of their contingency plans. The emphasis is on ensuring that all the resources needed for mission and business critical functions will be available for timely replacement if required. This includes people, communications, support equipment, services, and many other resources in addition to computing power.

9. **Public Access** - Federal agencies are encouraged to provide public access to information. Organizations should reduce their risks by separating public access systems or records from agency internal systems. (This can be done using a firewall with a demilitarized zone, which is explained elsewhere in this book.)

10. **Assistance** - This Circular provides for assistance to agencies in implementing the revised policy. *NIST* is tasked with helping agencies with security planning, interconnectivity, incident handling, training, and information sharing, as well as providing general assistance. The *Department of Justice* is tasked with helping agencies with legal issues surrounding incidents. The *General Services Administration* is tasked with providing guidance on including security in the acquisition process and providing or making available security services.

3.10   Common Threats

This section is a summarization of the statements concerning security threats that appeared in the chapter on threats in NIST's Handbook on Security (NIST 1995). Some of the statements are outdated but much is still relevant. Only the relevant parts are retained.

Computer systems are vulnerable to many threats that can inflict various types of damage resulting in significant losses. This damage can range from errors harming database integrity to fires damaging equipment in computer centers to catastrophic events causing the loss of the entire system. Losses can originate from the actions of trusted employees defrauding a system, from outside hackers, from terrorist attackers, or from careless data entry clerks. Precision in estimating computer security-related losses is not possible because many losses are never discovered, and others are ignored to avoid unfavorable publicity. The effects of all threats vary considerably from the diminution of confidentiality or data integrity, to the lack of availability of the system or its information, to the total destruction of a subsystem or the entire system.

This list is probably not exhaustive, and some threats may combine elements from more than one area. This overview should be useful to commercial and government organizations because its perspective is very broad. Threats against particular systems might be different from those discussed here.

To control the risks of operating an information system, managers and users need to know the vulnerabilities of the system, the values of the systems assets, and the probability of occurrence of the threats that may exploit them. Knowledge of the threat[34] environment allows the system manager to implement the most cost-effective security measures. In some cases, managers may find it more cost-effective to simply tolerate the expected losses (when the cost of a countermeasure is greater than the risk). Such decisions should be based on the results of a formal risk analysis.

3.10.1 Errors and Omissions

Errors and omissions are an important threat to data and system integrity. Many programs, especially those designed by users of personal computers, usually lack quality control measures. However, even the most sophisticated development programmers cannot detect all types of input errors or omissions. A sound awareness and training program can help an organization reduce the number and severity of errors and omissions.

Users, data entry clerks, system operators, and programmers frequently make errors that contribute directly or indirectly to security problems. In some cases, the error is the threat, such as a data entry error or a programming error that crashes a system. In other cases, the errors create vulnerabilities. Errors can occur during all phases of the systems lifecycle. A long-term survey of computer-related economic losses conducted by Robert Courtney, a computer security consultant and former member of the Computer System Security and Privacy Advisory Board, found that 65 percent of losses to organizations were the result of errors and omissions. This figure was relatively consistent between both commercial and government sector organizations.

Programming and development bugs can range in severity from benign to catastrophic. In a 1989 study for the House Committee on Science, Space and Technology, entitled *Bugs in the Program,* the staff of the Subcommittee on Investigations and Oversight summarized the scope and severity of this problem in terms of government systems as follows (NIST 1995):

---

[34] If a vulnerability is such that no threat exists to take advantage of it, then nothing is gained by removing or minimizing the vulnerability, see Chapter 6, *Risk Assessment*.

"As expenditures grow, so do concerns about the reliability, cost and accuracy of ever-larger and more complex software systems. These concerns are heightened as computers perform more critical tasks, where mistakes can cause financial turmoil, accidents, or in extreme cases, death."

Since the study's publication, the software industry has changed considerably, with measurable improvements in software quality. Yet software "horror stories" still abound, and the basic principles and problems analyzed in the report remain the same. While there have been great improvements in program quality, as reflected in decreasing errors per 1,000 lines of code, the concurrent growth in program size often seriously diminishes the beneficial effects of these program quality enhancements.

Installation and maintenance errors are another source of security problems. For example, an audit by the President's Council for Integrity and Efficiency (PCIE) in 1988 found that 10 percent of mainframe computer sites studied had installation and maintenance errors that introduced significant security vulnerabilities.

3.10.2   Fraud and Theft

Computer systems can be exploited for both fraud and theft both by "automating" traditional methods of fraud and by using new methods. For example, individuals may use a computer to skim small amounts of money from a large number of financial accounts, assuming that small discrepancies may not be investigated. Financial systems are not the only ones at risk. Systems that control access to any resource are targets (e.g., time and attendance systems, inventory systems, school grading systems, and long-distance telephone systems).

Computer fraud and theft can be committed by insiders or outsiders. Insiders (i.e., authorized users of a system) are responsible for the majority of fraud. A 1993 *InformationWeek*/Ernst and Young study found that 90 percent of Chief Information Officers viewed employees who have no-need-to-know-information as threats. The U.S. Department of Justice's Computer Crime Unit contends that "insiders constitute the greatest threat to computer systems." Since insiders have both access to, and familiarity with, the organization's information system (including what resources it controls and its vulnerabilities), authorized system users are in a better position to commit crimes.

Insiders can be both general users (such as clerks) or technical staff members. An organization's former employees, with their knowledge of an organization's operations, may also pose a threat, particularly if their continued access is not terminated promptly upon their departure.

3.10.3   Employee Malicious Actions

Malicious activities perpetrated by employees (i.e., insiders or intranet users) include at least the seven following examples:

1) Destroying hardware or facilities,
2) Embedding logic bombs in applications or data,
3) Entering data incorrectly (intentionally or unintentionally),
4) Crashing systems (intentionally or unintentionally),

5) Deleting data (intentionally or unintentionally),
6) Holding data hostage, and
7) Altering data (intentionally or unintentionally).

Employees are familiar with their organization's computers and applications, including knowing what actions might cause the most destruction. The downsizing of organizations in both commercial and government sectors has created a group of individuals with organizational knowledge, who may retain potential system access (e.g., if system accounts are not deleted in a timely manner). The irrational scare created over the Y2K issue (namely, that computers would malfunction because of the IBM-compatible computer software that limited the year to only two characters and commonly referred to as the Year 2000 (i.e., Y2K) problem). This issue allowed many foreign and domestic programmers (many of whom were aliens with no security clearances or background investigations) with little computer knowledge to have access to the critical software of many Governmental and commercial organizations where they could plant logic-bombs, back doors, and learn about potential vulnerabilities that could be exploited later.

The number of incidents of employee sabotage is believed to be much smaller than the instances of theft, but the cost of such incidents can be quite high. The motivation for software malicious activities can range from altruism to revenge. "As long as people feel cheated, bored, harassed, endangered, or betrayed at work, sabotage will be used as a direct method of achieving job satisfaction—the kind that never has to get the bosses' approval (NIST 1995)."

3.10.4   Loss of Physical and Infrastructure Support

The loss of supporting infrastructure includes power failures (outages, spikes, and brownouts), loss of communications, water outages and leaks, sewer problems, lack of transportation services, fire, flood, civil unrest, and strikes. These losses include such dramatic events as the 1993 explosion at the World Trade Center and the 1992 Chicago freight tunnel flood, as well as more common events, such as broken water pipes. A loss of infrastructure often results in system downtime, sometimes in unexpected ways. For example, employees may not be able to get to work during a winter storm, although the computer system may be functional.

3.10.5   Malicious Hackers

Malicious hackers can include both outsiders and insiders. Much of the rise of hacker activity is sometimes attributed to increases in Internet connectivity. One study of a particular Internet site found that hackers attempted to break in once at least every other day.

The hacker threat should be considered in terms of past and potential future damage. One example of malicious hacker activity is that directed against the public telephone system. Studies by the National Research Council and the National Security Telecommunications Advisory Committee show that hacker activity is not limited to toll fraud. It also includes the ability to break into telecommunications systems (such as switches), resulting in the degradation or disruption of system availability. While unable to reach a conclusion about the degree of threat or risk, these studies underscore the ability of hackers to cause serious damage.

The hacker threat often receives more attention than more common and dangerous threats. The U.S. Department of Justice's Computer Crime Unit suggests three reasons for this:

1) The hacker threat is a more recently encountered threat.
2) Organizations do not know the purposes of a hacker—some hackers only browse (i.e., innocuous hackers), some steal, some do damage.
3) Malicious hacker attacks make people feel vulnerable, particularly because the hacker's identity is unknown.

3.10.6   Industrial Espionage

Industrial espionage is the act of gathering proprietary data from Government or commercial organizations for the purpose of aiding some commercial organization. Industrial espionage can be perpetrated either by companies seeking to improve their competitive advantage or by foreign Governments seeking to aid their domestic industries. Foreign industrial espionage carried out by a Government is often referred to as economic espionage. Since information is processed and stored on computer systems, computer security can help protect against such threats. However, it can reduce the threat of authorized employees selling that information by limiting the capability of users to copy information (onto disks or into information systems on the Internet) or carry this information (as stored on disks) from the company's premises.

Industrial espionage is on the rise. A 1992 study sponsored by the American Society for Industrial Security (ASIS) found that proprietary business information theft had increased 260 percent since 1985. The data indicated 30 percent of the reported losses in 1991 and 1992 had foreign involvement. The study also found that 58 percent of thefts were perpetrated by current or former employees. The three most damaging types of stolen information were 1) pricing information, 2) manufacturing process information, and 3) product development or specification information. Other types of information stolen included customer lists, basic research, sales data, personnel data, compensation data, cost data, proposals, and strategic plans.

Within the area of economic espionage, the Central Intelligence Agency has stated that the main objective is obtaining information related to technology, but that information on U.S. government policy deliberations concerning foreign affairs and information on commodities, interest rates, and other economic factors is also a target. The FBI concurs that technology-related information is the main target, but also lists corporate proprietary information, such as negotiating positions and other contracting data, as a target.

3.10.7   Malicious Code

Malicious code refers to viruses, worms, Trojan horses, logic bombs, and other code intended to do malicious damage or for malicious purposes.

Actual costs attributed to the presence of malicious code have resulted primarily from system outages and staff time involved in repairing the systems. Nonetheless, these costs can be significant.

3.10.8   Foreign Government Espionage

In some instances, threats posed by foreign government intelligence services may be present. In addition to possible economic espionage, foreign intelligence services may target sensitive systems to further their intelligence missions. Some sensitive information that may be of interest includes travel plans of senior officials, civil defense and emergency preparedness, manufacturing technologies, satellite data, personnel and payroll data, and law enforcement, investigative, and security files. Guidance should be sought from the pertinent security office regarding such threats.

3.10.9   Threats to Personal Privacy

The accumulation of vast amounts of electronic information about individuals by governments, credit bureaus, and private companies, combined with the ability of computers to monitor, process, and aggregate large amounts of information about individuals have created a threat to individual privacy. The possibility that all of this information and technology may be able to be linked together has arisen as a specter of the modern information age. To guard against such intrusion, Congress has enacted legislation, which defines the boundaries of the legitimate uses of personal information collected by the commercial and government organizations.

The threat to personal privacy arises from many sources. In several cases federal and state employees have sold personal information to private investigators or other "information brokers." One such case was uncovered in 1992 when the Justice Department announced the arrest of over 24 individuals engaged in buying and selling information from Social Security Administration (SSA) computer files. During the investigation, auditors learned that SSA employees had unrestricted access to over 130 million employment records. Another investigation found that 5 percent of the employees in one region of the IRS had browsed through tax records of friends, relatives, and celebrities. Some of the employees used the information to create fraudulent tax refunds, but many were acting simply out of curiosity (even this is illegal).

More and more, Americans are becoming increasingly concerned about their personal data. While the magnitude and cost to society of the personal privacy threat are difficult to gauge, it is apparent that information technology is becoming powerful enough to warrant public fears.

REFERENCES AND BIBLIOGRAPHY

3.1   NIST (1995). "An Introduction to Computer Security: The NIST Handbook," NIST Computer Policy, Special Publication 800-12, October 1995.

3.2   NSA (1998). "Network Security Framework," Security Solutions Framework, Network Security Group, National Security Agency, December 1998.

3.3   PDD 62 (1998). "Combating Terrorism," Presidential Decision Directive, May 1998.

3.4   PDD 63 (1998). "Protecting America's Critical Infrastructures," Presidential Decision Directive, May 22, 1998.

3.5   PDD 67 (1998). "Enduring Constitutional Government and Continuity of Government Operations," Presidential Decision Directive, 21 October 1998.

3.6   Public Law 100-235, "Computer Security Act of 1987," January 8, 1998.

3.7   Bernstein, Terry, et al. (1996). "Internet Security for Business," Wiley Computer Publishing, 1996.

3.8   Russell, Deborah and G. T. Gangemi, Sr. (1991). "Computer Security Basics," O'Reilly, 1991.

3.9   Barrett, Randy (2001a). "Trust Me!" Interactive Week, August 20, 2001.

3.10 Smith, Sr., Charles L. (1998). "Computer-Supported Decision Making: Meeting the Demands of Modern Organizations," Ablex Publishing Corporation, Greenwich, Connecticut and London, England, 1998.

3.11 OMB (1998). "Information Security Policies For Changing Information Technology Environments," IT Bulletin, OMB, 1998.

3.12 Dorobek, Christopher (2001). "The Ripple Effect," Federal Computer Week, p. 36, September 24, 2001.

3.13 Telelogic (2001). "Telelogic DOORS," http://www.telelogic.com/products/doors/, 2001.

3.14 Andress, Mandy (2001). "Security Policy In a Box," InfoWorld, page 54, October 22, 2001.

3.15 UCITA (2001). "Uniform Computer Information Transactions Act," National Conference of Commissioners on Uniform State Laws, http://www.law.upenn.edu/bll/ulc/ucita/ucita01.pdf, July 1999.

3.16 ITAA (2001). "ITAA Asks for Removal of Legal Obstacles to Information Sharing to Protect Against Cyber Threats," http://www.itaa.org/infosec/, ITAA, 2001.

3.17 HIPAA (2001). "The Health Insurance Portability and Accountability Act of 1996 (HIPAA) Page," http://www.hcfa.gov/hipaa/hipaahm.htm, 2001.

3.18 FSMA (2001). "The Financial Services Modernization Act of 1999," http://banking.senate.gov/conf/, 2001.

CHAPTER FOUR

A COMPREHENSIVE SECURITY ANALYSIS PROCESS

4.1   Introduction

A complete process for performing a security analysis is shown in Figure 4.1, *A Comprehensive Security Analysis Process*. Each of the steps in the process is depicted and will be briefly presented and discussed in this chapter. Since many of these steps are part of a risk assessment, the steps will be briefer in this chapter since they will be explained in more detail in Chapter Six, *Risk Assessment*. The process begins with the formulation of a security policy for the organization, continues through the risk assessment process, and ends with the development of a migration plan and the performance of a security test and evaluation of the implemented countermeasures.

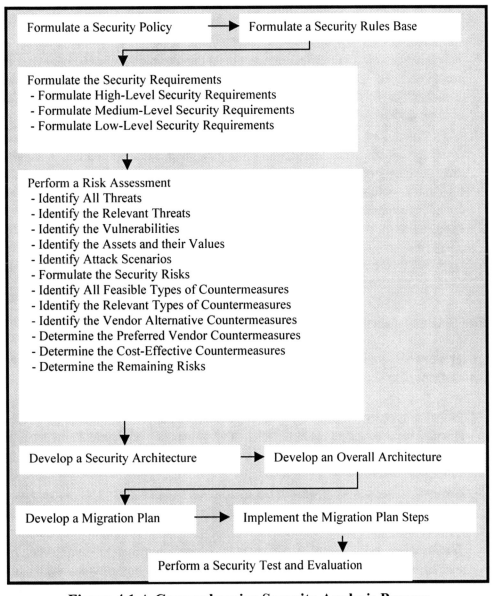

Figure 4.1 A Comprehensive Security Analysis Process

Some basic rules for performing a security analysis are:

1) *Focus on the client/analyst relationship:* Identify and understand the client, such as understanding the incentives, culture, important past events, and goals of both the employees and the organization they represent. Your success as an understanding analyst has much more bearing on the success or failure of your engagement than the technical disciplines involved.

2) *Define your role:* Determine what you are expected to accomplish, what tasks you are making a commitment to perform, what tasks you expect the client to perform, and where the boundaries of the analyst/client relationship are.

3) *Visualize the process:* Create a mental picture of the desired results of the security analysis. Visualizing a successful result creates a subconscious motivation for you and verbalizing a common goal that both you and the client can agree upon is paramount.

4) *You analyze; they make decisions:* The security analyst should remove emotional attachment to their own advice. Remember that the client understands the complexities of their environment and that the clients must live with the results of their decisions.

5) *Make beneficial results the primary goal:* Analysis is more than advising; it is assisting clients who wish to reach a goal. Most clients want the analyst to not only recommend solutions but also to help implement them. A good definition for results-oriented security analysis is to achieve the possible at a reasonable cost. By considering implementation issues throughout the lifecycle process, such as corporate culture, readiness to change, training requirements, and corporate communications channels, the analyst can keep an eye on the realm of possibility, by avoiding sidetracks into the theoretical, and preparing the client for the real-world issues of implementation and system operation.

4.2 The Security Analysis Process

Computer network security is divided into three categories: physical, behavioral, and technical situations. These aspects of security are explained below:

1) *Physical security* includes physical access limitations to the secure area, such as guards, fences, gates, cyber locks, and protected equipment, as well as water damage control, power outage control, earthquake damage control, and fire damage control,

2) *Behavioral security* includes training and awareness capabilities for ensuring that all relevant personnel are knowledgeable of the correct security procedures to follow, and

3) *Technical security* includes all aspects of the computer systems (viz., hardware and software), as well as network or communications devices and backup and recovery systems.

Each of the steps in a complete security analysis process is presented in the figure above and is explained below.

4.2.1 Formulate a Security Policy

The beginning of a security analysis is the organization's security policy. If the organization has a documented security policy, then this is the starting point and this step can be omitted. However, if the organization does not have one, then some persons in management should create one. This can be accomplished using a set of questions from the security analyst directed to high-level managers and/or their named representatives.

The sample security policy in Appendix D1, *Security Policy: An Example*, can be used to stimulate the process of *eliciting*[35] the security policy and procedures from persons representing organizational management. The elicitation process can be performed by:

1) An automated system using a computer with an embedded question and answer process that provides questions to the manager(s) (via the monitor) who then inputs the answers to these questions (via the keyboard), which are then stored in a database, or

2) A manual system using a person (e.g., the security analyst) who asks questions of the manager(s) who then responds with the answers, which are then recorded.

The process should continue until satisfactory answers to all the questions have been obtained.

4.2.2 Formulate a Security Rules Base

The second phase of a security analysis is to use the organization's security policy to create a Security Rules Base. If the organization has a rules base, then this step can be omitted. However, if the organization does not have a rules base, then the analyst can create one from the security policy with the assistance of some high-level managers or their named representatives.

In some cases, the security rules base can be developed in a facilitator[36] environment that is attended by the following: a facilitator, a scribe, and technical people who are knowledgeable about the security needs of the information system.

The sample security rules base in Appendix D2, *Security Rules Base: An Example*, can be used as a stimulus for a brainstorming session to create a particular rules base. The rules base should follow from the organization's security policy. However, the rules base should be more detailed but not to the level of identifying any vendor products.

---

[35] Eliciting is the process of obtaining specialized information (usually for input to a computer system) from a knowledgeable person.

[36] The security analyst is the facilitator. A scribe is needed to record all relevant statements. The facilitator is not the knowledgeable person, the technically knowledgeable employees are. The facilitator opens the meeting, presents the guessed security rule, and asks for comments, and keeps the discussion on track and moving forward.

4.2.3   Formulate the Security Requirements

The security requirements should follow directly from the security policy and rules base. The high-level security requirements, often called the user requirements, can be determined given the security rules base. This activity can be accomplished by security analysts, although the user requirements are more formal if they come from the organization's management.

The next three phases of the security analysis are to identify the three levels of security requirements:

1) Formulate the high-level security requirements,
2) Formulate the medium-level security requirements, and
3) Formulate the low-level security requirements.

It is not mandatory to obtain the low-level security requirements (the most detailed) to complete the security analysis.

4.2.3.1   High-Level Security Requirements

The *high-level security requirements* (the least detail) are the general requirements that should come from the organization's information system's owners and users.

These requirements can be obtained by a security analyst using a question and answer process in a facilitator environment. The facilitator should have a good idea of what the requirements are and these should follow from the Security Policy and the Security Rules Base.

1) Identify the categories for the security requirements. The categories should be complete (i.e., they include all of the possible requirements). Preferably the categories should be mutually exclusive[37], but they are still useful even if they are not. Later, the categories can be made to be mutually exclusive by moving the requirements around and even by redefining the categories.
2) Select a category and invite the managers who are knowledgeable of the processes that are in the selected category.
3) Present a set of requirements for the selected category to the managers and request their comments.
4) Various issues need to be addressed. Do we need to make any changes? Do we need to add any requirements? Are any stated requirements unnecessary?
5) After the meeting has been adjourned, the facilitator, the scribe, and any additional personnel required, should make the needed corrections to the set of requirements for the stated category and send them (usually via e-mail) to the

---

[37] Categories are *mutually exclusive* if the contents in any category are such that no element in that category is in any other category. In Venn diagram terms, if A and B are any two categories, then $A \wedge B = \Phi$, or A intersect B is the empty set.

attendees asking for their feedback. Make the suggested changes that appear in the feedback if the feedback is correct.

6) Select the next category and repeat the process (steps 2 through 5) until all the requirements have been identified.

7) When all of the requirements have been identified, create a complete set of security requirements. Then another meeting is required where as many of the managers are invited to meet once more to go over the complete set of security requirements.

8) Subsequent to the final meeting, the requirements are finalized (until some event causes them to change once more) and placed in a living database and also documented for issue to the relevant personnel.

The requirements, as they are identified, should be placed in an object-oriented database where each of these child-elements can be easily traced back to a prior parent-element in the security requirements database. The security policy should form the parent-elements of the database.

The process of obtaining the requirements is an iterative process because following each session, where the requirements are identified and installed in the requirements database, at the next session the analyst should present the latest version of the requirements to the owner/users for their review and comment. All appropriate changes should be made and the analyst should end the process when the requirements are sufficient to proceed to the next phase.

These high-level security requirements may be sufficient for inputs to the creation of the Security Architecture, however, they may not be and in this case, a more detailed set is needed.

4.2.3.2   Medium-Level Security Requirements

The *medium-level security requirements* can be identified by decomposing the high-level requirements to more detail. When sufficient detail is attained, the security architecture can be created. This may be an iterative process with the analyst attempting to create the security architecture and having to return to the medium-level requirements database to create more detail in order to continue.

The medium-level security requirements are complete when they are sufficient for developing a security architecture. The completeness can be determined in an iterative process of attempting to form the architecture from the requirements and if there is insufficient detail, return to the requirements and decompose some more. In some cases, the security rules base is sufficient for creating an architecture and hence this step may not be required. The detail in the requirements is dependent upon the detail in the analyst's architecture model.

4.2.3.3   Low-Level Security Requirements

Although it is not required to complete the security analysis, the analyst may wish to identify the most detailed security requirements, called the *low-level security requirements*.

These requirements are the decomposition of the high-level or medium-level security requirements. Often technicians and programmers, as assistants to the security analysts, are involved in this step.

The low-level security requirements can be used as the specification for software (either for purchasing it if a COTS package can be found, or for writing the programs) and for hardware acquisitions. In fact, the low-level security requirements are often referred to as the "security specifications."

For a security analysis, it is not necessary to create the detailed low-level security requirements database since none of the products of a risk assessment depend on these specifications.

4.2.4 Perform a Risk Assessment

The risk assessment process is explained in great detail in Chapter Six, *Risk Assessment*, but the risk assessment process's character is explained here. The primary objective of a risk assessment is to identify the cost-effective countermeasures that will minimize risks due to the threats and existing vulnerabilities.

To accomplish this objective, the following twelve actions or steps are required:

1) Identify all threats,
2) Determine the relevant threats,
3) Identify the system vulnerabilities,
4) Identify the organization's assets and their values,
5) For each threat, identify a gamut of attack scenarios,
6) Formulate the security risks,
7) Identify all feasible types of countermeasures,
8) Cull out the irrelevant types of countermeasures,
9) List the alternative countermeasures by vendor for each type of countermeasure,
10) Determine the preferred vendor countermeasure (for each type),
11) Determine the cost-effective countermeasures, and
12) Determine the remaining risks.

The process begins with identifying all of the threats.

4.2.4.1 Identify All Threats

The objective of this task is to identify all of the threats that could possibly be employed to attack this organization's information system network.

A *threat analysis* first should identify all of the potential threats to the organization's information system network. A good starting point is to have a list of all of the potential threats that could be launched against any information system network. This is a list that the analyst keeps as a living document upgrading it as new threats appear and old threats disappear. Categorizing the threats into the three areas of physical, behavioral, and technical is recommended.

Call this list the T_0 table. For any given information system, the relevant threats, T_R, can be denoted by the following equation:

$$T_R = T_0 - \delta T$$

where δT is the set of threats that are considered irrelevant so that the list T_R represents a culling of the original list T_0.

The list δT can be formulated by the analyst who should examine the total list T_0 and with knowledge of the customer organization's information system network, delete those threats considered to be irrelevant. These deletions represent the sublist δT.

4.2.4.2   Determine the Relevant Threats

Since many of the threats are not relevant to this organization, cull out those that do not apply. A formal approach to the culling would require that a comparison elimination decision approach be taken:

    a) disjunctive comparison,
    b) conjunctive comparison,
    c) comparison across attributes (Pareto dominance), and
    d) comparison within attributes (elimination by aspects) (Smith 1998).

See section 4.3, *Comparison Methods*, for a discussion of the decision options for comparison techniques. When using the formal comparison method, the decision-maker should begin with any two alternatives and eliminate one of them, then compare the winner with another alternative, until only one is left.

A quicker and easier method is to have the decision-maker select the desired countermeasure from among the alternatives based on which one seems to be the best; this is a *judgmental comparison*. The "best" one can be determined from the performance specifications, the cost, the vendor's reputation, past history with this vendor, all of these, and all other attributes that are important and known to the decision maker.

4.2.4.3   Identify the Vulnerabilities

A *vulnerability analysis* uses the relevant threats identified in the earlier step and using a scanning tool, e.g., **nmap** or SATAN, to identify the various vulnerabilities of the information system. The process to identify the cost-effective countermeasures is shown in Figure 4.2, *An Illustration of the Process to Identify the Cost-Effective Countermeasures*. The initial step is shown on the left side of the figure and the vulnerability analysis is shown in the center.

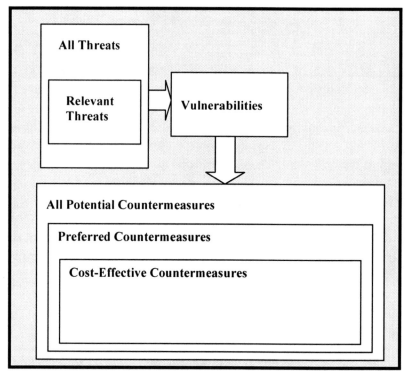

Figure 4.2 An Illustration of the Process to Identify the Cost-Effective Countermeasures

This process can be represented by the following formula:

$$V = \mathbf{g}\,(T_R)$$

where V is the set of vulnerabilities for this information system and **g** is a function represented by a scanning analysis, and a systems analysis, and any other analysis that might be used to ascertain the system's vulnerabilities.

4.2.4.4   Identify the Assets and Their Values

The organization's information system assets need to be identified and the value of each of these to the organization (dollar value of each asset) needs to be determined. These assets should be prioritized. The prioritization can be based on the value given to each asset.

For example, suppose that the organization is very dependent on the Internet for its business. Then the availability and the customers' information are very important. The security of the customers' information is paramount. If the organization's 24-hour business is an average of $1 M, and a maximum denial-of-service duration is 6 hours, then the loss of business of that period is $250 K (assuming that persons who call in and cannot get a connection will not call back).

For this organization, the information system assets are listed in Table 4.1, *Organization Information Assets: An Example,* shown below.

Table 4.1 Organization Information Assets: An Example

Asset	Cost	Potential Loss
Availability of system (Web server)	$50 K	$ 250 K
Availability of information (Web server)	$100 K	$ 100 M
Existence of customer information (Hard disk)	$10 K	$ 10 M
Privacy of customer information (Hard disk)	$10 K	$ 10 M
Organizational reputation (Info system)	$10 M	$ 100 M
Etc.	Etc.	Etc.
Etc.	Etc.	Etc.
Etc.	Etc.	Etc.
Etc.	Etc.	Etc.
Etc.	Etc.	Etc.
Etc.	Etc.	Etc.
Etc.	Etc.	Etc.

4.2.4.5   Identify Attack Scenarios

For each threat and system vulnerability, develop a scenario for the threat that takes advantage of a system vulnerability. This process should yield a set of *attack scenarios* from which the analyst can determine the probability of attack ($P_A (S_k)$) (I will assume also that the attack, if it occurs, will be successful because of the system vulnerability to this threat) and also formulate the potential loss ($C (S_k)$) that might result from such an attack.

An example of an attack scenario is the case of hacker (the threat) who can easily enter an organization's information system network as an Internet external user who sends a virus as an attachment to an innocuous-seeming e-mail message. The receiver, who is not aware of the rule to not open e-mail attachments from sources that are not recognized, opens the attachment which then infects the system and the receiver's PC subsequently crashes following the alteration of the user's operating system. The system owners had not instituted a policy that advised users to not open e-mail attachments from sources that they did not recognize (the vulnerability).

4.2.4.5   Formulate the Security Risks

For each attack scenario (S_k), formulate the concomitant security risk using the equation:

$$Risk (S_k) = P_A (S_k) * C (S_k), \text{ for } k = 1, 2, …, n,$$

where $P_A (S_k)$ is the probability of successful attack for the kth scenario, $C (S_k)$ is the potential organizational loss that could occur for the kth scenario, and n is the number of scenarios.

After computing each of the risks, Risk (S_k), for all k scenarios, continue to the next step.

4.2.4.6 Identify All Feasible Types of Countermeasures

For each risk that is computed (for a particular threat and its associated threat scenario that takes advantage of a system vulnerability), identify a set of alternative types of countermeasures that can minimize the effects of this type of attack.

These types of countermeasures are vendor-independent so that just the type of countermeasures is identified (e.g., firewall). A list of possible countermeasures should be used to stimulate the analysis.

The objective of this phase of the process is to obtain or create a list of all of the potential countermeasures that could be applied to any information system to decrease the identified risks to the system. This list will be called the C_0 list and can be obtained from the list given in this book or by forming a living document list developed from forming and then adding or deleting the various countermeasures available from commercial vendors.

If it is the case that a single type of countermeasure will minimize the effects of several scenarios (i.e., threats), then the analyst should select the scenario that has the greatest losses and discard the others (that apply to this specific type of countermeasure). Then select all of the alternative countermeasures (i.e., vendor alternatives for this type of countermeasure) that are applicable to minimizing the effects of this particular threat/scenario.

For the remaining risks, repeat the process until all of the risks have been matched with at least one type of countermeasure.

4.2.4.7 Identify the Vendor Alternative Countermeasures

The identification of the vendor alternative countermeasures requires the following process:

1) Identify some type of countermeasure,

2) Either through the analyst's knowledge or by using the Internet or some other convenient reference, identify alternative vendor products for this countermeasure,

3) Fill in the indicated rows using Table 4.2, *Vendor Products and Attributes,* shown below, and

4) Repeat this process for each type of countermeasure until all of the types are exhausted.

Table 4.2 Vendor Products and Attributes

Vendor	Product	Attributes
Firewall		
Axent	Raptor	Speed. Memory. Ease of use. Cost. Etc.
Cisco	PIX Firewall	Speed. Memory. Ease of use. Cost. Etc.
Check Point	Firewall-1	Speed. Memory. Ease of use. Cost. Etc.
Etc.	Etc.	Speed. Memory. Ease of use. Cost. Etc.

Anti-virus Software		
Norton	AntiVirus 2000	Speed. Frequency of update. Cost. Etc.
Etc.	Etc.	Etc.
Etc.	Etc.	Etc.
Etc.	Etc.	Etc.
Etc.	Etc.	Etc.
Etc.	Etc.	Etc.

4.2.4.8   Determine the Preferred Vendor Countermeasures

A preferred list of countermeasures can be determined using one of the formal analyses, namely, a *multi-attribute utility analysis* to rank order the countermeasures (e.g., prioritized best (i.e., most preferred) to worst), or some form of *comparison analysis*, which can be used to cull certain countermeasures from the original list C_0 until the "best" one is left (Smith 1998). If, for some reason, the analyst does not have time to use one of the formal analyses, the analyst can use an *intuitive evaluation* based on their feelings about the alternatives (Smith 1998). The preferred list, C_P, (i.e., preferred by either the owners/users and/or the analyst(s)) is then formed and can be represented as:

$$C_P = C_0 - \delta C$$

where the items in the list $\square C$ are identified by scanning the information system using the relevant threat list T_R. The list $\square C = \mathbf{h}\,(V)$, can be deduced by examining the capabilities of the list of countermeasures and determining which of these are capable of diminishing the system's vulnerabilities. This examination process is represented by the function $\mathbf{h}$.

Of course, in general, some of the risks may require a physical countermeasure, some may require a behavioral countermeasure, and most will require a technical countermeasure.

For example, suppose the threat scenario is a malicious hacker who wishes to gain electronic entry to the system so as to perform some malicious act. The potential success of this scenario can be minimized through the use of a firewall (type of countermeasure). There are many firewall vendors, for example, Firewall-1 by Check Point, Raptor by Axent, or Cisco PIX Firewall, and any other firewall that the decision-maker wishes to add to the list of alternatives. For each of these vendor firewalls, the decision-maker can list the attributes (e.g., cost, performance specifications, and vendor reputation). The last will likely be rather subjective. The decision maker can then eliminate the alternatives until one is left, or just select the preferred one based on judgmental comparison (the option that gives the decision maker the warm fuzzies).

Note that firewalls can also perform the encryption/decryption for virtual private networks (VPNs), can perform as an IDS, and also can perform identification and authentication, access control, and other duties.

4.2.4.9 Determine the Cost-Effective Countermeasures

Finally, a list (or set) of the cost-effect countermeasures (C_{CE}) can be determined using the total value[38] of the assets protected and comparing this value with the total cost[39] for each countermeasure, and can be represented by the following formula:

$$C_{CE} = \{C: Tot_i\,(C) < Tot_i\,(A_P)\}$$

where A_P is the set of assets protected by the countermeasures (C_i, i = 1, 2, ... m), $Tot_i\,(A_P)$ is the total value of the assets A_P, $Tot_i\,(C)$ is the total cost of the ith countermeasure C_i, and {C: X} is read "the set of countermeasures such that the total cost of the countermeasure is less than the total cost of the relevant assets that it protects."

If C_{CE} is not the empty set (i.e., nonexistent), then the most cost effective countermeasure C_{CE} (min) is the minimum cost one, which is:

$$C_{CE}\,(min)\ such\ that\ C_{CE}\,(min) \le Tot_i\,(C_{CE}),$$

where $Tot_i\,(C_{CE})$ is the total cost of each countermeasure that is cost-effective.

4.2.4.10 Determine the Remaining Risks

Some of the risks will have countermeasures that are not cost-effective so these risks will not be countered, physically, behaviorally, or technically. Even those risks (i.e., threats) that are countered, there may still be some residual risk associated with this threat-countermeasure pair.

The security analyst should evaluate the remaining risks (both countered and uncountered) to ascertain the level of residual risk for the system.

4.2.5 Develop a Security Architecture

Once the risks and the set of countermeasures have been determined, the Security Requirements for the information system network can be updated to reflect the identified needs for the security upgrades to the system. The security architecture will consist of the complete set of countermeasures (legacy plus new).

4.2.6 Develop an Overall Architecture

The security architecture, when embedded in the reference architecture, becomes a part of the overall architecture. The particular location of each technical countermeasure is

---

[38] The total value of an asset is the worth in dollars to this organization, even if the analyst or owner has to quantify the value given some qualitative remark, such as "the loss of this particular information would jeopardize our business."

[39] The total cost of a countermeasure is the cost required to purchase, install, unit test, security test, modify if required, and operate the countermeasure.

important and is reflected in the overall architecture. The overall architecture is the target architecture. What is needed now is a plan for migrating from the legacy architecture to the target architecture. This plan is called the migration plan.

4.2.7   Develop a Migration Plan

The *migration plan* should be created based on a tradeoff of risks (technical, schedule, and performance) associated with the specific upgrades, the availability of funds, the difficulty of security testing and evaluating the upgrades, the time to install and test the upgrades, and any other aspect of the upgrade process that is important to the organization. The migration plan should be based on a *security lifecycle process* that includes the following (in the order listed):

1) Update the organization's security policy and security rules base (if needed),
2) Update the security requirements for the information system,
3) Update the risk analysis,
4) Update the security and overall architectures,
5) Update the threats and vulnerabilities lists,
6) Update the remaining countermeasures needed to achieve the target architecture (if any, i.e., install the recommended additional or improved countermeasures),
7) Update the security test and evaluation plan for the implemented countermeasures,
8) Update the test results and evaluation of the implemented security tests,
9) Update the certification and accreditation status of the system vis-à-vis the Designated Approving Authority, and
10) Update the Security Certification and Authority Process document.

Once these considerations have been identified and used as inputs to an analysis and evaluation of the upgrade process, a set of discrete upgrades will be identified and a roadmap can be constructed. This roadmap can consist of a Gantt chart that shows:

1) The different tasks (that make up the total effort),
2) The planned schedule (start and stop times),
3) The level of effort for each month for each task (so that a costing analysis can be done), and
4) Any deliverables that are required at the completion of a task.

Associated with the chart should be all the verbiage needed to clarify the chart and the migration process.

4.2.8   Implement the Migration Plan Steps

The next phase of the effort will be implementing the migration plan to proceed from the current system to the target system.

It is important to properly manage each of the ten-roadmap steps. This can be done using the POSDCORB process (Smith 1998). For each of the ten steps or phases presented above for the lifecycle process, management should commit to the activities shown in Table 4.3, *Management Activities and Products*.

Table 4.3 Management Activities and Products

Mgt. Activity	Description	Product
Planning	Make preparations for the particular phase that is next.	Planning Document
Organizing	Identify and structure the tasks and create organizational positions to perform the identified tasks.	Task Descriptions
Staffing	Evaluate personnel and fill the positions.	Task Assignments
Directing	Provide leadership and delegate authority.	Authorities
Coordinating	Coordinate activities and handle any changes to the task descriptions.	As Required
Reporting	Document decisions and prepare information reports.	Decision Reports
Budgeting	Ensure that the phase is properly funded.	Phase Budget

A *quality assurance process* can be used to ensure that the lifecycle phases are performed properly, are completed within the assigned budget, are completed within schedule constraints, reflect high quality, and are based on the appropriate assignments of personnel. *Total quality management* can be incorporated throughout the process by ensuring that a proper relationship is maintained with the identified user security requirements and the product of each lifecycle phase.

4.2.9 Perform a Security Test and Evaluation

After each upgrade, there should be a Security Test and Evaluation (ST&E) performed to ensure that the countermeasure upgrades are operating as they should.

If any countermeasures are not performing as they should, then some modifications are required. Subsequent to these modifications, further ST&E is required to make sure that the modifications were successful.

After the ST&E has been performed and the interim upgrade is declared successful, the next upgrade can begin according to the schedule.

4.2.10 Change Management

Change management is a formal process for ensuring that any changes made to the existing software are performed properly. The formal process usually involves the following steps: review of the proposed change, implementation of the change if the change is approved, and testing and evaluation of the implemented change to ensure that it is operating properly.

Change management usually involves a Configuration Control Board (CCB), which convenes periodically and reviews the various changes for a particular system.

The comparison decision methods are described next.

4.3   Comparison Methods

Five methods for comparing alternatives are presented below (Smith 1998). The first is Disjunctive Comparison.

4.3.1   Disjunctive Comparison

Disjunctive comparison is when the decision-maker identifies minimally acceptable standards for each relevant attribute, so that alternatives that pass the critical standard of one or more attributes are retained. Alternatives are rejected only if they fail to exceed all of the critical standards. This type of comparison can be easily computed as follows:

If $x_j \geq c_j$ (or $x_j \leq c_j$ as the specific case may be) for at least one $j = 1, 2, ..., m$; then accept the alternative,

where x_j is the value for the jth attribute and c_j is the value for the jth reference standard. The case is that $x_j < c_j$ (or $>$ as the specific case may be) for all values of j for the alternative to be rejected. This rule is noncompensatory[40]. Choose ">" or "<" as the comparison to be used depending on whether "bigger is better" or "smaller is better," respectively.

4.3.2   Conjunctive Comparison

Conjunctive comparison is when the decision-maker identifies minimally acceptable standards for each relevant attribute, so that alternatives that pass the critical standard of all attributes are retained. Alternatives are rejected if they fail to meet or exceed a single minimum standard. This type of comparison can be easily computed as follows:

If $x_j \geq c_j$ (or $x_j \leq c_j$ as the specific case may be) for all $j = 1, 2, ..., m$; then accept the alternative,

where x_j is the value for the jth attribute and c_j is the value for the jth reference standard. The case is that $x_j < c_j$ (or $>$ as the specific case may be) for just one value of j for the alternative to be rejected. This rule is also noncompensatory. Choose ">" or "<" as the comparison to be used depending on whether "bigger is better" or "smaller is better," respectively.

4.3.3   Comparison Across Attributes

Comparison Across Attributes (sometimes called Pareto Dominance) can be described as follows. Choose alternative a_1 over a_2 if a_1 is better than a_2 on at least one aspect and not worse than a_1 on any other aspect, where an aspect is the score of a specific attention on a

---

[40] A compensatory rule is one where an attribute with a low value can be compensated by an attribute with a high value. A noncompensatory rule is one that is not compensatory.

specific attribute. That is, if the attributes of the alternative are x_1, x_2, and x_3, and the values of the attributes, v (x_j) for a_2 (v^{a2} (x_j)), are such that

$$v^{a2} (x_1) \geq v^{a1} (x_1) \text{ and } v^{a2} (x_2) \geq v^{a1} (x_2)$$

but

$$v^{a2} (x_3) > v^{a1} (x_3)$$

then a_2 is selected because it dominates a_1. In this case, "$\geq$" means "is at least as good" and "$>$" means "is strictly better than." This rule is also noncompensatory.

4.3.4   Comparison Within Attributes

Comparison Within Attributes (sometimes called Elimination by Aspects) also requires using option attributes. Attributes are assumed to have different importance weights. For some alternative, select an attribute with a probability that is proportional to its weight. Alternatives, which do not have attribute scores above this reference value, are eliminated. For a remaining alternative, select an attribute with probability proportional to its weight and perform the evaluation again. Continue the evaluation process until a single alternative is left. This rule is similar to a lexicographic comparison (i.e., it is like the alphabetic method used to find a word in the dictionary except the selection method is based on a probabilistic analogy instead of its spelling). This rule is also noncompensatory.

4.3.5   Additive Difference

A compensatory method is the *Additive Difference* method. For alternatives a_1 and a_2, compute the valued differences U_j (a_1) - U_j (a_2) where U_j is the value for the jth attribute for j = 1, 2, ..., m. Elicit the weight, f [U_j (a_1) - U_j (a_2)], for each difference from the decision maker, where—$1.0 \leq f_j [.] \leq 1.0$. This means that the decision maker likes a_1 better if U > 0 and likes a_2 better if U < 0, where,

$$U = \Sigma j \, f_j \, [U_j \, (a_1) - U_j \, (a_2)].$$

If U = 0, the method is indeterminate, or the two alternatives can be considered as being equally acceptable so it doesn't matter which one the decision maker selects.

In formulating a definition of the issue, taking different views of the problem will contribute to a better issue definition, and the concept of an inquiring system will assist the analyst in finding a solution of the problem.

4.4   Perspectives and Inquiring Systems

Alternative *perspectives* (e.g., political, hostile, self-critical, and financial) can be used to view a situation in different ways [Linstone 1975] so as to get a more accurate view of the relevant events occurring in the environment. These perspectives can be used to create a more accurate hypothesis to be used in the evaluation of an information system's needs for adequate security.

It is useful to formulate the hypothesis (i.e., the issue or problem that needs to be resolved) with a variety of perspectives (Smith 1992). In this manner, the hypothesis can be formulated to encompass most, if not all, of the political and financial constraints (and even schedule timelines that can be revealed in a Gantt chart), the gamut of threats, the vulnerabilities, and the various assets and their values that need to be considered when attempting to find an appropriate countermeasure for responding to the threats.

That is, when evaluating an information system, it is useful to view the security issue considering the politics of the situation, the potential threats that could be applied to attack the system, the various vulnerabilities of the system to these threats, the costs to acquire and install potential countermeasures, and the potential losses by the organization because of attacks that are based on these threats that can take advantage of the system's vulnerabilities. To determine the likelihood of a particular attack, the cost to the attacker to carry out the attack should be considered. Attacks that are extremely expensive (to the attacker) may be considered to be highly unlikely. However, for terrorist attacks that are funded by some country or a very rich backer, the upper boundary or limitation on the attacker's cost may be considerable.

The following are useful perspectives for properly defining an information system issue:

1) *Political view* (What Government mandates (e.g., PDD 63 and OMB Circular A-130) and Government agency directives (e.g., NIST directives such as the Special Publication series) are relevant to this system?)

2) *Threat view* (How might internal and external users attack this system?)

3) *Vulnerability view* (How good are the existing countermeasures, even if there are none, and what threats can get through the existing countermeasures?)

4) *Financial view* (What are the relevant assets that are connected to, or dependent upon, the information system, and what is each asset's value in dollars?)

Once the hypothesis has been formulated, an *inquiring system* [Churchman 1971] approach can be used to guide the process of analyzing and evaluating the statement of the issue. The primary purpose of an inquiring system is to provide a pragmatic and structured systemic framework for evaluating a statement of an issue. The statement can be a hypothesis, or a design concept, or any idea that can be verbally and/or mathematically described. Four types of inquiring system have been defined, namely: Lockean, Leibnitzian, Kantian, and Hegelian (Churchman 1971).

These four inquiring systems yield four different structured approaches to evaluating a particular hypothesis. The four inquiring systems are:

1) The *Lockean inquiring system* is best for analyzing and evaluating a statement based on *empirical data inputs*. The empirical data elements are defined in terms of the user's needs for confirming or disconfirming the statement. Data elements are *primary* in this inquiring system. Alternative statements of the issue, if formulated, are complementary. That is, the alternative hypotheses should be mutually exclusive so that no part of one is a part of another. For the evaluation of the information system situation, only a single hypothesis is recommended, else the analytical

situation would get overly complicated. However, this inquiring system is not the best one for evaluating information systems.

2) The *Leibnitzian inquiring system* is best for analyzing and evaluating a statement that is based on a *model or analytical concept* and requires an evaluation that is both logical and analytical. Technical data elements that may be required for confirming or disconfirming the statement are defined in terms of the user's needs. The technical data must support the analytical foundation of the model or formula. For data that does not support the statement, either the statement is rejected or the hypothesis must be modified to better match the data. Ideally, the problem is resolved based on a mathematical analysis or what serves as a mathematical proof. Theory is primary in this inquiring system, which makes this inquiring system inappropriate for evaluating information systems.

3) The *Kantian inquiring system* is best for analyzing and evaluating a statement that has *both empirical and analytical aspects*. The empirical data elements are deduced and collected and analytically evaluated to ascertain if they match the analytical model. This evaluation is based on an empirical and an analytical evaluation where the user compares the empirical data to ascertain whether it matches the analytical result. Both data and theory are of equal value in this inquiring system. This inquiring system is best for evaluating information systems.

4) The *Hegelian inquiring system* consists of developing a thesis and a conflicting antithesis based on an early formulation of the problem. By arguing the position and the counterposition with a *dialectic evaluation*, a synthesis is eventually formed. The conflicting arguments can be based on empirical, analytical, or both empirical and analytical evaluations. This approach is not recommended for evaluating an information system hypothesis because the analysis would be too extensive and complex.

The statement of the information system security issue encompasses both empirical data (i.e., the threats, vulnerabilities, and values of the identified assets) and analytical data (viz., the risk estimates). Thus, a Kantian inquiry is the most appropriate process to use. Proceed as follows:

1) For a particular risk value (as computed using a threat scenario, vulnerability, and relevant asset value or assets values, i.e., potential losses), compare this analytically computed risk with the threat effect reduction capability of alternative countermeasures to identify the appropriate countermeasure or countermeasures for this particular risk.
2) If several countermeasures will perform sufficiently well, then identify the total cost (acquisition, installation, testing, and operation) of each one and rank order them on the basis of performance and cost to identify the most cost-effective one. This can be done with a multi-attribute utility analysis.
3) Repeat this process (steps 1 and 2) for all of the identified threat scenarios.
4) When completed, list all of the cost-effective countermeasures.

5) Create a Remediation Plan for implementing these listed countermeasures in a manner that minimizes the technical, schedule, funding, and performance risks of the countermeasures.

Another approach, using the Common Criteria (CC), see http://csrc.nist.gov/cc/, was originally developed for *security testing of products by an independent organization* for serving as security mechanisms to protect any information system (NIAP 1997, NIAP 1999, NIST 1999). The CC approach later became a useful process for evaluating the entire information system. However, the CC approach is complex and involved and difficult to understand, so many organizations and analysts do not wish to use it. Moreover, the CC criteria are not embedded in a total security analysis process as presented here and therefore the CC process is not recommended.

The decision process as presented here is to formulate a hypothesis, find a solution to this issue, and then implement[41], test, and operate the identified solution.

I.   The Hypothesis Process:

1) Identify all of the potential threats.
2) Identify those that are relevant to this organization's system.
3) Identify the vulnerabilities using a scanner and/or architecture model.
4) Identify the assets that could be attacked and estimate their values.
5) Identify the threat scenarios (an exhaustive set of size n) that could exploit the identified vulnerabilities.
6) Formulate the Hypothesis.

For each threat, identify the vulnerabilities that could be exploited by the threat. Determine the scenarios that could be used to perform an attack for this threat. For each scenario, estimate the potential losses that might occur. Of these, select the largest loss (i.e., perform a worst case analysis). Thus, perform an exhaustive analysis by creating scenarios for all of the relevant threats, identified vulnerabilities, and identified assets (and their values) that are affected by the threats, so that the complete hypothesis will consist of the following statements (n of them):

1) Describe threat scenario 1 (TS_1) that exploits an identified vulnerability and could result in a loss to the organization of X_1 dollars.
2) Describe threat scenario 2 (TS_2) that exploits an identified vulnerability and could result in a loss to the organization of X_2 dollars.
3) ...
4) Describe threat scenario n (TS_n) that exploits an identified vulnerability and could result in a loss to the organization of X_n dollars.

---

[41] In some cases, the countermeasure is implemented in an offline system, then tested, and modified if required, and then it is implemented in the online system after it has been proven to be a reliable mechanism. Most organizations cannot afford to have an operating (or offline) backup system.

123

II. The Resolution Process:

1) Identify a comprehensive list of the alternative countermeasures (required to counter each scenario) and attributes of these countermeasures (including performance attributes, vendor attributes, and total cost).

2) Compute the risk (for each of the n scenarios and associated maximum loss).

3) For each computed risk (R_i), identify the relevant countermeasures that could reduce the success probability of the ith threat scenario (thereby lessening the ith risk).

4) Rank order the identified countermeasures (for the ith risk) if there are more than one countermeasure (and repeat for all of the n risks).

5) Formulate a list of the cost-effective countermeasures.

For each threat scenario (TS_n), compute the risk (R_i). For each risk (R_i), determine the countermeasures that could be used to reduce the ith threat scenario's success probability. If there is more than one countermeasure for this threat scenario, rank order them using a multi-attribute utility analysis. Select the one with the highest utility value. Do for all of the n threat scenarios. Delete any replications. Place all of the identified unique cost-effective countermeasures in a table. Identify the locations in the organization's network where each type of countermeasure must be located (e.g., firewalls may be required at each entry point in the organization's network, and there may be two of them at each point to ensure no single point-of-failure).

III. Implement the Solution:

1) Create a remediation or migration plan that minimizes the upgrade risks (technical, schedule, and performance).

2) Follow the plan and upgrade the information system, accordingly.

For each of the upgrade risks, determine an upgrade process (remediation) that minimizes these risks. For example, if a particular countermeasure has been difficult to implement in other environments (which means that it probably should not have been selected in the first place), then this countermeasure should be implemented in an offline location and tested prior to installation in the online environment. Using this process, develop a plan for installing all of the countermeasures. After the plan has been approved, follow the plan during the implementation process.

IV. Perform Security Test and Evaluation of the Upgrade:

1) Create a Security Test and Evaluation (ST&E) Plan.

2) Implement the ST&E Plan.

3) Modify or replace, if required, the security mechanisms that are not operating properly.

Create an ST&E plan. Get the ST&E plan approved. After all of the countermeasures have been installed, implement the ST&E plan and document the results. If any countermeasure did not work properly, either modify the device or replace it. If it is replaced, then the replaced mechanism must be tested and evaluated.

V. Operate the System:

1) Operate the system in its environment.
2) Maintain and upgrade the system, as required.
3) Respond properly to all security incidents.

If the tested system is an offline system that is a replica of the actual online system, then subsequent to the test and evaluation process, the offline system should be installed as the online system with, perhaps, the original online system remaining as a hot backup in case any aspect of the new system is deemed non-operable.

REFERENCES AND BIBLIOGRAPHY

4.1   Smith, Sr., Charles L. (1998). "Computer-Supported Decision Making: Meeting the Demands of Modern Organizations," Ablex Publishing Corporation, Greenwich, Connecticut and London, England, 1998.

4.2   NIAP (1997). "Common Evaluation Methodology for Information Technology Security, Part 1: Introduction and General Model," Version 0.6, National Information Assurance Partnership (NIAP), CEM-97/017, January 1997.

4.3   NIAP (1999). "Common Methodology for Information Technology Security Evaluation, Part 2: Evaluation Methodology," Version 1.0, National Information Assurance Partnership (NIAP), CEM-99/045, August 1999.

4.4   NIST (1999). "Information Technology Security Testing—Common Criteria," Version 1.1, National Voluntary Laboratory Accreditation Program (NVLAP), U.S. Department Of Commerce, National Institute of Standards and Technology (NIST), Draft Handbook 150-20, April 1999.

4.5   Smith, Sr., Charles L. (1992). "A Theory of Situation Assessment," PhD Dissertation, George Mason University, Fairfax, VA, May 1992.

4.6   Linstone, H. (1984). "Multiple Perspectives for Decision Making," North-Holland, Amsterdam, 1984.

4.7   Churchman, C. (1971). "The Design of Inquiring Systems: Basic Concepts of Systems and Organizations," Basic Books, Inc., New York, 1971.

CHAPTER FIVE

SECURITY ARCHITECTURES

5.1   Architectures

A *system* is the actual set of elements used for creating an information system, whereas an *architecture or architecture model* is a representation of the system.

An *architecture* (or architecture model) provides a structured representation of an information system for use by analysts who can achieve some understanding of the manner in which information is communicated, processed, and presented, for some particular organization. This architecture is an abstraction of the information system and provides a *definition* of the system components and a *description* of how these components will interact with one another, and a *description* of the information that is communicated from one component to another. In a normal lifecycle process, the creation of this model (of an architecture) will precede the acquisition, installation, and testing of the actual information system components.

Cisco (2001) states that, "An architecture provides both a coherent framework that unifies disparate solutions onto a single foundation, and a roadmap for future network enhancements." That is, whenever an organization is able to create an architecture for its information system, it will enable the organization to identify those customer-preferred vendors who can provide the "best" products, which enable customers to meet their needs with the most cost-effective products available. These products will enable the organization to create an open architecture, that is, an architecture that is composed of products (hardware and software) that are portable, scalable, and interoperable. In real life, it is quite difficult to create a truly open architecture, and this is the reason many customers will purchase most of their equipment from a very few vendors.

The three types of architecture, as presented in Chapter One, are:

1) *Reference* Architecture,
2) *Security* Architecture, and
3) *Overall* Architecture.

To create a security architecture for some reference system, it is necessary for the analyst to understand the architecture of the reference system. The reference system plus the security system will be referred to as the "overall system" (or overall architecture), which can be described with an architecture model. In most cases, the reference architecture (or legacy system) will have some security mechanisms, but usually will require some improvements to meet the security requirements.

A *reference architecture* should follow from a set of fairly detailed requirements for the system and if security were considered when generating these requirements, then the reference architecture will contain the countermeasures needed to protect the system. However, many reference systems were not generated with security in mind and thus the reference architecture will contain little, if any, security countermeasures. In this case, the security architecture will be defined as a result of a set of security requirements, which should be generated following a risk assessment.

Ideally, an architecture for an information system should graphically and verbally present the following system ingredients:

1) Each subsystem element (viz., hardware devices (e.g., firewalls and IDSs) and software products (e.g., DBMSs, anti-virus software, and auditing)),

2) The interfaced information (viz., the data that must flow among the network elements),

3) The system functions (i.e., the tasks that each element must perform),

4) The communications connections (i.e., the lines connecting the various devices),

5) The network devices (e.g., routers, firewalls, WANs, and LANs),

6) The cabling devices, and

7) Any other element needed to complete the system.

Usually an architecture model consists of a graphical part and a verbal part that together depict what the system should look like and how it should operate or work. If the system is yet to be built, the architecture is a *normative model* (i.e., it describes how the system *should look* and operate from a purely analytical perspective) and if the system exists, then the architecture is a *descriptive model* (i.e., it describes how the system *actually looks* and operates). In most cases, organizations do not maintain a current architecture model for its existing information systems, so the security analyst must use the latest version of the model, and with verbal inputs from users and technicians, can deduce what the actual architecture should be and make the corrections accordingly. A model that is recommended by an analyst, say an information system architect, for how the system *should be built* is a combination of the normative and descriptive models and is called a *prescriptive model*. Prescriptive models can simplify performing a security analysis, if they are properly prepared. Some of these can be based on templates[42]. However, a template that does not provide a good set of directions for guiding the analyst in the preparation of a security analysis can be worse than not having any model at all.

Ideally, a reference architecture, for an existing system, is a descriptive or prescriptive representative model that includes illustrations of the generic devices (but not vendor specific) that are contained in the actual information system. This can be checked by comparing the actual components with the illustrated devices shown in the architecture. If they don't match, then the analyst will have trouble implementing the security architecture as developed using the architecture unless it is corrected to reflect the actual existing system.

It is always a good idea to compare the architecture with the actual system to make sure that the architecture is correct and up-to-date. If they do not match, then the analyst should modify the reference architecture accordingly.

The benefits of an approach that includes the creation of reference and security architectures are (Cisco 2001):

1) *Speed* - Defining a framework with a created architecture and its concomitant consistent services allows rapid deployment of new applications and will enable an organization to quickly implement changes without re-engineering the network.

2) *Reliability* - Increased network uptime is due to a consistent architectural approach to network design.

---

[42] A *template* is a formatted system, sometimes a single page, that describes some object or how to prepare some product.

3) Response to change - Using an architecture-based network will decrease the time to test new solutions. Adaptation to new organizational requirements can quickly take place.

4) Interoperability - Alternative solutions work together based on a common architectural approach.

5) Simplification - Processes are streamlined because products are strategically deployed in alignment with the defined architectural framework.

6) Cost savings - Using a defined architecture, resource and time requirements are minimized, thereby reducing the cost to design and implement new networking technologies and solutions.

In some cases, an organization may create a document titled "Concept of Operations (CONOPS)" that usually describes the following elements:

1) What functions the operational personnel need to perform,
2) What functions the information system needs to perform,
3) What the network nodes are and what they do,
4) How the system should behave in normal and extreme circumstances,
5) What information is displayed and what information is transmitted to/from others, and
6) What information is needed by each node in the network.

In general the CONOPS adds operational details needed to better understand the mechanisms operating in the architecture, but an analyst can get by without this document. In fact, much of this information in a CONOPS should be provided in a well-written reference architecture.

5.1.1 Web Architectures

Web-based applications and connections to the Internet or through some ISP are often referred to as Web Architectures.

Many organizations now provide Web-based applications for users so that they can use the Internet for communicating their inputs and outputs (usually to save money on the required communications links). This requires that the various applications that will make use of the Web must be programmed in one of the different languages that offer Web capabilities, namely, HTML, XML, Cold Fusion, or the new language Curl.

Much of a customer's LAN (and PC) security for Web-based communications is, or can be, provided by the customer's ISP. A list of the attributes and their preferential value (1 meaning the least preferred and 5 meaning the most preferred) is contained in Table 5.1, *ISP Attributes and Preferential Values*, (Wetzel 2001). The values were obtained from a questionnaire mailed to 75,000 ISP customers with only 4,175 responses. Only three of the 26 attributes correlate with security parameters and are the following items:

1) Network Reliability (availability), 2) Managed Security Services (security services), and 3) VPNs. The most desired services not currently provided by ISPs, (based on 2,980

responses to the question, "What additional services would you like to receive from your ISP?"), in order of priority for the top three, are: 1) virus scanning, 2) voice-over-IP, and 3) online backup and recovery, two of which (1 and 3) are security services.

Table 5.1 ISP Attributes and Preferential Values

No.	Attribute	Preferential Value
1	Network Reliability	4.73
2	Value for Price	4.57
3	Network Performance	4.55
4	Customer Service Responsiveness	4.48
5	Time to get Service Up and Running	4.21
6	Technical Support	4.20
7	Availability of Broadband Services	4.14
8	Network Capacity (Scalability)	3.98
9	Network Reach	3.96
10	Reputation	3.86
11	Ease of Setup/Startup	3.81
12	Knowledge/Expertise of Sales Force	3.79
13	Managed Security Services	3.72
14	Quality of Services	3.65
15	Breadth of Services	3.63
16	Performance Monitoring and Tools	3.56
17	Billing	3.54
18	Service-Level Agreements	3.44
19	Managed Access	3.43
20	Value-Added Services	3.40
21	Web-Based Interface	3.39
22	Web Hosting	3.34
23	Turnkey Solution	3.31
24	Virtual Private Network Services	3.14
25	Data Center Services	2.95
26	Brand Awareness	2.81

5.1.1.1 Hypertext Markup Language

The Hypertext Markup Language (HTML) was developed as the initial Web language for creating Web pages. However, programmers rely on applets and scripts based on the Java and C^{++} object-oriented (OO, or OO-like) languages to accommodate new user-interaction models and to add functionality that is not in the HTML language. A markup language is a mechanism to identify structures in a document.

Providing animation for Web pages requires a programming language unrelated to HTML that can encapsulate the desired capabilities.

5.1.1.2   Extensible Markup and Simple General Markup Languages

The *Extensible Markup Language (XML)* is a markup language for documents containing structured information that extends the capabilities of the HTML language (Walsh 1998).

"Structured" means that the document information contains both content (words, pictures, etc.) and some indication of what role that content plays (e.g., content in a section heading has a different meaning from content in a footnote, which means something different than content in a figure caption or content in a database table). The XML specification defines a standard way to add markup to documents (Castro 2001).

"Document" refers not only to traditional documents, like this book, but also to all other XML data formats including vector graphics, e-commerce transactions, mathematical equations, object meta-data, server application programming interfaces (APIs), and many other kinds of structured information.

In HTML, both the tag semantics and the tag set are fixed. An **<h1>** is always a first level heading and the tag **<ati.product.code>** is meaningless. Changes to HTML are always rigidly confined by what the browser vendors have implemented and by the fact that backward compatibility is a necessary requirement. For those who wish to transmit information widely, features supported by only the latest releases of Netscape and Internet Explorer are not useful.

Unlike HTML, XML specifies neither semantics nor a tag set. XML is a meta-language for describing markup languages. In other words, XML provides a facility to define tags and the structural relationships between them. Since there is no predefined tag set, there cannot be any preconceived semantics. All of the semantics of an XML document will either be defined by the applications that process them or by style sheets.

XML is defined as an application profile of the *Simple General Markup Language (SGML)*. It is a restricted form of SGML. SGML has been the standard, vendor-independent way of maintaining repositories of structured documentation for over ten years, but it is not well suited to serving documents over the Web. Defining XML as an application profile of SGML means that any fully conformant SGML system will be able to read XML documents, but using and understanding XML documents do not require a system that is capable of understanding the full generality of SGML.

XML was created so that structured documents could be used over the Web (Castro 2001). The only viable alternatives, HTML and SGML, are not practical for this purpose. HTML comes supplied with a set of semantics and does not provide arbitrary structure. Similarly, SGML provides arbitrary structure, but this language is too difficult to implement just for a Web browser. Full SGML systems are capable of solving large, complex problems that justify their expense. Just viewing structured documents sent over the Web usually is not a good justification.

Perhaps XML can be expected to completely replace SGML, but while XML is being designed to deliver structured content over the Web, some of the very features it lacks to make this practical, make SGML a more satisfactory solution for the creation and long-time storage of complex documents. In many organizations, filtering SGML to XML will be the standard procedure for Web delivery (Walsh 1998).

5.1.1.3   Cold Fusion

The *Cold Fusion* language is another extension of the HTML and XML languages. Its most recent version is Cold Fusion 5, a tag-based language with visual tools. The language is produced by Macromedia, a subsidiary of Allaire Corporation. Its components include the following three capabilities: 1) server technology, 2) tools for development, and 3) an environment for programming the Web pages.

Cold Fusion offers the capability to connect with a full range of backend systems, including databases, mail servers, directories, and packaged applications. It can generate software that integrates with key Internet and enterprise technologies, including Common Object Request Broker Architecture (CORBA), Enterprise Java Beans (EJB), XML, C/C++, and Java. With Cold Fusion, organizations can develop Web pages quickly (Macromedia 2001).

5.1.1.4   Curl

In 1995, the Defense Advanced Research Projects Agency (DARPA) granted funds for the Massachusetts Institute of Technology (MIT) for a study of a new Web language called *Curl*. Tim Berners-Lee, inventor of the World Wide Web was one of three who helped start the company Curl Corporation. There are three products (Korzeniowski 2001):

1) *Curl* - the new programming language,
2) *Curl Surge* - a browser plug-in that works with either Microsoft's Internet Explorer or Netscape's Navigator browsers, and
3) *Curl Surge Lab* - a Web page development environment.

Both HTML and XML have limitations that Curl either now overcomes or is pledged to overcome. The objective of the Curl language is to avoid the use of Java and C++ languages to attain desired Web page capabilities. Curl is compatible with the Windows NT 4.0, Windows 2000, Windows 95, and Windows 98 operating systems. The newer Microsoft Corporation OSs are the Windows Millennium Edition (Me) and the next version of Windows, the Windows XP Professional, which is a 64-bit version for technical workstations (Microsoft Windows 2001). Support for the Macintosh and the Linux OSs is currently in process.

Curl uses a syntax that is consistent with semantics for expressing Web content at levels that cover simple format to contemporary OO programming capabilities. However, the Curl language is missing middleware components, so complete development of Web page products requires using an additional language but the company is attempting to add this needed capability.

By teaming with other software development companies, Curl is being assisted in the development of Web-based software, such as dynamic applications and services to improve or replace traditional, static Web applications, especially for banking software (Curl1 2001, Curl2 2001). This capability solves many disadvantages of Web development and usage.

5.1.1.5   Web Software Security

Unfortunately, the various languages now available for programming Web pages do not offer any special security capabilities. One of the associates of Curl Corporation, *adisoft AG*, offers its "banking customers and other companies secure, rapid, and cost-optimized data communications using the Curl programming language," (Curl2 1001). But how this security is provided is not described. Banks and health institutions usually have the most stringent security requirements of any U.S. organizations. Patient information is protected by information requirements regulations stipulated in the federal Health Insurance Portability and Accountability Act (HIPAA).

The principal capability for Web security seems to be to provide Web pages from servers that are located in the organization's DMZ.

5.2   Communications Networks

Communications networks are the links that provide for subsystems to communicate with other subsystems. In the past, computers and communications have been separated because computers are based on silicon technology and optical communications systems are based on the technology of light-emitting compounds. Recently, some Motorola scientists have discovered how to grow light-emitting semiconductors on a silicon wafer, thus combining the best qualities of both computing and communications technologies on a single chip (Scanlon 2001).

5.2.1   Wide Area Networks

The most popular *wide area network (WAN)* is the Internet. This network is based on the Transport Control Protocol/Internet Protocol (TCP/IP) suite of protocols that computers use to find, access, and communicate with each other. There are many other networks that are provided by different vendor called public WANs.

The TCP/IP architecture consists of four layers:

1) Application layer,
2) Transport layer,
3) Internet layer, and
4) Physical or network interface layer.

Information transmissions over the Internet make use of the TCP/IP suite of protocols for encoding, transmitting, and decoding the information. The data sent is dissembled and follows a particular path so that when it arrives at the destination it can be reassembled so that it can be read, understood, and used without any problems (assuming that there is no attempt to modify or delay the message).

Data to be sent over the Internet starts at the *Application layer*, that is, at the layer where the application program resides. This layer also contains the various services required to ensure that the transmitted data can properly interface with other devices and software products on the network. These services or utilities provide the user with connectivity, file transfer capabilities, utilities for remote administration, and Internet utilities, such as Telnet, PING, and FTP. The data is then formatted into a message for the Transport layer.

The two major components of the transport layer are the TCP and the User Datagram Protocol (UDP). This layer provides the functionality for ensuring that there is error checking, flow control, and verification of the message. TCP is a connection-oriented[43] protocol and UDP is a connection-less[44] protocol. UDP is much faster than TCP but has only rudimentary error checking and flow control. It also has reliability problems. The Transport layer formats the message for the Internet layer.

The Internet layer comprises three protocols: 1) the IP, 2) the Address Resolution Protocol (ARP), and 3) the Internet Control Message Protocol (ICMP). Each has a particular purpose. In addition, there are two other less-used protocols: 4) the Reverse Address Resolution Protocol (RARP) and 5) the Internet Group Management Protocol (IGMP).

IP addressing, a scheme for standardizing how devices are identified and differentiated from one another, occurs within the Internet layer. As long as both devices are using TCP/IP, then these devices can communicate with one another. The task of the ARP is to resolve a logical IP address into its physical equivalent address. ICMP is used by routers to send information back to a source device about the transmission. The PING (for identifying the connection) utility is transmitted using the ICMP. The message is formatted for the physical layer.

The Physical layer is the actual wire or fiber link between the communicating devices. This layer contains a collection of services and specifications that provides and manages access to the network hardware with the following responsibilities:

1) Interfacing with the network devices,
2) Checking of errors in incoming message packets,
3) Tagging outgoing messages packets with error-checking information,
4) Acknowledging the receipt of a message packet, and
5) Resending that packet if no acknowledgement is returned by the receiver.

The government is giving serious thought to implementing a private network for all government agencies in order to move aggressively on cybersecurity. This approach will require a closed IP network to be called *Govnet* (Barrett 2001b). The desire is to acquire a network that is totally separate from the Internet or other public or private networks so that, hopefully, it will be immune to functional disruptions that plague other shared networks, as well as malicious code from external networks. Proposals for Govnet are due by November 24, 2001, so the government (General Services Administration) is moving fast on this issue. The private network approach for government agencies should provide a much more secure network than moving to integration with public networks or other private networks. Govnet plans are to provide the following capabilities (Verton 2001):

---

[43] A connection-oriented protocol is one that establishes a connection with a target computer system and maintains that connection for the entire duration of message transmission.

[44] A connection-less protocol does not establish a connection with the target computer system.

1) Operate only in the US and Canada,
2) Have dedicated hardware and personnel,
3) Have no connections to the Internet or other networks,
4) Provide voice and video capabilities,
5) Use NSA encryption, and
6) Provide bandwidth-on-demand.

However, Bradner offers several arguments against Govnet in favor of an Internet approach, one of which is the need for two computers for each user who also wishes to be connected to the Internet so that they can access e-mail and various information sources (Bradner 2001). If, for any reason, the Govnet (if it ever actually exists), is connected to the Internet, then whatever security level it has is immediately reduced to the security level of the Internet, which is none. In addition, Govnet will not protect against internal attackers.

5.2.2 Metropolitan Area Networks

Metropolitan area networks (MANs) are networks that provide a central connection point for a set of local area networks as an interim network between the local networks and a WAN. Many competitive local exchange carriers (CLEC) and incumbent local exchange carriers (ILEC) are trying to eliminate the bandwidth bottleneck in the metropolitan area by extending optical technologies from the LAN and/or the WAN core into the MAN. These new MANs can potentially replace all the connectivity provided by the traditional local loop by creating MAN architectures that include (McClellan and Metzler 2001):

1) Site-to-site in the metropolitan area,
2) Access to the Internet or ISP, and
3) Access to WAN services between metropolitan areas.

MAN CLECs are using a much simpler and more homogeneous MAN architecture based on long-haul Ethernet and Ethernet Layer 2/Layer 3 switches. In the example (Figure 5.1, *Point-to-Point Ethernet-Based MAN*), the service interface units (SIUs) at one site could be dedicated to a single subscriber or shared by a number of subscribers in a large multi-tenant building.

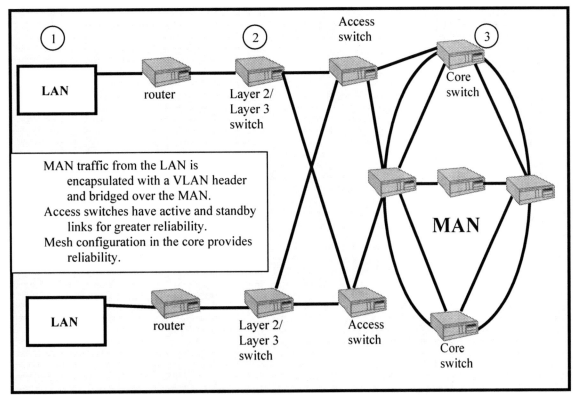

Figure 5.1 Point-to-Point Ethernet-Based MAN

In either case, the MAN traffic of each subscriber would be encapsulated with a unique 802.IQ virtual LAN (VLAN) header and bridged over the MAN. As with the transparent LAN service, additional routed ports would be configured for Internet access and WAN services. Figure 5.1 is an illustration of a simple, homogeneous MAN architecture that is built on long-haul Ethernet and Layer 2/Layer 3 Ethernet switches.

MAN architectures can also be designed based on Coarse Wavelength Division Multiplexing (CWDM) and Dense Wavelength Division Multiplexing (DWDM) technologies (McClellan and Metzler 2001).

5.2.3   Local Area Networks

Local area networks (LANs) are networks that provide connectivity for the personal computers, workstations, and local servers. LANs are frequently protected by a firewall and are connected to the Internet or other WAN or a MAN through a router. A sample network consisting of LANs, MANs, and a WAN is shown in Figure 5.2, *A Sample Network*. LANs are formed by one of the following systems: 1) Token Ring, 2) Ethernet (the most popular), and 3) Fiber Distributed Data Interface (FDDI) networks. One of the favorite architectures is the peer-to-peer architecture.

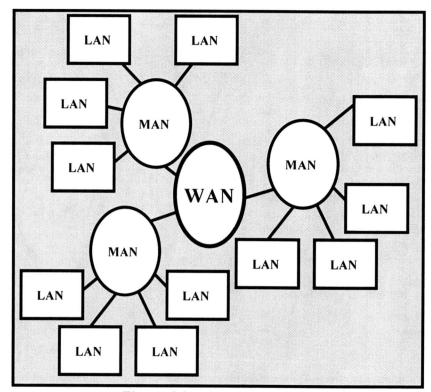

Figure 5.2 A Sample Network

5.3   Peer-To-Peer Architectures

At least 30 years ago companies were working on architectures that would now be labeled peer-to-peer but today, three factors have caused a heightening interest in the peer-to-peer movement:

1) Inexpensive computing power,
2) Network bandwidth, and
3) Increased cheap storage.

Peer-to-peer (P2P) computing is the sharing of computer resources and services by direct exchange between systems. These resources and services include the exchange of information, processing cycles, cache storage, and disk storage for files. Peer-to-peer computing takes advantage of existing desktop computing power and networking connectivity, allowing clients to economically leverage their collective power to benefit an entire organization (P2P Org 2001).

In a *peer-to-peer architecture*, computers that traditionally have been used solely as clients can communicate directly among themselves and can act as both clients and servers, assuming whatever role is most efficient for the network. This reduces the load on servers and allows them to perform specialized services (such as mail-list generation, billing, etc.) more effectively. At the same time, peer-to-peer computing can reduce the need for information technology organizations to grow parts of its infrastructure in order to support certain services, such as backup storage.

P2P computing promises to become a model capable of promoting a number of interesting distributed computing technologies into the spotlight (Sundsted 2001). In modern organizations, peer-to-peer is about more than just the universal file-sharing model popularized by Napster. These peer-to-peer capabilities include the following:

1. *Collaboration.* Peer-to-peer computing empowers individuals and teams to create and administer real-time and off-line collaboration areas in a variety of ways, whether administered or not, across the Internet, or behind the firewall. Peer-to-peer collaboration tools mean that teams have access to the latest information. Collaboration increases productivity by decreasing the time for multiple reviews by project participants and allows teams in different geographic areas to work together. As with file sharing, it can decrease network traffic by eliminating e-mail and decreases server storage needs by storing the project locally.

2. *Edge services.* Peer-to-peer computing can assist organizations by delivering services and capabilities more efficiently across varied geographic boundaries. Edge services move data closer to the point at which it is actually used by acting as a network caching mechanism. For example, an organization with sites in multiple continents has a requirement to provide the same standard training across multiple continents using the Web. Instead of streaming the database for the training session on one central server located at the main site, the organization can store the video on local clients, which act essentially as local database servers. This speeds up the session because the streaming happens over the local LAN instead of the WAN. It also utilizes existing storage space, which is a saving since it can result in the elimination of the need for local server storage.

3. *Distributed computing and resources.* Peer-to-peer computing can help organizations with large-scale computer processing needs by using a network of computers. Peer-to-peer technology can use idle computer capability and unused disk space, allowing organizations to distribute large, computationally intensive jobs across multiple computers. Results can be shared directly among peers participating in the computation. The combined power of previously unused computational resources can easily surpass the normal available power of an enterprise system without distributed computing. The results of this cooperation are: 1) faster completion times and 2) lower cost because the technology takes advantage of unused computational power available on client systems.

4. *Intelligent agents.* Peer-to-peer computing also allows networked computing subsystems to dynamically work together using intelligent agents. These agents reside on peer computers and communicate various kinds of information back and forth among the peers. The agents may also initiate tasks on behalf of other peer systems, e.g., intelligent agents can be used to: 1) prioritize tasks on a network, 2) change traffic flow, 3) search for files locally, or 4) determine anomalous behavior

(as an IDS might do) and stop the activity (e.g., a virus or worm) before it can affect the network.

5.3.1 Architectures for Advanced Peer-to-Peer Networking

One of the first commercial network architectures was IBM's System Network Architecture (SNA). Many information systems still use the SNA, but IBM has developed another architecture called the *Advanced Peer-to-Peer Networking (APPN)* architecture. The APPN is based closely on the SNA concept, which has a connection model quite similar to the seven-layer OSI model used today. Even though SNA has evolved over the years, a new architecture concept was needed and this new concept is the APPN, which supports peer-based communications, directory services, and routing between two or more network systems that are not directly attached (IBM 2001).

An illustration of the APPN architecture is shown in Figure 5.3, *An Architecture Using an Advanced Peer-to-Peer Network*. This network comprises three basic types of peer nodes: 1) *low-entry nodes*, 2) *end nodes*, and 3) *network nodes*, as shown in the figure.

A low-entry node functions as a funnel for the network services that are provided by an adjacent node depicted as a server in the figure. An end node (two are shown), accesses the network through an adjacent network node (or router). To communicate on the network, an end node establishes a control point-to-control point (CP-CP) with an adjacent network node and uses the CP-CP session to register resources, request directory services, and request routing information.

Network nodes contain full APPN functionality. The control point in a network node manages the resources of the network node, as well as the attached end nodes and low-entry nodes. Also, the control point in a network node establishes CP-CP sessions with adjacent end nodes and network nodes and maintains the network topology and directory databases created and updated by gathering information dynamically from adjacent network nodes and end nodes.

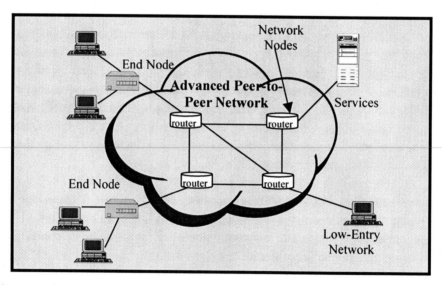

Figure 5.3 An Architecture Using an Advanced Peer-to-Peer Network

A *system configuration* is a description of how the parts or elements of a system are arranged. For example, in a LAN, certain printers, routers, firewalls, or computer systems may be disconnected or disabled. Similarly, a *software configuration* is a description of how particular elements of the software system are arranged. For instance, some applications may be disabled. *Configuration management* is the process of ensuring that a system is appropriately defined; see Figure 5.4, *Configuration Management,* below.

Configuration management is a collection of processes and tools that promote network consistency, track network change, and provide up-to-date network documentation and visibility. By building and maintaining configuration management best-practices, you can expect several benefits such as improved network availability and lower costs. These include:

1) *Lower support costs* due to a decrease in reactive support issues,

2) *Lower network costs* due to device, circuit, and user tracking tools and processes that identify unused network components, and

3) *Improved network availability* due to a decrease in reactive support costs and improved time to resolve problems.

APPN configuration services are responsible for activating connections to the APPN network. Connection activation involves three separate actions:

1) Establishing a connection,

2) Establishing a session, and

3) Selecting an adjacency option.

The minimum requirement is a single pair of sessions to one adjacent network node, which ensures proper topology updating. Reducing the number of adjacent nodes increases the time required to synchronous routers.

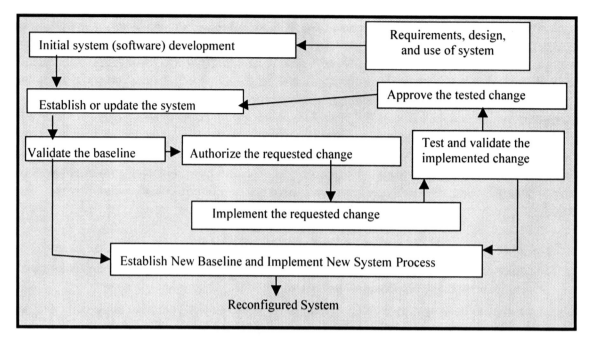

Figure 5.4 Configuration Management

Directory services assist the network devices in locating service providers. Establishing a session is dependent on these services. A central directory or database can be used to maintain a centrally located directory for an entire network because it contains configured, registered, and cached entries.

Topology management in the APPN topology is accomplished using a complete picture of all elements or devices in the network architecture. Whenever the configuration is modified, the picture is also modified to reflect the current system configuration.

5.3.2 Architectures for Voice, Video, and Integrated Data

Cisco created the *Architecture for Voice, Video, and Integrated Data (AVVID)* framework, which is a standards-based (as required for openness) design to avoid the use of proprietary networks for system architectures. This architecture leads to a roadmap for organizations wishing to develop an information system (local or distributed) that has the advantages of an open architecture, namely portability, scalability, and interoperability.

This architectural approach offers the various end-to-end networking services that enable an organization to take advantage of the rapidly changing elements of the modern information technology environment. The newer architectures comprise three levels or tiers:

1) Clients,
2) Applications servers, and
3) Database servers.

Because of the open nature of the AVVID, this approach speeds up both the acquisition and implementation of emerging technologies and cost savings through the capability to promote competition. Many products, such as the *de facto* standard Microsoft Windows operating systems, have well-documented application programming interfaces (APIs) that enable interaction among hardware devices and software applications. The AVVID relates well to the evolved networks, such as the TCP/IP[45] Internet systems, the UNIX systems, and the Ethernet, Token Ring, and FDDI networks.

The AVVID architecture approach advances the state-of-the-art in voice, video, and data solutions that take advantage of improvements in PC processing power and networking standards so that voice, video, and data can be transmitted along a single IP-based network infrastructure.

The AVVID architecture provides the capability to connect the various client systems (PCs, laptops, PDAs, and cellular phones) to network platforms (routers, LAN switches, gateways, servers, and other devices). It also can provide quality-of-service (QoS) through security (i.e., availability options and intrusion detection systems), multi-protocol label switching (MPLS), and intelligent network services (Cisco 2001).

Service control and communications services are key parts of any networking architecture because they provide the software and tools to break down the barriers of complexity arising from new and emerging technologies. These services enable integrators and customers to tailor their organizational network infrastructure and customize intelligent network services to meet specific application requirements. These packages can perform the following actions:

1) Perform perimeter security (i.e., firewall software),
2) Perform access control,
3) Identify and provide user privileges (i.e., identification, authentication, and authorization),
4) Call setup and teardown configuration management software, and
5) Prioritize and allocate communications bandwidth.

These services (service control and communications services) are the glue that connects the layers of the AVVID framework with the Internet solutions. Vendor product solutions (such as Database Management Systems (DBMSs) and Decision Support Systems (DSSs)) that abound in the information technology environment are available through network connections can be enabled, accelerated, and delivered through utilization of the AVVID architecture.

5.3.3 Asynchronous Transfer Mode

Asynchronous Transfer Mode (ATM) is a means of digital communications for voice, video, images, and data that is capable of very high speeds (up to 10 Gigabits/second (Gbps) today with up to 40Gbps in the future). ATM uses short, fixed-length packets called cells for transport. Information is divided among these cells, transmitted, and then re-assembled at their final destination using packet numbers. ATM is a standard from the International Telecommunications Union-Telecommunication Standardization Sector (ITU-T) for

---

[45] "TCP/IP" is an industry standard suite of protocols that computers use to find, access, and communicate with each other over a transmission medium. It is the standard protocol suite for the Internet.

information that is conveyed in small, fixed-size cells, and is sometimes called *cell relay* (ATM Forum 2001).

A typical ATM network consists of a collection of different types of ATM switches that are interconnected by ATM switch-to-switch links, (or network-to-network interfaces, public or private) into an ATM network (Alles 2000). ATM connections can be of two types, permanent virtual connections (PVCs) or switched virtual connections (SVCs). PVCs are connections that are set up through administrative actions. SVCs are set up by signaling protocols operating either between the end systems[46] and the ATM network or within the ATM network.

ATM is compatible with the telecommunications infrastructure for use as the backbone for other networks and possesses key network interfaces and protocols. ATM is the world's most widely deployed backbone technology. It is used at the perimeter and as internal networks for many organizations. It is best known for being easy to integrate with other technologies and for its sophisticated management features that allow alternative carriers (i.e., vendors) to guarantee quality of service (QoS). These features are built into the different layers of ATM, giving the protocol an inherently robust set of controls (ATM Forum 2001). ATM networks can provide connections with a guaranteed QoS, which can facilitate the operation of networked multimedia and continuous bit rate applications.

Any telecommunications network is designed in a series of layers so that a typical configuration may have used a mix of time division multiplexing (TDM), Frame Relay, ATM, and/or IP. Within a network, carriers often extend the characteristic strengths of ATM by blending it with other technologies, such as ATM over Synchronous Optical Network/Synchronous Digital Hierarchy (SONET/SDH) or Digital Subscriber Link (DSL) over ATM. By doing so, vendors extend the management features of ATM to other platforms in a cost-effective manner. A blend of ATM, IP, and Ethernet options can be found in the WAN. When carriers expand to the WAN, most do so with an ATM layer.

ATM consists of a series of layers. The first layer (the adaptation layer) holds the bulk of the transmission, which is a 48-byte payload, and divides the data into different types. The ATM relay information contains five bytes of additional information, referred to as *overhead* so that the total packet contains 53-bytes of information. The overhead information directs the transmission. The final layer (the physical layer) connects the electrical elements with the network interfaces. It can act as a unique bridge between legacy equipment and the new generation of operating systems and platforms. ATM easily communicates with both legacy and new equipment, allowing carriers to maximize any organization's investment in its information technology infrastructure.

The LAN environment of a campus or building appears sheltered from the headaches associated with high-volume traffic that is transported on larger networks. However, the changes of LAN interconnection and performance are no less critical. ATM is a proven technology that is now in its fourth generation of switches. Its strength is in its capability to anticipate the market and quickly respond.

Distance is not a problem with ATM since the integrity of the transport signal is maintained even when different kinds of traffic are traversing the same network. Because of

---

[46] In the OSI model, an end system is the computer containing application processes that can communicate through all seven layers.

its capability to scale up to very high data rates (e.g., 2.5 Gigabits/second or higher), different services can be offered at varying speeds and at a range of performance levels.

The Metropolitan Area Network (MAN) is one of the fastest growing areas in the communications sphere. Traffic may not travel more than a few miles within a MAN, but this information is generally doing so over leading edge technologies and at very high speeds. A typical MAN configuration is a point of convergence for many different types of traffic that are generated at many different sources. The appeal of ATM as a MAN is that it easily accommodates divergent transmissions, often connecting legacy equipment with very high-speed networks (ATM Forum 2001).

Not all ATM switches are alike. Although all ATM switches will perform cell relay, these switches differ in terms of capabilities such as the variety of interfaces and services supported, redundancy, the depth of ATM networking software supported, and the sophistication of traffic management mechanisms (Alles 2000).

Enterprise switches are sophisticated multi-service devices designed to produce the core backbones of large WANs thus complementing the duties of high-end multi-protocol routers that are commonly in use today. Enterprise switches are capable of supporting LAN switching, packet WAN interfaces such as frame relay, ATM-connected servers, and ATM routers. Enterprise switches can also serve as the single point of integration for all of the various services and technologies employed in modern enterprise backbones.

By merging all of these services onto a common platform and ATM transport infrastructure, modern network planners can achieve greater manageability and eliminate the need for complex overlay networks.

Just as for any transmission medium, security over an ATM network is accomplished using encryption of message traffic.

5.3.4   Point-To-Point Tunneling Protocol

The Point-To-Point Tunneling Protocol (PPTP) is an extension of the Point-To-Point Protocol (PPP). PPP is the choice for connecting dial-up serial mode to ISPs (R. Smith 1997). PPTP acts like a single, coherent link-level protocol that allows it to carry a variety of Internet and non-Internet networking protocols between the devices that provide support to it.

The PPTP and PPP protocols do not provide encryption services but can be provided by extension protocols to provide confidentiality and data integrity protection between sites. These are the building blocks of VPN.

Often, dial-up connections make use of the Serial Line Internet Protocol (SLIP) and the PPP, that provide IP connectivity over standard telephone lines. SLIP and PPP offer almost full connectivity to the Internet. The only real difference between a PPP/SLIP connection to the Internet and a dedicated connection to the Internet, other than the difference in bandwidth, is the somewhat lower availability of PPP/SLIP links.

5.3.5   Frame Relay and Multi-Protocol Label Switching

Frame relay is a high-speed communications technology developed in response to the expressed need for cheaper lines than leased lines for connecting LAN, SNA, Internet, and voice applications. It is a way of transmitting information over a WAN that divides the information into frames or packets. Each frame has a label that the network can use to decide where the packet should go. The frame can carry multiple network layer protocols (including IP) and because it is a connection-oriented approach, it becomes a simple reference to a virtual connection (Hopkins 2001).

Frame relay is also capable of providing WAN connectivity, but it does so across a shared public network. It is a widely implemented packet-switching protocol that offers an alternative to virtual private network lines or leased lines and is primarily used for data communications and is not recommended for voice links.

Frame relay and X.25 are packet-switched technologies. Both must be configured as permanent virtual circuits (PVCs), meaning that all data entering the cloud at point A is automatically forwarded to point B. For large WAN environemnts, frame realy can be far more cost effective than dedicated circuits. The WAN connection point is defined through the use of a unique *data link connection identifier (DLCI)*. An attacker can divert traffic to their network by using the same local exchange carrier and the same physical switch, but must know your DLCI.

Frame relay is a telecommunication service designed for cost-efficient data transmission for intermittent traffic between local area networks (LANs) and between end-points in a WAN. It places data in a variable-size unit called a frame and leaves any necessary error correction (retransmission of data) up to the end-points, which speeds up overall data transmission. For most services, the network provides a PVC, which means that the customer sees a continuous, dedicated connection without having to pay for a full-time leased line, while the service provider figures out the route each frame travels to its destination and can charge based on usage. An enterprise can select a level of service quality - prioritizing some frames and making others less important.

Frame relay is offered by a number of service providers, including Worldcom/MCI, Sprint, Quest, and AT&T. Frame relay is provided on fractional T-1 or full T-carrier system carriers. Frame relay complements and provides a mid-range service between Integrated Services Digital Network (ISDN), which offers bandwidth at 128Kbps, and ATM, which operates in somewhat similar fashion to frame relay but at speeds from 155.520Mbps or 622.080Mbps or higher rates. Frame relay:

1) Is based on the older X.25 packet-switching technology, which was designed for transmitting analog data such as voice conversations. Unlike X.25, which was designed for analog signals, frame relay is a fast packet technology, which means that the protocol does not attempt to correct errors. When an error is detected in a frame, it is simply "dropped" (i.e., thrown away). The end points are responsible for detecting and re-transmitting dropped frames. (However, the incidence of error in digital networks is extraordinarily small relative to analog networks.)

2) Is often used to connect local area networks with major backbones as well as on public WANs and also in private network environments with leased lines over T1 lines. It requires a dedicated connection during the transmission period. It's not ideally suited for voice or video transmission, which requires a steady flow of transmissions. However, under certain circumstances, it is used for voice and video transmission.

3) Is based on relay packets at the data link layer of the Open Systems Interconnection (OSI) model rather than at the Network layer. A frame can incorporate packets from different protocols such as Ethernet and X.25. It is variable in size and can be as large as a thousand bytes or more.

Multi-Protocol Label Switching (MPLS) is a development by the Internet community and seeks to combine the flexibility of the IP network layer with the benefits granted by a connection-oriented approach to networking. MPLS, like Frame Relay, is a label-switched system that can transport multiple network layer protocols, and similar to Frame Relay, MPLS sends information over a WAN in packets (or frames) (Hopkins 2001). MPLS addresses requirements for:

1) Efficient and simplified high-speed forwarding IP packets,
2) Provision of scalable networks,
3) Control over quality of service (QoS), and
4) Traffic engineering and the control of traffic routing.

MPLS networks can use Frame Relay, ATM, and PPP as the link layer. In MPLS networks, either an explicit path can be defined or the IP routing mechanism can determine the path. The use of explicit paths enables MPLS to perform traffic engineering by considering required bandwidth as a constraint for the computed path.

Since Frame Relay is a label-switching system, using Frame Relay as one of the underlying layer 2 (layer 2 is the data link layer that uses frames for message transmission) technologies for supporting MPLS is reasonable because Frame Relay switches become Label Switched Routers (LSRs). However, the more likely scenario is to use ATM or router-based MPLS core networks with Frame Relay forming a highly appropriate access to these MPLS core networks.

A layer-3 approach requires that the MPLS network do IP lookups. An IP lookup is performed by the LSR and the packet is encapsulated in a Frame Relay frame for delivery to the destination. However, no firm definition of how Frame Relay and MPLS should interact has been identified, since both the layer 2 and layer 3 options both have certain advantages and disadvantages. MPLS can allow an organization to set up VPNs through service provider networks (Giacalone 2001).

5.4   Architecture Creation

Architectures are created from the following three inputs:

1) The Security Policy,
2) A Security Rules Base, and
3) The Security Requirements (architecture level).

and have five principal components,

1) WANs and/or the Internet,
2) Local backbones or communication rings,
3) Client and server computers,
4) Intranets, and
5) Extranets.

The *WAN or Internet* is the communications method for transferring information among the intranets and extranets. The *backbone communication rings* are the connections of the intranet or extranet elements. The *client and server computers* are the computers that support the users (i.e., clients) and the system database and management, applications, middleware, and services (i.e., servers, which may be mainframes). *Intranets* provide LANs for the enterprise at specific sites. *Extranets* are the LAN sites for organizations outside the enterprise.

Client systems can consist of desktop PCs, laptop PCs, and PDAs or cellular phones that are connected to one another and to the various servers via ISPs, modems, and wireless networks.

A simple overview of a generic enterprise architecture, without any security countermeasures, is shown in Figure 5.5, *Generic Enterprise Architecture: Without Security*.

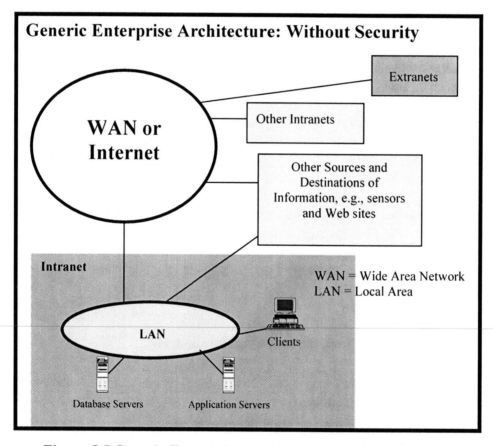

Figure 5.5 Generic Enterprise Architecture: Without Security

Architectures can have many different forms, but the most important one is a graphical depiction of the entire enterprise network showing the major system components, communications, computers, network devices, and information flow. Architectures can be of varying detail so that they can represent an overview of the system or a very detailed presentation of the system, in which case the representation of the components will be very small on regular-sized graphical paper or regular sized on large graphical paper. General architectures can be created using the user requirements whereas a very detailed architecture will require either the architecture requirements or the system specifications (for hardware and software component selection or development).

In some cases, an organization may have an existing information system but no architecture model; perhaps it was lost or never existed. In any case, before a security architecture can be developed, an architecture of the reference system must be obtained or created. The architecture, in this case, can be created from the actual existing information system itself. By identifying each component and seeing how it is connected to other components, a technical person can eventually develop the graphical and verbal architecture models.

Assume that there is physical security at the intranet sites, that is, persons entering the facility must have a badge and that there are badges, guards, fences, gates, and cyber clocks to ensure that only authorized persons can enter the facility. Because of local area physical security, the security needs for the intranet mostly apply to information coming in or going out on the intranet.

5.5   Single Sign-On

Most users who have several applications to which they must access, prefer a Single Sign-On (SSO) process (Murray 1999) to minimize their memory and activity requirements as well as improving security, since users tend *not to paste* their passwords on their monitor when there is just one password to remember. However, if they have two or more passwords to remember, they may have written copies of the passwords pasted on their monitor. Also the need to transmit passwords over the links is minimized because users log on to a local server, thereby increasing security. Thus, an SSO capability contributes to system security. Some recommendations for implementing an SSO capability are offered in Table 5.2, *Single Sign-On Recommendations*.

Table 5.2 Single Sign-On Recommendations

No.	Recommendation
1	Implement an RBAC capability for access control and this will facilitate the implementation of the SSO system. This is because the identification of roles and role privileges is important to the SSO system.
2	Select the appropriate SSO software. For example, the WebLogic server and concomitant software is applicable to most situations and is used by 90% of commercial customers.
3	Select an SSO system that can provide extensibility later and can adapt to the organization's needs (such as Web servers, email servers, and portal servers).
4	Keep the RBAC tables up-to-date by quickly removing any user who no longer needs access and promptly adding new users.

No.	Recommendation
5	Ensure that the SSO system handles user authentication efficiently so that the response time is not noticeable to the users.
6	Avoid duplication of user access control lists when implementing the SSO system.

SSO is the capability for any authorized user to sign on once, and then access any application or database (to which they have access privileges), for the duration of a session (after which the user logs off) or for up to 24 hours.

SSO can be trivial or complex depending on the vendor selected and the type of network that the organization has. A suggested SSO capability is the one provided by BEA Systems, Inc using their WebLogic server and concomitant software. Information on the WebLogic server[47] can be found at the following URL address: http://www.weblogic.com/docs51/resources.html.

A soon to be released standard, the Security Assertion Markup Language (SAML) based on XML, will provide an SSO capability across the multitude of Web access management platforms available today (Ploskina 2001). However, Microsoft is not a player in the SAML standard because Microsoft already has an SSO capability called Passport, which is provided with the Kerberos capability in the Windows operating systems (Ploskina 2001).

5.6   Example of a Quick-Look Security Analysis

Generally, security protection consists of the following capabilities:

1) *Identification and authentication* - ensuring that an entity is identified and is who or what they say they are,

2) *Availability* - ensuring that the system or its information is accessible by any user when they want or need it,

3) *Confidentiality* - ensuring that information is not accessed and understood by unauthorized entities,

4) *Data integrity* - ensuring that information is not modified by unauthorized entities,

5) *Access control* - ensuring that all authorized users have proper access to information and applications,

6) *Nonrepudiation* - ensuring that the sender (receiver) cannot later deny have sent (received) a message that was actually sent and received, and

7) *Accountability* - ensuring that each network entity can be held responsible for any detrimental action (either intentional or unintentional) they cause.

A quick-look risk sample analysis for each security category for a generic system might reach the following conclusions for threats and requirements:

---

[47] In many cases the word "server" actually means "software."

Security Issue #1 (Identification and Authentication) - The *threat* is "unauthorized persons acting as users," which **implies** the *requirement* for a proper identification and authentication capability such as implementing a User Identification and Password Authentication, and for strong authentication, a biometric authenticator, e.g., fingerprint. (One can argue that because of the physical security at each intranet facility, there is no need for system identification and authentication. But is this physical security absolutely perfect? To ensure that users who log on are really who they say they are will require strong authentication, such as a biometric identifier.)

Security Issue #2 (System Availability) - The *threat* is "potential down time for a device or no access to information," which **implies** the *requirement* for clustered computers (or hot backups) with load balancing and automatic failover software. (Some may argue that this is not a security issue, however, single-point-of-failure and loss-of-information-availability are usually included as security issues.) Another aspect of this requirement is the blocking of virus (or malicious) codes or denial-of-service (DoS) attacks. Malicious code blocking software is generally called anti-virus software.

Security Issue #3 (Confidentiality) - The *threat* is "falsification or corruption of critical information, either intentionally or unintentionally (such as a bit error during transmission)," which **implies** the *requirement* for encryption of messages transmitted over the WAN or Internet. Implementation of confidentiality measures can also satisfy the *data integrity* aspect of security. All information that is considered to be sensitive, for any reason, should be transmitted in a manner that ensures confidentiality.

Security Issue #4 (Data Integrity) - The *threat* is "unauthorized modification of stored or transmitted information," which **implies** the *requirement* for a method of identifying any changes in the information from its original form.

Security Issue #5 (Access Control) - The *threat* is "unauthorized access to sensitive information," which **implies** the *requirement* for an access control method, such as Role-Based Access Control.

Security Issue #6 (Nonrepudiation) - The *threat* is "a user who transmits a message and may later deny having sent the message," which **implies** the *requirement* for a digital signature included in the information transmitted.

Security Issue #7 (Accountability) - The *threat* is "a user may perform some malicious action (either intentionally or unintentionally) and, in this case, the specific user should be identified as the perpetrator of a malicious action," which **implies** the *requirement* for auditing user (or application) actions. (Because of supervisor activities and the retention of strip recordings, (for FAA environments) it may be that the user and the associated action can always be identified without an audit trail. But again, is this manual process absolutely foolproof?)

151

In addition, there is a security issue of *no-object-reuse*, which means that any sensitive information left in the computer system by some user who is logging off should be erased prior to another user logging on. In some cases no-object-reuse is not a requirement, such as for Federal Aviation Agency (FAA) flight controllers who must be able to take over as a replacement flight controller and pick up the various tasks where a previous flight controller left off. Erasing all previous information would be very detrimental to the assumption of the tasks (viz., monitoring and advising aircraft pilots during the duration of their flight, from gate departure for takeoff, to en route, to landing and gate arrival).

The security countermeasure devices that can meet the security requirements are as follows:

1) *Identification and Authentication* - Software for obtaining and checking user IDs and passwords are required. For strong authentication, some type of biometric capability may be required. Also, firewalls that can exclude unauthorized personnel and include authorized personnel are required.

2) *Availability* - Clustered servers with load balancing and automatic failover software are required. In addition, there may be a requirement for anti-virus software and anti-DoS software that perhaps can be installed in the firewall.

3) *Confidentiality* - Encryption of transmitted messages can be performed by software on the firewall through the use of VPN software. Encryption can also include data integrity capabilities.

4) *Access Control* - Role-based access control can be provided by some database management systems (e.g., Oracle 8i), but also can be provided by vendor commercial off-the-shelf (COTS) software.

5) *Nonrepudiation* - This requirement, if needed, can be provided by the use of digital signatures, which requires that each user have a Security Certificate. However, at present, nonrepudiation appears not to be a requirement.

6) *Accountability* - This can be provided by auditing all system activities and making sure that each user is properly identified. In addition, IDSs may be required to identify intrusions into the LAN and WAN links.

Physical security requires that users have badges and can enter a secure facility through cyber-locked doors (requiring a password code for entry) or pass through a fenced-in enclosure with gates that are manned by guards.

The behavior aspect of security requires that training and awareness capabilities be provided. Thus, each user, manager, or other personnel should be provided with a training program appropriate to their needs as well as awareness training to ensure that all personnel are knowledgeable of the various security situations such as:

a) Protect portable computers,
b) Do not give out information to callers who are unable to identify themselves as authorized personnel,
c) Do not open e-mail attachments from sources whom you do not know, and
d) Other things that may enable a malicious action.

All critical information should be protected with a *backup and recovery system*. This type of capability is facilitated if all data resides in a single location on a storage area network. The data should be backed up periodically (at a remote site that would not be affected by a catastrophe at the primary site) and there should be recovery procedures for restoring the information in case some sort of catastrophe causes a loss of critical data. A Contingency Plan should be developed to ensure that the system is minimally affected by the loss of information or its information storage capability. This contingency plan should be periodically practiced. An organization that provides an outsourcing capability for storage is Sanrise Incorporated, a storage service provider (SSP), at http://www.sanriseinc.com/.

An illustration of a generic enterprise system with the security devices added is shown in Figure 5.6, *Generic Enterprise Architecture: With Security*. Protection of the local area network equipment and applications at the intranet can be provided by a firewall.

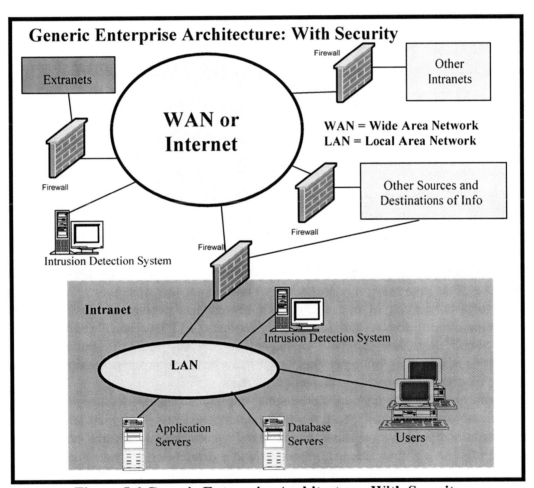

Figure 5.6 Generic Enterprise Architecture: With Security

Firewalls require a set of rules to ensure that authorized personnel and messages are allowed to pass through to the LAN and that unauthorized personnel and messages are blocked (i.e., disallowed). In addition, message protection (confidentiality and data integrity for information traveling over the WAN or Internet) can be provided by a VPN encryption capability at the firewall. It is assumed that message security is not required for information traveling inside the LAN.

The servers for encoding the data can be reduced to just encryption if the information going to the destination is not considered sensitive. Intrusion detection systems (IDSs, on the LANs and WAN links) can create audit trails of all LAN or WAN activity, and then, either online or offline, analyze the audit trials and take appropriate actions, such as notifying the System Security Officer (SSO).

The security architecture suggested here is compatible with the various COTS products (hardware and software) that are currently available as well as those that will be issued in the future assuming that the newer emerging technologies conform to the standards defined for open architectures. This suggested security architecture, when its subsystems are acquired, installed, joined as a unified enterprise system, and tested will be composed of components that are interoperable, scalable, and portable. In cases where proprietary products must be used, the other security architecture components recommended will be minimally affected by the non-standard product. The case may be that through the use of specifically generated products, security may be enhanced.

A list, in matrix form, of the vulnerabilities for the three areas of security control, is shown in Table 5.3, *General Vulnerability Area/Security Control Area Matrix.*

5.7   Client/Server Architectures

The most popular information system design is the client/server architecture. In its modern form, the architecture consists of three tiers:

> Tier I - The client or personal computer tier (front-end tier).
> Tier II - The application servers (middle tier).
> Tier III - The database management system servers (backend tier).

This design is just an extension of the client/server two-tier architecture and two improvements that have made this progression possible are the: 1) higher reliability and 2) lower costs of servers. In fact, the cost of servers has become so low that many different functions of the system (e.g., Web services, FTP services, electronic mail services, portal[48] services, database management services, domain name services, firewall services, and encryption services) are often placed on separate server platforms to enhance the throughput and responsiveness of the system.

SSP Solutions, Inc has developed a personal computer system (i.e., client) that contains an entire repertoire of security features (viz., smart card and PIN for identification and authentication and protection against unauthorized access; secure Internet communications for confidentiality and data integrity; intranet and extranet access; and access control to sensitive information) to ensure that the PC is properly protected against a gamut of threats (Catrerinicchia 2001). However, in many cases the primary security protection is contained in servers, such as identification and authentication, access control, encryption/decryption

---

[48] *Portal services* are provided to external users who wish to obtain information about an organization by accessing the organization's portal servers over the Internet.

for confidentiality and data integrity, firewall protection, and anti-virus and anti-DDoS capabilities. Thus, client protection might be redundant or even not needed.

Table 5.3 General Vulnerability Area/Security Control Area Matrix

No.	Vulnerability Area	Physical	Behavioral	Technical
1	System access control			X
2	External connectivity and telecommunications			X
3	System integrity monitoring and reporting (e.g., virus protection)			X
4	Session control			X
5	System auditing			X
6	Account management			X
7	System backup			X
8	System contingency planning/disaster recovery			X
9	Maintenance			X
10	Configuration management			X
11	Labeling and data control			X
12	Sanitization and disposal			X
13	Information system security documentation		X	
14	Information security roles and responsibility		X	
15	Training and awareness		X	
16	Facility physical access	X		
17	Physical environment	X		
18	Personnel security		X	
19	Contingency plan and disaster recovery	X		

In any architecture, load balancing of the servers can be used to improve security. For example the Cisco operating system has a server load-balancing scheme that relies on a site's firewalls to protect the site from most attacks. In general, the operating system is no more susceptible to direct attack than is any switch or router but any site can take the following three steps to increase its security capabilities:

1) Configure the firewalls in a private network to protect against attackers trying to access real or nonexistent IP addresses in the firewall subnet.

2) Deny all access attempts from unexpected addresses by configuring input access lists on the access router or on the server operating system device to deny

accesses from the external network aimed directly at the interfaces on the server device.

3) Configure firewalls to deny all unexpected access attempts targeted at the firewalls, especially those originating from the external network.

As mentioned above, the system's firewalls can have a load balancing capability to ensure maximum throughput and with an automatic failover, so that the system can gracefully degrade should a firewall server suddenly fail.

Because of the increasing use of distributed systems, the desirability of integrating middle-tier application servers with backend database management servers is becoming quite compelling (Biggs 2001). Some benefits of this approach are:

1) Security measures are improved (e.g., SSO is simplified and authorization can be defined to the database row level),
2) New applications can be deployed faster,
3) Time required to access data is decreased (i.e., improved),
4) Advanced options (e.g., clustering, load balancing, and automatic failover) improve reliability and scalability,
5) Applications are more reliable,
6) Data exchanges are more reliable,
7) Applications are easier to manage,
8) Administrative overhead is reduced,
9) Applications can be more easily expanded to meet increased workloads, and
10) The adoption of emerging technologies, such as Web services, is easier.

The integration of application and database servers is totally dependent upon the XML technology (Biggs 2001).

REFERENCES AND BIBLIOGRAPHY

5.1   Smith, Sr., Charles L. (1998). "Computer-Supported Decision Making: Meeting the Demands of Modern Organizations," Ablex Publishing Corporation, Greenwich, Connecticut and London, England.

5.2   Murray, William Hugh, (1999)."Enterprise Security Architecture," *Handbook of Information Security Management*, Micki Krause and Harold F. Tipton, Editors, Auerbach, pp. 371-386.

5.3   Korzeniowski, Paul (2001). "A New Web Language Is Born," *eAI Journal*, pp. 31-32, August 2001.

5.4   Curl1 (2001). "Curl Corporation and BTexact Technologies Announce Software Agreement," Curl Corporation, http://www.curl.com/html/, 2001.

5.5   Curl2 (2001). "Adisoft AG to Develop Future Online Banking Software in Curl," Curl Corporation, http://www.curl.com/html/, 2001.

5.6   Macromedia (2001). "Macromedia Cold Fusion 5," http://www.macromedia.com/software/coldfusion/productinfo/why_use_cf/, 2001.

5.7   Walsh, Norman (1998). "A Technical Introduction to XML," http://www.xml.com/pub/a/98/10/guide0.html, October 1998.

5.8   IBM (2001). "IBM Systems Network Architecture (SNA) Protocols," http://www.cisco.com/univercd/cc/td/doc/cisintwk/ito_doc/xtocid151470, 2001.

5.9   Cisco (2001). "White Paper: Cisco AVVID—The Architecture for E-Business," http://www.cisco.com/warp/public/cc/so/cuso/epso/avpnpg/vvaid_wp.htm, 2001.

5.10   Sundsted, Todd (2001). "A new-fangled name, but an old and useful approach to computing," http://www-106.ibm.com/developerworks/java/library/j-p2p/, Chief Architect, PointFire, Inc., March 2001.

5.11   P2P Org (2001). "What is peer-to-peer?" http://www.peer-to-peerwg.org/whatis/index.html, P2P Organization, 2001.

5.12   ATM Forum (2001). "About ATM," http://www.atmforum.com/pages/aboutatmtechfs1.html, 2001.

5.13   Scanlon, Bill (2001). "Motorola Solves 30-Year Optical-Silicon Chip Puzzle," Interactive Week, p. 18, September 10, 2001.

5.14   Hopkins, Harman H. (2001). "Frame Relay + MPLS and How They Fit Together," Frame Relay Forum http://www.frforum.com/4000/whitepapers/MPLSwhitepaper.html), February 2001.

5.15   Alles, Anthony (2000). "White Paper: The Next-Generation ATM Switch: From Testbeds to Production Networks," Cisco Systems, http://www.cisco.com/warp/public/cc/so/neso/vvda/atm/ngatm_wp.htm, 2000.

5.16   Microsoft Windows (2001). "Windows Operating Systems," Microsoft, http://www.microsoft.com/windows/default.asp, 2001.

5.17   Castro, Elizabeth (2001). "XML for the World Wide Web," Peachpit Press, 2001.

5.18   Biggs, Maggie (2001). "Putting App Servers to Work," Federal Computer Week, pp. 30-31, August 20, 2001.

5.19   Wetzel, Rebecca (2001). "Getting What You Pay For," Interactive Week, pp. 34, 37, September 10, 2001.

5.20   Ploskina, Brian (2001). "Sign-On-And-Go Security," Interactive Week, p. 12, October 1, 2001.

5.21   Ptacek, Thomas H. and Timothy N. Newsham (1998). "Insertion, Evasion, and Denial of Service: Eluding Network Intrusion Detection," Secure Networks, Inc., http://secinf.net/info/ids/idspaper/idspaper.html, January 1998.

5.22   Giacalone, Spencer (2001). "MPLS VPNs Improve WAN Connectivity," Network World, p. 43, October 15, 2001.

5.23   Barrett, Randy (2001b). "Bush Wants Separate Net," Interactive Week, Page 14, October 15, 2001.

5.24   Bradner, Scott (2001). "Does Going It Alone Make Sense?" Network World, Page 26, October 22, 2001.

5.25   Catrerinicchia, Dan (2001). "SSP Unveils Secure PC," Federal Computer Week, http://www.ssp.com/, p. 34, November 5, 2001.

5.26   McClellan, Rolf and Jim Metzler (2001). "Designing the New Metropolitan Area Network (MAN)" Network World, pp. 48-50, November 5, 2001.

5.27   Verton, Dan (2001). "Bush Plan to Unplug Feds From Internet Draws Criticism," Computer World, p. 7, November 5, 2001.

PART II. DEFINING THE PROBLEM

This section contains a presentation of the process for defining the problem. It is called a Risk Assessment. It also contains a discussion of the threat, vulnerabilities, threat scenarios, risk computations, and alternative countermeasures. There is also a discussion of the return on investment (for monies spent on countermeasures versus the security benefits provided).

Charles L. Smith, Sr.

CHAPTER SIX

RISK ASSESSMENT

6.1 Overview

There are many types of risk, but the primary type is security risk and is often referred to simply as "risk." Some other types of risk are technical, schedule, and performance. The primary type of risk is the most important and is explained in detail in this book. The other types of risks are also addressed but in a more limited manner.

Risk, for some specific threat, is the potential loss that could occur to the organization should its information system be attacked (by this specific threat) and hence is a probabilistic cost. Thus, the total risk, for any organization's information processing system vis-à-vis the gamut of relevant threats (and concomitant scenarios[49]), is the sum of all the risks (i.e., costs) computed for each of the threats. Some of the risks will be much larger than others and the probabilities for each particular type of attack can also vary from miniscule (very near zero) to highly likely (near one).

Attempts to *quantify* risk probabilities and costs (i.e., numerical estimates) are not recommended because the numbers will be very subjective and most possibly quite inaccurate since they are guesses based on little or no information. Instead, the recommended method here is to *qualify* risk probabilities (e.g., low, average, and high) and losses (minimal, medium, or maximal) in terms that can be reasonably estimated. These in turn can be quantified, such as low = 0.1, average = 0.5, and high = 0.8, and minimal = $10,000, medium = $50,000, and maximal = $1,000,000. Thus, it is possible to quantify the risks even when they were first estimated qualitatively.

Some organizations are interested in both risk and safety. *Safety*, on the other hand, is the likelihood that persons or property may suffer physical harm (including death) or physical damage should the organization's information system not function in an appropriate manner. Safety is not an issue that is addressed here.

At a fundamental level, a *risk assessment* is a procedure to estimate potential losses that may result from system vulnerabilities (and concomitant threats) and to quantify this potential damage so that a set of cost-effective countermeasures can be identified (Russell 1991). A risk assessment begins with identifying the information system assets and their values to the organization and comparing these values with the costs of relevant countermeasures to identify those that are cost-effective because of a desire to protect these assets against identified threats.

Risk management is the process of dealing with risks through the use of countermeasures and attempting to limit damage from the various threats to the information system. Thus, risk assessment is a part of risk management. Risk management is only briefly covered in this book.

6.1.1 Risk Assessment Process Considerations

A *risk assessment process* should be simple and user-friendly for it to be advantageous to the typical organization, commercial or government. Software products for performing a risk assessment are available, however they are usually not very good and there is not one that I would recommend. The most recent addition to the list of risk assessment tools is

---

[49] A *scenario* is a verbal description of a situation in which some particular set of events results from some particular cause.

Carnegie Mellon's OCTAVE^{SM}, which was presented in Section 1.1.2, *Overview*. One of the reasons that I do not recommend these tools is that these products are so complicated that I do not believe that the typical organization would wish to embark on a risk analysis based on using such a complex product. Also, many of these products are quite expensive. However, these opinions are a subjective assessment and some organizations may in fact *want* to use one of these products. An exhaustive search for a risk assessment tool was not undertaken so there is a possibility that a good one exists, but the author is not aware of it.

Appendix III of the Office of Management and Budget (OMB) Circular A-130, states that a risk assessment should be a part of a risk-based approach to determining adequate, cost-effective security for an information system network. This is why most, if not all, government agencies require a risk assessment as part of their security analyses. Unfortunately, it is often the case that the people charged with doing the risk assessment are not knowledgeable of how one should be performed nor are the management personnel aware of what a proper risk assessment should look like. The result is that risk assessments are often not prepared properly.

Whether your organization is a commercial one or a government one, a proper risk assessment should be performed prior to purchasing and installing security mechanisms or methods (i.e., countermeasures). Unfortunately, Appendix III also states that a formal risk assessment is not required, an approach that I find is inappropriate, especially for large organizations.

"The Appendix no longer requires the preparation of formal risk analyses. In the past, substantial resources have been expended doing complex analyses of specific risks to systems, with limited tangible benefit in terms of improved security for the systems. Rather than continue to try to precisely measure risk, security efforts are better served by generally assessing risks and taking actions to manage them. While formal risk analyses need not be performed, the need to determine adequate security will require that a risk-based approach be used. This risk assessment approach should include a consideration of the major factors in risk management: the value of the system or application, threats, vulnerabilities, and the effectiveness of current or proposed safeguards. (OMB A-130, Appendix III)"

A properly performed formal risk assessment should be a major ingredient of every security analysis. The risk assessment should consider the threats, vulnerabilities, the identity and value of each of the information system's assets, threat scenarios, alternative security countermeasures, selection of the preferred countermeasures, remaining risks, and definition of a target system that includes an infrastructure that comprises the selected countermeasures as defined in an overall system architecture. It should determine the mathematically-computed risks, identify the countermeasures to cost-effectively minimize these risks, and should be followed by the creation of a migration (i.e., remediation) plan to upgrade the legacy system to the target system based on a roadmap that includes the tasks and schedule for the upgrades (i.e., a Gantt chart).

Risk assessment is the analysis effort required to identify the preferred countermeasures associated with protecting an organization's information system. Countermeasures are the security mechanisms (e.g., firewalls) or methods (e.g., fences and guards) that are required to protect the information system and its data.

Since performing a risk assessment is the same thing as the process of asking the right questions, it is essential that the analyst perform a risk assessment prior to determining the countermeasures required to protect the organization's information system. Bernstein et al. (1996) state "Failure to assess the value of information (regardless of whether that information is or is not accessible via the Internet) is one of the major shortcomings in corporate information security programs today!"

Because the information system supports the organization's mission, there are at least three primary objectives for the risk assessment of an information system:

1) It should *reflect the security requirements* appropriate for the given state of the environment[50] into which the information system will operate.

2) It should clearly *convey these security requirements* to the information system's product team and the developers.

3) It should *recommend a set of countermeasures* that should sufficiently reduce both the risks to the organization's information system and the risks to the organization's mission, in a cost-effective manner.

Countermeasures should provide the level of protection required for managing threats to the organization with the objectives that:

1) No information subsystem (hardware or software) will be used to attack other information subsystems.

2) No information subsystem will be used to introduce malicious code or information into the network.

3) No information subsystem will be used to decrease the security posture of the set of all information subsystems (i.e., the total information system network) and specifically, no information subsystem will be used to decrease the availability of other information subsystems.

4) No untrusted information subsystem (i.e., another organization's information system network) will be allowed to connect with this organization's information system network without a sufficient set of countermeasures in between (e.g., a firewall).

A process for a complete risk assessment is shown in Figure 6.1, *A Risk Assessment Process.*

---

[50]The environment for an information system includes the organization's mission and security policy, the communications links (viz., private or public), the size of the information system, and the type of support provided (e.g., organizational administration, payroll, analyses and evaluations, response to customer requests, etc.).

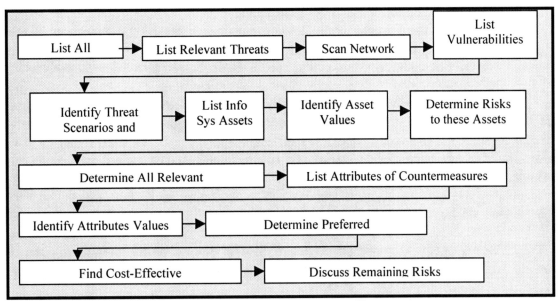

Figure 6.1 A Risk Assessment Process

A Risk Analysis is principally a risk assessment and should include the following steps:

Pre-Risk Assessment
    Identify the Mandated Requirements (see section 3.1.2)
    Identify a Security Policy and create a Security Rules Base
    Identify the Security Requirements
Risk Assessment
    Identify all the Threats
    Identify the Relevant Threats
    Identify the Prioritized Information System Assets and their Values
    Identify the Vulnerabilities—(Scanning results and analysis)
    Identify Threat Scenarios and Potential Losses
    Compute the Risks Associated with the various Threat Scenarios
    Determine the Alternative Security Mechanisms and Methods for Mitigating the
        Vulnerabilities (sometimes called "remediation")
    Determine the Total Costs and Performance of each Security Mechanism
    Select the Stakeholders' Preferred Security Mechanisms and Methods
    Determine the Remaining Risks and the Concomitant Vulnerabilities and the
Potential Effects of these Vulnerabilities
Post-Risk Assessment
    Revisit the Security Requirements, and possibly the Security Policy and Security
Rules Base
    Propose a Security Architecture (devices and locations; information flow;
        concept of operations; and software and locations)
    Propose a Migration (or Remediation) Plan for moving from the Legacy System
        to the Target System
    Propose a Security Test and Evaluation (ST&E) Process according (or similar) to
        the outline suggested in the DITSCAP

167

Perform the Security Test and Evaluation

Document the Results of the Implementation of the ST&E Process—(Security Performance; Upgrade needs)

Modify the countermeasures that need to be mitigated to operate properly

A Facilitated Risk Analysis Process (FRAP) is often performed by government agencies (DITSCAP 1999). However, the FRAP process is not recommended and a somewhat different approach is presented here. This different approach is based on experience with performing risk assessments in a variety of environments and knowledge of what a risk assessment should produce.

6.1.2   Other Risks

The risk assessment discussed above has to do with the probability that a particular type of attack might successfully occur and the potential harm (i.e., damage) that it might cause. That is, it is a probability of loss.

There are other kinds of risk, namely, technical, schedule (or managerial or cost), and performance (Smith 1998). *Technical risk* means that there is a possibility that the product or service may not be ready for implementation, for example, there may be a flaw in the design that has as yet been undiscovered or the technological support required may be beyond the state-of-the-art. *Schedule risk* means that the roadmap as defined may not have considered some part of the implementation situation and the product or service takes longer to acquire, install, test, and get into operational condition than originally planned, thus causing a management problem and increasing the cost of the item. *Performance risk* means that the product or service does not perform as well as advertised. This is sometimes a function of the vendor; some can be trusted more than others can. Clearly, all of the different risks are interdependent.

6.1.3   Risk Management

Some organizations have a full-time staff devoted to studying all types of security risks that might affect their information system network (NIST Handbook 1995). Potential situations for risk-oriented policies include the following:

1) Risk management and risk management strategy,
2) Contingency planning,
3) Protection of sensitive information,
4) Guarding against unauthorized software,
5) Acquisition of software,
6) Telecommuting,
7) Bringing in disks from outside the workplace,
8) Access to other employees' files,
9) Encryption of files and e-mail,
10) Rights of privacy,

11) Responsibility for correctness of data,

12) Suspected malicious code, and

13) Physical emergencies.

Risk management is the process of assessing risk, taking steps to reduce risk to an acceptable level, and maintaining that level of risk. Management is concerned with many types of risk, technical, schedule, performance, as well as security. Computer security risk management addresses risks that arise from an organization's use of information technology. One can use a particular strategy when performing risk management, such as a) protecting the high-value information assets or b) minimizing overall risks to all of the information system assets.

Both commercial and government organizations manage these various risks. For example, to maximize the return on their investments, commercial organizations must often decide between aggressive tactics (but high-risk) versus slow-growth activities (but more secure). These decisions require an analysis of risk, relative to: a) potential benefits and b) consideration of alternatives, and then, c) implementation of the options that management has selected to be the best courses of action.

A risk management effort should focus on those areas that result in the greatest effects on the organization (i.e., those threats that can cause the most harm). A risk management methodology does not necessarily require the analysis of each of the components of risk separately. For example, assets/consequences or threats/likelihoods may be analyzed together.

Risk management can assist a manager in selecting the most appropriate security controls; however, it is not a silver bullet that instantly eliminates all difficult issues. The quality of the output depends on the quality of the input and the type of analytical methodology used. For all practical purposes, complete and accurate information is never available which means that uncertainty is always present. Despite these disadvantages, risk management can provide a powerful tool for analyzing (and responding to) the risks to the information system network.

The two primary functions of risk management are:

1) The interpretation of those risks that are not cost-effective to counter and the level of these remaining uncountered risks that must be accepted (NIST Handbook 1995), and

2) The identification of those cost-effective countermeasures that should be purchased, installed, and operated.

6.2   Risk Assessment

Typically, for security of an information system network, a *Risk Assessment* is performed, not a safety analysis. The risk assessment should identify the following objects (see Table 6.1, *Risks, Activities, and Responsibilities*):

169

1) Threats,
2) Vulnerabilities,
3) Threat scenarios,
4) Information system assets (prioritized) and their values to the organization,
5) Risks,
6) Alternative countermeasures (i.e., security mechanisms or methods to counter the computed risks and thereby provide protection for the organization's assets),
7) Countermeasures that are cost-effective,
8) Implementation plan (i.e., or remediation plan which is a migration plan to progress from the legacy (as is) system to the desired target (to be) system), and
9) Remaining risks that are not countered because it is not cost-effective to counter them.

Risk is the probabilistic cost of potential damage to the system, given the threats, vulnerabilities, threat scenarios, and values of the system's assets. It is estimated based on available approximate information and hence is somewhat subjective. By applying probabilities to the various threats and identifying the potential approximate damage costs in case of a successful attack, the risks can be determined. However, it is not a good idea to attempt to quantitatively identify damage costs since the result most likely will be a lot of guesswork. Moreover, the probability of a certain type of attack is also just guesswork. But the more realistic and knowledgeable the analyst, the better will be the estimated information, namely probabilities and loss estimates.

However, one can decide, in a qualitative way, what the probabilities or losses might be. So, it is best to categorize the probabilities and potential damages using a three- or five-level qualitative assessment. For example, probabilities can be assessed as high, medium, and low or very high, high, medium, low, and very low. Similarly, the potential cost of an attack (or loss) can be assessed similarly. With the categorizations of the probabilities and losses, a multiplication table must be devised for risk, where:

Risk ($) = Probability of Successful Attack x Potential Cost of That Attack ($).

Table 6.1 Risks, Activities, and Responsibilities

Parameter	Activity	Responsibility
Pre-Risk Assessment		
Identify Mandated Requirements	Identify and Analyze	Owners, Named Representatives, and Security Analysts
Security Policy and Rules Base	Identify and Analyze	Owners, Named Representatives, and Security Analysts
Risk Assessment		
Threats	Identify and Analyze	Security Analysts, Users
Vulnerabilities	Identify and Analyze	Security Analysts, Users
System Assets/Values	Identify	Owners, Named Representatives, and Users
Threat Scenarios	Create	Security Analysts

Parameter	Activity	Responsibility
Countermeasures	Identify	Security Analysts
Selection of Countermeasures (General and Specific)	Identify and Analyze	Security Analysts, Owners, and Users
Remaining Risks	Identify and Analyze	Security Analysts
Post Risk Assessment		
Migration Plan	Identify and Analyze	Security Analysts
Countermeasures Implementation	Implement	Vendor and Organizational Technicians
Performance of Countermeasures	Security Test and Evaluate	Security Analysts
Remaining Risks	Identify and Analyze	Security Analysts

Since probabilities are dimensionless, risk has the dimension of Dollars (or whatever the monetary system is that is being used). Sometimes an annualized probability (i.e., the probability that the information system will be successfully attacked during a one-year duration) is used and the result is Annualized Loss Expectancy (ALE) so that Risk, in this case, is the ALE.

Thus, the multiplication table can be described as shown in Table 6.2, *Multiplication Table for Risk Computation*. For example, if the Probability of Attack were *High* and the Potential Loss were *Low*, then the Risk would be *High* x *Low* = *Medium* (from Table 6.2).

Table 6.2 A Multiplication Table for Risk Computation

Probability Of Attack	Potential Loss				
	Very High	*High*	*Medium*	*Low*	*Very Low*
Very High	Very High	Very High	High	Medium	Medium
High	Very High	High	High	Medium	Medium
Medium	High	High	Medium	Low	Low
Low	Medium	Medium	Low	Low	Very Low
Very Low	Medium	Medium	Low	Very Low	Very Low

The analyst has to have some idea of what these values mean, for example, *Very High Cost* could be at least $10 Million, *High Cost* could be $1 Million to $10 Million, *Medium Cost* could be $100 Thousand to $1 Million, *Low Cost* could be $10 Thousand to $100 Thousand, and *Very Low Cost* could be less than $10 Thousand. Similarly, a *Very High Probability* would be at least 0.9, a *High Probability* would be 0.7 to 0.9, a *Medium Probability* would be 0.3 to 0.7, a *Low Probability* would be 0.01 to 0.3, and a *Very Low Probability* would be no more than 0.01.

Of course, these values could be modified to suit the organization's owners or users, according to the organization's size, the amount of business the organization does each year (or month or week or day, and time of year), the average business per customer transaction, and any other pertinent parameter that might be relevant to the values of the organization's information system's assets based on the opinion of the owners or users.

Since probabilities are non-negative numbers that are always less than or equal to one (i.e., P (x) is in the closed interval [0,1]), the resulting risks will be less than or equal to the costs, usually less than. For example, if the probability of a particular attack is *Very Low* (i.e., say an average of 0.01) and the potential loss is *Low* (i.e., say an average of $50,000), then the Risk for that particular attack would be an average of $500, a value in the *Very Low* range for Risks, assuming that Risks are categorized the same as Costs, namely ($0 \leq$ Very Low Risk < $10 K, $10 K < Low Risk < $100 K, $100 K < Medium Risk < $1 M, $1 M < High Risk < $10 M, Very High Risk $\geq$ $10 M).

This is the type of risk ($500) that it would not be cost-effective to counter.

If the qualitative values for the probabilities and losses are the same, as is the case in Table 6.2, then the resulting matrix of multiplicative values should be symmetrical and the main diagonal should be a replication of the gamut of qualitative values and the counter-diagonal should be a middle cost value, if both the Costs and Risks are categorized into an odd and equal number of values (e.g., five).

Some analysts who wish to perform a risk assessment are nontechnical and do not have a mathematical background nor are they able to understand the technical rigor of a true risk assessment. For these analysts, there is a method for identifying risks that avoids the mathematical formulas that accurately reflect the strict definition of a risk. One nonmathematical approach to determine risk is the following (this method is simplistic and arithmetic, i.e., simple addition and multiplication only):

$$R = L * (C + I + A)$$

where L is the likelihood of an attack and is a number 1, 2, or 3 where 1 means "not too likely," 2 means "quite likely," and 3 means "very likely." The variable C is the potential cost of an attack on the Confidentiality of the system and may be a number 0, 1, 2, or 3 where 0 means "negligible cost," 1 means "low cost," 2 means "medium cost," and 3 means "high cost." The variables I and A mean (Data) Integrity and Availability, respectively. Their costs are rated in the same way. This means that for the nonmathematical approach, Risk has an integer value in the range [0 to 27], where 0 means no risk and 27 means the maximum risk.

The values for L, C, I, and A are very subjective and must be estimated by the analyst. However, this method does not require the knowledge of probability mathematics nor of the value of losses due to various threat scenarios. Still, the analyst must provide values for L, C, I, and A for the gamut of risks for the variety of threat attacks considered feasible.

The analyst must also decide what the threshold point is for the computed risks so that any risk above this point is considered to require a countermeasure and any below the point can be ignored (for now, but must be considered in the analysis of the remaining risks). Again, this is a nonmathematical or non-rigorous approach and cannot be argued in an analytical manner.

It is assumed for this type of analysis, that there are three subsets of risk:

1) Risks to the *confidentiality* of information (C),
2) Risks to the *integrity* of information (I), and
3) Risks to the *availability* of information (A).

This approach has no mathematical or physical meaning but some analysts want to use it because they feel that it is simpler, and hence easier to accomplish, and certainly may be. This nonmathematical form of risk computation does have the correct dimension, namely "cost." However, this approach is not recommended because it suffers from many of the same maladies of the formal risk assessment, namely an inability to accurately estimate both the probability of attack and the cost of the potential damage to the system. It also has the disadvantage identified above, namely, it cannot be supported with a mathematical analysis.

In determining the cost-effective countermeasures to combat the relevant threats, the formal risk assessment does provide probabilistic costs (i.e., the risks) with which the total costs of the countermeasures can be compared. The nonmathematical approach does not offer this capability, which is essential to a proper risk assessment. However, with the identified threshold for risk values, there is a method (albeit nontechnical) that allows the analyst to identify those risks that require some sort of countermeasure for each risk.

6.2.1 List Of Threats

The principle threats to any information system network are the following:

1) Loss of system or information availability,
2) Destruction of information,
3) Loss of privacy of information,
4) Loss of information accuracy, or
5) Damage to the organization's reputation.

There are many ways in which these events can happen and these are shown in Figure 6.2, *A Taxonomy of Security Threats*, as a taxonomy[51] for the various threats to the security of an information system network.

---

[51] A taxonomy is a structured representation of a way of decomposing some topic.

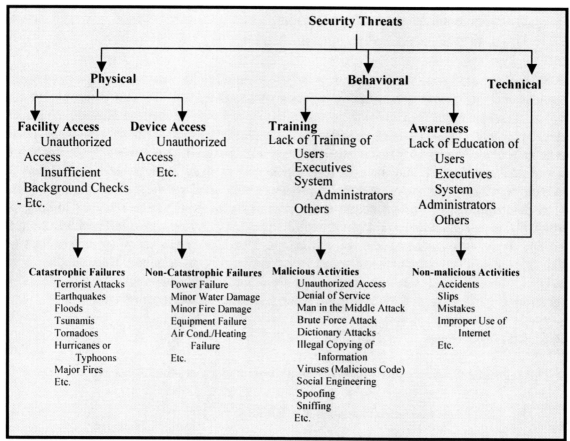

Figure 6.2 A Taxonomy of Security Threats

This list of threats can be used as a stimulus in a brainstorming session to ascertain the relevant threats. However, this list should be a living table since the list is dynamic and should be updated as the gamut of threats changes because new threats emerge, old threats disappear, and some threats change.

The threats are categorized into three groups:

1) physical (facility versus device access),
2) behavioral (training versus awareness or education), and
3) technical, which is divided into four parts,
   a) catastrophic failures,
   b) non-catastrophic failures,
   c) malicious activities, and
   d) non-malicious activities.

Some of these threats may not be what some might refer to as threats (e.g., random system failures); however, they are included here because they represent situations that can result in a decrease in system and information availability. Security should address the following issues:

1) identification and authentication,
2) access control,
3) authorization,
4) confidentiality,
5) data integrity,
6) availability, and
7) non-repudiation.

Although availability does have certain aspects that might not be considered a part of the security area, all aspects of availability are included here under the security umbrella.

Some Web addresses where information system network security threats can be located are the following:

http://www.symantec.com/
http://www.netscape.com/
http://www.cisco.com/
http://www.sun.com/
http://www.yahoo.com/

The organizational losses that could occur due to a specific scenario are shown in Table 6.3, *Losses versus Threat Scenario Type*. A malicious action that could occur due to the threat is also shown in the table. Each of the threats to the information system network is listed and explained in Table 6.4, *Threats and Explanations*.

Repudiation is sometimes viewed as a potential threat. This occurs when the sender (or receiver) denies having sent (or received) a message or information that they indeed sent, or the receiver denies having received a message, that they indeed received.

Table 6.3 Losses versus Threat Scenario Type

Organizational Loss	Threat Scenario Type
Loss of information (stolen or damaged) due to unauthorized user access	1) System administrators creating access holes for important users. 2) System administrators not removing former user names (IDs) from system access control lists. 3) System administrators not installing software patches as soon as they are available. 4) System software not enforcing User ID and Password (or other authentication mechanism) for granting user, server, or program application and database information access. 5) System administrators not enforcing organizational security policy due to a lack of knowledge of, understanding of, or access to, the policy.
Loss of system or information availability	Denial-of-service, direct damage to equipment, random subsystem failures, catastrophic failures

Destruction of equipment	Improper access to equipment (by internal or external person)
Destruction of information	Internal or external user destroying information (either intentionally or unintentionally), random subsystem failures, catastrophic failures
Loss of privacy of information	Intruder viewing or stealing information, internal user stealing information
Loss of information accuracy	Intruder modifies stored or transmitted information, internal user accidentally modifies information in a database
Damage to organization's reputation	Any intrusion that can cause embarrassment to the organization (such as loss of credit card numbers to an intruder)
Improper entry into the system	Lack of proper identification and authentication
Improper access to information	Lack of proper authorization and access control
Sender (or receiver) denies having sent (or received) message	Sender (or receiver) denies having sent (or received) message
Loss of relevant information that can be used in a malicious attack	Use of sniffing device to discover information that can be used to attack the system
Loss of capability to locate the true source of a malicious message or action	Use of false address to fool the system into thinking that a message came from another source

Table 6.4 Threats and Explanations

Threat	Action	An Explanation
Physical		
Facility Security	Unauthorized Entry	Use of fences, gates, guards, patrols, cyber locks, and user badges to gain entrance to the organizational facility. Use of server rooms with cyber locks.
System Hardware Security	Unauthorized Entry	Prevention of Authorized or authorized persons doing physical damage to a subsystem.
System Information Security (data removal from the premises by internal users)	Removing Sensitive Information	Sensitive Information Stored on Physical Disks (Floppies (up to 1.44 Megabytes), Super Disks (up to 120 Megabytes), Zip Disks (up to 250 Megabytes), Compact Disks (up to 680 Megabytes), etc.).
Behavioral		
Training	Improper Personal Behavior	Training courses designed for Managers, System Administrators, Security Administrators, Technical Personnel, System Users, etc.).

Threat	Action	An Explanation
Awareness Education	Lack of Pertinent Knowledge	Awareness knowledge courses designed to convey proper facts to organizational personnel.
Standard Operating Practices	Lack of Pertinent Knowledge	E.g., Ensure that system users are removed from access lists when they depart the organization.
Technical		
Catastrophic Failures		
Any Massive Failure	Massive Failure	Failures due to an Earthquake, Flood, Tsunami, Tornado, Hurricane or Typhoon, Terrorist Bombing, Aircraft Crashes, Act of War, or Major Fire event.
Non-Catastrophic Failures		
Power Failures	Random Failure	Use of Uninterruptable Power Supply and Power Generators.
Minor Water Damage	Random Failure	Use of Water damage control subsystems.
Minor Fire Damage	Random Failure	Use of Fire control subsystems.
Climate Control	Random Failure	Use of Air Conditioning Systems and Heating Systems.
Dual Routers, Firewalls, Bridges, Servers, etc.	Random Failure	Use of dual subsystems and clustered servers with load balancing and automatic failover software to avoid a single-point-of-failure situation.
Malicious Activities		
External Hacker/Cracker[52] Attack	Attack by an External Source	External persons who get into the system through an identified vulnerability (using a scanner such as **nmap**). Denial-of-Service attacks Acquisition of sensitive information (e.g., private customer or patient info, intellectual property, etc.) Exploit well-known defects in COTS software Unauthorized access, then increase privileges.
Unauthorized user attack	Attack by Unauthorized Users	1) System administrators creating access holes for important users. 2) System administrators not removing former user names (IDs) from system access control lists. 3) System administrators not installing software

---

[52] A "cracker" is usually a malicious hacker but can be an entity that attempts to enter (electronically) an information system by guessing User IDs and passwords.

Threat	Action	An Explanation
		patches as soon as they are available. 4) System software not enforcing User ID and Password (or other authentication mechanism) for granting user, server, or program application and database information access. 5) System administrators not enforcing organizational security policy due to a lack of knowledge of, understanding of, or access to, the policy.
Internal User Attack	Attack by an Internal Source	Internal persons who use their access to, and knowledge of, the network to perform malicious activities, such as damaging information in a critical database.
Malicious E-Mail	Malicious Attachments	External or internal persons who send e-mail with malicious attachments (viruses or worms) that can do damage to the system.
Social Engineering	Phone Calls	Use of phone calls asking for information from authorized internal users.
Man-in-the-middle Attack	External Attack	Use of personal computer to intercept message traffic and routing traffic to an unintended destination.
Brute Force Attack	External Attack	Use of computer power to decrypt traffic using the decoding algorithm and every possible key of the correct length.
Dictionary Attack	External Attack	Use of a dictionary to identify passwords.
War Dialing	External Attack	Use of computer to incrementally dial phone numbers seeking a specific type of answerer.
Dumpster Attack	Person Attack	Delving into the garbage of a targeted organization seeking pertinent information.
Replay Attack	External Attack	Use of a personal computer to intercept traffic and then replaying it at a later time or to another destination address.
Misrouting Attack	External Attack	Use of a personal computer to intercept traffic and then rout it to a different destination address.
Malicious Subsystem Configuration Change	External Attack	Unauthorized modification of firewall rules set. Countermeasure - Use of Out-of-band subsystems
Sniffing	External Attack	Use of information collection devices such as plaintext User IDs and Passwords or sensitive information.
Spoofing	External Attack	Intentional use of false address to fool the receiver.

Threat	Action	An Explanation
Spamming	External Attack	Use of the e-mail capability to send unsolicited messages.
Spoofing	External Attack	Providing an incorrect IP address so that the source of a message will be misidentified.
Non-Malicious Activities		
Accidents	Internal Action	Performing some unintended non-malicious operation.
Slips	Internal Action	Performing a keyboard operation that is based on an inadvertent error.
Mistakes	Internal Action	Performing a keyboard operation that is based on a misunderstanding.
Improper Use of the Internet	Internal Action	Use of the Internet to access a pornographic site or other forbidden site.

6.2.1.1   Physical Threats

Physical threats consist of those activities where an unauthorized person gains physical access to the information system facility or its components (e.g., a critical server). These threats can be prevented through the use of fences, gates, badges, cyber locks, and other such entities, and security guards and their operational procedures.

Physical threats can be perpetrated by individuals who may decide to do physical damage to the information system, such as hitting a server with a hammer or stealing a personal computer or workstation. Both unauthorized persons and authorized persons with no need for access can perpetrate a physical attack on the information system.

Physical threats also consist of improper behavior of a prospective employee prior to employment, which can be countered with background checks for access to a facility or its sensitive areas. Employees should have background checks that can assure that the employee has a level of trustworthiness sufficient for their job functions.

6.2.1.2   Behavioral Threats

Behavioral threats consist of those situations where a user, manager, or administrator is unable to properly respond to a situation because of a lack of knowledge. This dearth of knowledge can be mitigated through the use of training and awareness courses created to present the information that a particular employee or relevant information system person requires to ensure that they can properly handle most situations that arise due to some sort of threat to the organization's information system. For example, social engineering attacks are included in behavioral threats because the counter to these attacks requires training and awareness courses on the proper actions by an employee should they become involved.

If an employee intentionally violates a security mandate, such as copying sensitive information onto a disk (e.g., floppy, super, zip, or compact) and then removes it from the organization's premises for use elsewhere, then if caught, that person should suffer the penalties as clearly stated in the organization's security policy.

Social engineering is the situation where a caller pretends to be someone from the organization who needs some sensitive information, say a user ID and password. If successful, the caller will be given the information even though it is critical that the person called should not divulge such information over the phone. Proper training and education can prevent these types of lapses from occurring, or greatly diminish the likelihood of their occurring.

6.2.1.3   Technical Threats

Technical threats include threats to the information system hardware, software, and information by electronic means and minor failures. Technical threatening situations can be the result of actions by either internal or external users. Technical threats are divided into four categories:

1) Catastrophic failures,
2) Non-catastrophic failures,
3) Malicious actions, and
4) Non-malicious actions.

6.2.1.3.1   Catastrophic Failures

Catastrophic failures are those failures that occur due to some catastrophic event, such as natural disasters or war. These types of failures are often addressed in a separate plan called a Contingency Plan. However, in this context, these failures are addressed as a part of the risk assessment. The critical information, including operating systems, middleware, applications, and databases should be replicated in a remote location which would not be affected by a catastrophic failure at the primary site. The frequency of downloading (i.e., replicating) the information to a remote site depends on the organization and how often the information is modified.

Catastrophic failures can be due to acts of war, terrorists attacks (e.g., bombings or fully-fueled airplane crashes), earthquakes, mud slides, floods, tsunamis (tidal waves), tornadoes, hurricanes or typhoons, major fires, and other such events. A Contingency Plan is required for proactively dealing with the possibility of a catastrophe.

6.2.1.3.1.1   Contingency Plan

The primary method of dealing with the results of a potential catastrophe is through building a contingency plan and implementing a backup and recovery system. An important part of any threat analysis is the development of a contingency plan. A *Contingency Plan* is a set of procedures that, when carried out, will return an information system network that has been damaged from a catastrophe to its original state, or at least to an operational state.

A complete contingency plan should also include an alternative workspace area, that is, backup facilities. A computer system without people is not going to save the organization. Testing of disaster recovery plans (for those who practice their recovery plans) paid off during the World Trade Center attacks (Chen and Hicks 2001).

6.2.1.3.1.2   Contingency Plan Components

The contingency plan should contain backup and recovery operations that require the installation of a subsystem for replicating all critical information in the system to a site that would most likely not be affected by the primary system's demise. A Contingency Plan should include at least the following:

1) A description of an automated backup system (hardware and software),
2) A description of a process for replicating applications and information to this backup system, and
3) A prescriptive[53] recovery process that explains how to replace lost information or damaged equipment and get the information system back into its proper operation in the briefest possible time.

The critical information and replacement equipment (i.e., redundant devices) should be readily available and there should be employees who know how to quickly access and install these items.

6.2.1.3.2   Non-Catastrophic Failures

Non-catastrophic failures include failure or saturation of a subsystem such as throughput overload and/or crash of a server or firewall, local power outage, minor water damage such as leaking of an air conditioner or plumbing water pipe, minor fire damage, or heating/air conditioning outage.

6.2.1.3.3   Malicious Actions

Malicious actions can be caused by malicious code (viruses, worms, Trojan horses, back doors, etc.), sending e-mail messages with malicious code attachments, gaining unauthorized entry (electronic or physical) by an attacker, performing a social engineering phone call to obtain sensitive information, performing denial-of-service (DoS) or distributed DoS (DDoS) attacks, scanning the information system to gain vulnerability information such as identifying open ports, sniffing to gain information on plaintext information such as passwords, sniffing to collect encrypted information to be decoded later (if possible for the attacker), or spoofing to deny accountability.

One approach to combatting attackers is to use an identification and authentication process, such as one based on a User ID and Password. The User ID may not be the user's name (but can be) and the passwords are private information consisting of eight or more characters in length, some of which must be numerical and special characters, are not dictionary words, and are changed periodically. Also, the I&A software should break the connection if the user ID or password attempt fails after three tries. Once the user has successfully been identified and authenticated, a session can begin.

Generally, a *session* is when two hosts on a network are communicating some information between them. The session usually begins with a handshake and continues until

---

[53] A *prescriptive* process is one that explains how to accomplish some task.

one (or both) of the participants no longer has any information to transmit or receive. The *session layer* is the fifth layer of the OSI communications model. It establishes and maintains a connection between two or more systems. Identifying the IP address of the communicators is a part of the process.

IP *spoofing* involves providing false information about a person or host's identity to obtain unauthorized access to systems and/or the services they provide. The best defense against spoofing is to configure routers to reject any inbound packets that claim to originate from a host within the internal network. This simple precaution will prevent any external machine from taking advantage of trust relationships within the internal network. Cross network spoofing is more difficult to counter (Bernstein 1996). There are many types of spoofing and some of these are listed in Table 6.5, *Types of Spoofing Attacks*.

Table 6.5 Types of Spoofing Attacks

Spoof Type	Attack Scenario	Vulnerability	Countermeasure
IP Spoofing	Internal user sends a packet with a bogus external source IP address to a trusting host.	Source addresses are not checked.	Install a firewall that blocks packets with trusted internal addresses from entering the network.
DNS Spoofing	Send an unsolicited reply containing a bogus domain (name, address) pair to victim's DNS server.	The DNS has no authentication.	Use a modified DNS that doesn't cache entries.
Session hijacking	Attacker inserts bogus packets into an established session.	Authentication of players has already happened.	Encrypt the session traffic.
Email	Send bogus message with a fake "from" line to an SMTP server.	SMTP does not authenticate.	Check source IP address of row message. Digital signatures.
Logon	Use someone else's User ID and Password to gain entry to the network.	Doesn't check IP address at logon time.	Protect passwords. Implement strong authentication.
Routing	Send bogus Routing Information Protocol (RIP) or Internet Control Message Protocol (ICMP) redirect packets to a router.	No authentication in RIP. ICMP redirects source-routed packets.	Don't use RIP or ICMP with untrusted networks.
Anonymous Remailer	Attacker sends email via anonymous remailer account.	SMTP does not authenticate.	Digital signatures.
Web Spoofing	Attacker creates a shadow copy of the entire Web site, traffic is routed through the attacker's PC, allowing attacker to monitor victim's traffic.	The man-in-the-middle attack rewrites all of the URLs on some Web page so that they point to the	Disable JavaScript, make sure the browser's location line is always displayed, and pay heed to these displayed URLs.

Spoof Type	Attack Scenario	Vulnerability	Countermeasure
		attacker's server.	
Third Party	Send bogus email to the Network Interface Card (InterNIC) requesting bogus domain name change or alternate IP address.	InterNIC doesn't fully authenticate unless requested.	Install InterNIC authentication of changes to domain.

Sniffing is the use of a physical device to passively collect network information for analysis to determine such items as User IDs and passwords, if transmitted in plaintext. Countermeasures to sniffing include the following six options (Bernstein 1996):

1) *Strong authentication* - Install a one-time password tool. One-time passwords preclude the possiblility of an attacker using any password obtained from network sniffing.

2) *Physical inspection* - This countermeasure requires a personal inspection of the network in search of a sniffer and may not always be practical since some cabling runs beneath flooring.

3) *Twisted pair hub technology* - Some vendors implement a security feature in their twisted pair hub technology that prevents packets from traveling everywhere within a shared media network. This approach can control the threat of unauthorized physical sniffers and packet capture programs.

4) *Scanning* - There is a program that determines whether each UNIX host within your network is in promiscuous mode (i.e., can monitor traffic on the network). Unfortunately, it only identifies sniffers on individual hosts and does not provide protection against vampire taps[54] elsewhere on the network.

5) *Policy* - Include provisions that forbid employees from using sniffers with explicit penalties for such. (Of course, a malicious employee likely will intentionally violate this policy but it will be done with the knowledge that if caught, there will be a specific penalty.)

6) *Network encryption* - Cryptography can provide confidentiality that will prevent a malicious user using a sniffer from understanding the information it records.

An example of a malicious code is the Code Red Worm, in four variants, that was first transmitted around August 9, 2001 (iDefense 2001). This worm self-propagated through Web servers that operate using Microsoft Corporation's Windows NT 4.0 with Internet Information Services (IIS) 4.0 or 5.0 or Windows 2000 with IIS 4.0 or 5.0, or Cisco's Call Manager, Unity Server, uOne, ICS7750, Building Broadband Service Manager, and the unpatched Cisco 600 series DSL routers (all of which use the IIS code). It is not unusual for malicious code to be programmed to attack a popular software package, such as most any

---

[54] A *vampire tap* is a connection to a coaxial cable in which a hole is drilled through the outer shield of the cable so that a clamp can be connected to the inner conductor of the cable. A vampire tap is used to connect each device to a Thick Ethernet coaxial cable in the bus topology of an Ethernet 10Base-T LAN. A different connection approach, the British Naval Connector, is used for the thinner coaxial cable known as a Thin Ethernet.

operating system from Microsoft or Cisco, because these systems are numerous and sometimes vulnerable.

The Code Red Worm gains control by first sending a message that overflows the buffer within **idq.dll** (an unchecked buffer in the dynamic link library). There are patches to fix this problem. Possibly up to 750,000 servers have been infected, or about one in eight of the 6,000,000 servers worldwide that use the vulnerable software. Although the worm can write to the hard drive amd make some changes to the server, it remains in memory and propagates from there. Once infected, the worm spawns multiple copies that begin trying to infect other servers by going through a list of IP addresses checking on Port 80 to see if a Web server is vulnerable. For information on the Code Red Worm go to the URL address http://www.cert.org/advisories/CA-2001-19.html. Other URL addressess relevant to the Code Red Worm, such as the locations for Microsoft and Cisco operating system patches, are accessible from this Computer Emergency Response Team (CERT) Web page.

6.2.1.3.4 Non-Malicious Actions

Non-malicious actions include accidents, mistakes[55], slips[56], and improper use of the Internet. Accidents by authorized users can occur due to improper use of the system, such as unintentionally causing damage to a database. Often these types of error are due to a user who is not knowledgeable of the proper procedures, but sometimes they can be due to the user's incompetence or lack of computer abilities even though they may have much experience.

Improper use of the Internet can include such things as downloading or viewing pornographic material or any material that is not pertinent to the user's job. The non-malicious use of the information system should be addressed in the organization's security policy with the potential responses clearly defined so that there is no question as to the possible effects on the person who commits such events. Since each employee is aware of the organization's security policy (from reading the policy (in hardcopy form) or from having attended a training or education session or in some cases having it accessible online), there can be no logical impediment to a well-defined quick response to any transgression that is identified and supported by firm evidence (e.g., an audit trail).

6.2.1.3.5 List Relevant Threats

In most cases, the particular organization's information system network may not be the target of all of the potential threats. For example, if the organization is a small commercial establishment, then the threat of a hacker attacking the system is miniscule and they may not care to protect itself from such an unusual threat. However, should the organization be subject to a former employee's ire, it might become the target of such a threat. This is one reason why it is good practice to ensure that whenever a user leaves the organization, for any reason, that the former user is removed from the access control lists immediately upon departure. Procrastination or inaction will allow the former user to continue accessing the information system and to possibly perform some malicious action where sensitive

---

[55] *Mistakes* are due to the selection of inadequate approaches so that the desired outcome is not achieved.
[56] *Slips* are failures in execution or the execution of unintentional actions.

information could be stolen or damaged. The organization should have a good capability to enforce its identification and authentication, authorization, and access control polices, to ensure that unauthorized personnel cannot gain electronic access to the information system network.

The security analyst should list the considerations that perhaps may exclude this particular organization from some of the potential threats. Given the list of all threats to the system, the analyst should cull from the list any threats considered unusual (for this organization's information system), and create a new list of relevant threats with rationale for eliminating some from the list.

A list of some possible threats was presented in Table 6.4, *Threats and Explanations*.

6.2.2 Vulnerabilities

A *vulnerability* is a weakness in the information system that could be exploited by a malicious hacker. Two methods of determining system vulnerabilities are 1) through the use of an architectural analysis or 2) through the use of a scanner[57]. If a current, detailed architecture of the system is available, then it is possible to perform an analysis to identify some of the system's vulnerabilities.

To identify the system *technical vulnerabilities*, a scanner can be used. This requires performing an analysis using a system scanner such as one of the following:

"**nmap**" by Fyodor@insecure.org,

"strobe" by Julian Assange,

"netcat" by *Hobbit*,

"stcp" by Uriel Maimon,

"pscan" by Pluvius,

"ident-scan" by Dave Goldsmith,

SATAN tcp/udp scanners by Wietse Venema,

or other such devices or similar software systems. By properly hooking up the desired mechanism to, or installing the software in, your information system and scanning your system with the tool, it is possible to discover the various technical vulnerabilities in your information system. However, the user should be aware that operating a scanner may crash your system.

One of the scanners, **nmap**, is designed to allow system administrators and security analysts to scan an information system network to determine which hosts are up and what services they are offering. **Nmap** supports many different scanning techniques such as: UDP, TCP connect(), TCP SYN (half open), FTP proxy (bounce attack), Reverse-ident, ICMP (ping[58] sweep), FIN, ACK sweep, Xmas Tree, SYN sweep, IP Protocol, and Null scan. It also offers a number of advanced features such as remote OS detection via TCP/IP fingerprinting, stealth scanning, dynamic delay and retransmission calculations, parallel

---

[57] *Scanner* is this context is not the device for converting a hard copy image to electronic form but a device for scanning a network to identify any vulnerabilities in the system.

[58] A "ping" is a command to verify the existence of, and connection to, remote hosts over a network.

scanning, detection of down hosts via parallel pings, decoy scanning, port filtering detection, direct (non-portmapper) Remote Procedure Call (RPC) scanning, fragmentation scanning, and flexible target and port specification. The **nmap** program can be downloaded from the following URL address http://www.insecure.org/nmap/.

Running **nmap** usually results in a list of interesting ports on the port devices (such as routers or firewalls) being scanned. **Nmap** always gives the port's "well known" service name, number, state, and protocol. The device's state is "open," "filtered," or "unfiltered." *Open* means that the target machine will accept connections on that port. *Filtered* means that a firewall, filter, or other network obstacle is covering the port and preventing the **nmap** program from determining whether the port is open. *Unfiltered* means that the port is known by **nmap** to be closed and no firewall or filter seems to be interfering with **nmap's** attempts to determine this. Unfiltered ports are the common case and are only shown when most of the scanned ports are in the "filtered" state (Fyodor 2001).

Many of the identified vulnerabilities can be mitigated using software patches (such as those issued by Microsoft Corporation and Oracle) that should ensure that malicious users either cannot take advantage of your system or if they can, their intrusion process is more difficult. It is good practice to install all software patches that arrive from vendors who have products on your system and who have patches that can be downloaded from an Internet address.

The **nmap** program is distributed in the hope that it will be useful, but without any warranty; without even the implied warranty of merchantability or fitness for a particular purpose. It should also be noted that the **nmap** program has been known to crash certain applications, TCP/IP stacks, and even operating systems. **Nmap** *should never be run against a mission-critical system* unless the owners are prepared to suffer downtime. Thus, if it is at all possible, the system to be scanned should be an offline or backup system so that should it fail due to the **nmap** program, then no harm is done. Because **nmap** may crash your system or network, the vendor disclaims all liability for any damage or problems caused by **nmap** (Fyodor 2001). This disclaimer is a drawback for the **nmap** scanning system or for that matter, any scanning system that has such a policy.

The *physical and behavioral vulnerabilities* in your system must be identified by security analysts who, at least, may wish to:

1) Review relevant documentation (such as the policy statement, the security policy statement, the concept of operations, and the information system architecture model),
2) Question some of the owners, users, or other employees,
3) Inspect the information system facilities, and
4) Scrutinize other areas, as required.

Other sources of information may come from owners and named representatives who are aware of certain physical and technical intrusions that have occurred in the past.

There are organizations that will scan your network for you, for example, Rent-a-Hacker, which has the Web site: http://www.rent-a-hacker.com/. But care should be taken when writing a service level agreement that includes any concerns about your network crashing due to the scanner being used and the responsibilities of the service organization with respect to this event occurrence.

In some cases, *it is absolutely forbidden to scan the organization's network*. Often, the rationale for this ban is that the scanning package may cause the system to crash (as the warning offered in the **nmap** system description) and some organizations cannot tolerate their system going down for any reason and for even a short period. In these cases, an alternative process that is much less risky, for identifying the system vulnerabilities is to compare the identified relevant threats with a detailed description (viz., the implemented software and hardware products) of the system.

This description should identify the:

1) Specific products used,
2) Where they are located in the network,
3) What their capabilities are,
4) What their vulnerabilities might be, and
5) Any other attributes or disadvantages that the vendor might freely admit.

If an architecture model exists, then this model could be used. A knowledgeable analyst can identify many of the system's vulnerabilities using this method, though probably not as comprehensively as would be the case using a scanning tool.

For those who wish to hire a vendor to perform a vulnerability analysis, one is the Sword & Shield, which can be found at http://www.sses.net/. These types of vendor organizations can perform any or all of the following activities:

1) *Review Security Policy* - Review the formal statements of the organizational rules that regulate how to manage, protect, and use information system assets (information and devices). Unfortunately, many organizations do not have a security policy and in these cases a security policy needs to be developed before a comprehensive security analysis can begin.

2) *Vulnerability Assessment* - Perform an assessment of network vulnerabilities with a systematic review that: 1) determines the deficiencies of current security measures, and 2) evaluates the effectiveness of existing and planned countermeasures, against all relevant threats.

3) *Risk Assessment* - Provide an assessment to identify organizational assets and their worth, determine vulnerabilities, estimate the probability of exploitation and associated effects, associate these and compute potential losses, survey applicable countermeasures and associated costs, and perform a cost-benefit analysis to identify the preferred countermeasures.

4) *Identify Security Countermeasures* - Identify the potential countermeasures that could be employed to minimize the system's vulnerabilities.

5) *Security Awareness Training* - This activity is to encourage a culture of security consciousness, end-user awareness training, and security education through training and awareness courses.

6) Incident Response - Incident handling is an integral security activity that requires performing proactive analyses and planning for dealing with security-related events that may have an adverse effect on organizational assets.

6.2.2.1 List All Known Vulnerabilities

Sources of the knowledge required to answer questions about system vulnerabilities include:

1) Scanner output and analysis,
2) User testimonials, and
3) Analyses of your system using knowledge of threats and system vulnerabilities that usually exist in other similar systems, or analyses using the identified threats and a description of the system.

The primary objective of this task is to identify the vulnerabilities, that either have occurred or might occur, that could cause harm to this system should a particular type of threat attack be launched against the system by some attacker.

A vulnerability might be one that allows a hacker to enter the network and copy plaintext passwords. This in itself does not result in any losses, however, having these passwords will enable the hacker to then access the network and perform some malicious activity, such as erasing an important database (that may not be backed up) or access a sensitive database to gather information on a new product which the hacker then sells to a competitor. This latter action could result in the company losing potential profits, say $10M. Thus, when performing a risk assessment, it is the potential damage that could be done that should guide the analyst. The two primary situations for a vulnerability assessment are:

1) The situation of an existing information system and
2) The situation of an information system to be built for which there is only an architecture model.

For the first case, the owner can have a scanning of the information system performed, or the owner can have someone perform an analysis of the architecture and documentation to determine abstractly what the vulnerabilities might be.

For the second case, the owner can only determine abstractly what the vulnerabilities might be, based on an architecture of the new system and a sample list of vulnerabilities, such as Table 6.6, *List of Some Potential Vulnerabilities*. Any user should begin with this table and then expand it to include any other vulnerabilities that come to mind.

Identifying vulnerabilities can be accomplished in many different ways. A process for identifying the vulnerabilities is (NSA 1998):

1) Search through the graphical model of the system architecture for the absence of a particular type of countermeasure that will defeat some identified threat (and scenario). Do this for all of the identified threats.

2) Again, using the architecture model, examine the flow of information to see if there are any weaknesses in the information flow design by theorizing attack algorithms that might exploit these weaknesses.

3) Perform a code analysis of listings of legacy software to see if there are any weaknesses that could be exploited. (This step would be extremely labor intensive and is not recommended because it can be very expensive and unless the examiners are excellent coders, the task may not produce anything useable or reliable.)

4) Perform a security test and evaluation of the legacy system to identify any weaknesses that could be exploited. (This step is not recommended since a scanning process is the better method for identifying vulnerabilities.)

Table 6.6 List of Some Potential Vulnerabilities

Threats	Potential Vulnerabilities
Unauthorized Physical Entry	Physical entry by unauthorized person and subsequent theft or destruction of some portion of the premises and/or network equipment by that person
Non-catastrophic Event	Loss of power, minor water damage, minor fire damage, random server failure, and loss of heating or air conditioning
Catastrophic Event	Total loss of system due to terrorist bomb or aircraft crash, earthquake, volcanic eruption, major mud flow, flood or tsunami, wind damage from tornado or hurricane, or major fire
Social Engineering Attack	Divulging by employee of privileged information to malicious individual
Internal User Action	1) Compromise of sensitive information recorded on portable device (e.g., floppy disk, compact disk, zip drive disk, super disk, etc.) and then illegally removed from premises by employee 2) Intentional damage to sensitive information 3) Unintentional damage to sensitive information by user
Sneaker Attack	Loss of information that has been illegally recorded by attacker
Virus (Malicious Code) Attack	Loss of system due to damage from virus or other malicious code sent as an attachment through the e-mail system and opened by the victim
Denial-of-Service Attack	Loss of system or information availability due to a denial-of-service attack
Sniffer Attack	Loss of unencrypted information through use of sniffer by an attacker
Scanner Attack	Loss of information concerning system vulnerabilities to attacker using a scanning device and/or software
Hacker Attack	Unauthorized system entry (electronic) to perform malicious actions, such as stealing sensitive information, damage to applications or middleware or databases, deface Web page, loss of transmissions, and loss of system control (availability)

Threats	Potential Vulnerabilities
Man-in-the-middle attack	Loss of sensitive information, including user IDs and passwords, to attacker who intercepts communications between two users and acts as the source and destination of the messages
Brute Force Attack	Loss of encrypted information to attacker with computational capability to decrypt encrypted information by trying every key possibility
Ex-Employee Attack	Unauthorized entry and possible malicious attacks on, or loss of, assets to former employee who is knowledgeable of the organization's information system
Employee Accident	Damage to applications or database information due to accidents
Untrusted Connection	Connecting to an untrusted network without any protection in between such as a firewall followed by hacker entry through the untrusted network
Spoofing	Intentionally using incorrect IP addresses to fool the system

5) Perform a security test and evaluation of the upgraded system to identify any weaknesses that could be exploited.

6) Build a replication of the system in a laboratory where various tests can be performed to identify any weaknesses. (This is recommended for those who can afford it since it allows the client to investigate the capabilities of the implemented security mechanisms prior to a formal implementation.)

Unfortunately, many of these methods are quite time consuming and expensive. The use of scanning tools will perform essentially the same actions as the labor-intensive methods shown above. For these reasons, the first four are not recommended, but the fifth (that proposes an ST&E) is highly recommended. The sixth (system mock-up in a lab) can be too expensive for most organizations, but can lead to a most rigorous security evaluation.

6.2.2.2   List Relevant Vulnerabilities

Some of the vulnerabilities that will be discovered may not be pertinent to this particular system because the threats that would take advantage of a particular vulnerability are not considered credible. A sample list of relevant vulnerabilities is presented in Table 6.7, *List of Relevant Vulnerabilities*.

Table 6.7 List of Relevant Vulnerabilities

Threats	Potential Relevant Vulnerabilities
Employee unawareness	Using easily-guessed passwords
Employee unawareness	Non-installation of delivered vendor software patches
Employee lapses	Loss of laptops containing sensitive information
Social engineering	Giving out sensitive information over the phone to malicious callers
Vendor employee access	Lack of vendor attention to removal of back doors used during installation and testing

Threats	Potential Relevant Vulnerabilities
Employee unawareness	Lack of appropriate training and education of users, administrators, and managers
Administrative inabilities	Lack of proper system administration
Malicious hackers	Lack of appropriate firewall installations with appropriate rules
Malicious hackers	Lack of anti-virus software
Traffic monitoring and malicious hackers	Lack of encrypted transmissions and stored information
Administrative impropriety	Patching direct access to the system for some important user
Administrative unawareness	Not removing use access from access control lists when no longer needed
Unauthorized intrusions	Lack of appropriate identification and authentication, such as including biometrics or smart capabilities
Employee unawareness	Writing down passwords and leaving them in sight
Public malicious acts	Lack of use of a DMZ to protect e-mail, FTP, and Web-based servers
Administrative unawareness	Not turning off unneeded services such as Domain Name Service
Employee unawareness	Opening malicious code such as virus attachments to e-mail messages for unfamiliar sources
Administrative unawareness	Never inspecting networks to see if any scanners or sniffers are attached to the network
Administrative unawareness	Not limiting password entries to only three tries
Unauthorized physical entry	Lack of physical protection of premises and sensitive equipment
Malicious hackers	Hacker attacks against holes in the system

The primary inputs to the creation of a list of vulnerabilities, for this particular system, are the results of the scanning analysis based on the findings using the scanner software mechanism.

6.2.3   Information System Assets and Values

The *information system assets* are based on a subjective evaluation and are dependent upon the specific organization for which the security analysis is being done. The information system owners and principal users should be quizzed to properly identify those assets of most importance to this organization. If these persons are not aware of the prioritized assets, then the assets can be deduced from the organization's security policy or even its overall policy or its mission and strategic plan[59].

---

[59] The *mission* is a simple statement of what the organization wishes to achieve (i.e., its *raison d'etre*) and the *strategic plan* is a description of how the organization hopes to achieve its mission.

Once the important assets have been identified, then the values of these assets can be determined in much the same way by querying the owners and principal users. Because it is so difficult, if not impossible, to get accurate quantitative cost values it is recommended that qualitative values be used instead. However, if qualitative cost estimates are used (e.g., low, medium, or high cost), then it also may be quite difficult, if not impossible, to identify those countermeasures that are cost-effective. The process of identifying those countermeasures that are cost-effective becomes an *intuitive effect* process, which is decision making based on a *gut feeling* about the damage cost versus the countermeasure cost, for each countermeasure. An intuitive effect decision process is probably okay if the decision maker is someone who is very experienced and has much knowledge of the system attributes and their values and alternative countermeasure products, their performance, and costs.

For example, for a commercial organization that does all of its business on the Internet, the following types of attack concerns might include:

1) Denial-of-service attacks,
2) Stealing credit card information, and
3) Viewing or stealing customer lists.

These attacks could have serious consequences in this context. Thus, for this organization, obtaining and installing countermeasures (e.g., strong identification and authentication, firewalls, anti-virus software, and encryption of stored information) that protect against these types of attacks should be highly cost-effective.

For another organization that has private lines for its internal communications and does not deal directly with the public, and uses its system primarily for payroll, the principal types of attack concerns may include:

1) Viewing employee or customer private information (by internal employees), and
2) Lack of availability of the system to employees.

For this organization, encryption of employee or customer information and ensuring that system downtime is minimal such as installing clustered servers (with load balancing and automatic failover software), assigning and enforcing user authorizations (by using role-based access control, for example), and protecting against single-points-of-failure situations (e.g., using dual router and firewall architectures) likely would be cost-effective countermeasures.

Thus, it should be apparent that the selected countermeasures, to be cost-effective, are dependent upon the nature of the organization's business.

6.2.4   Create Threat Scenarios

The next step in the risk assessment process is to use the various threats and the specific vulnerabilities that have been identified and to theorize the scenarios that could occur, see Figure 6.3, *Scenario Creation*.

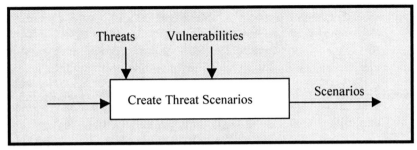

Figure 6.3 Scenario Creation

Some examples of threat scenarios, for some identified threats and vulnerabilities, are listed in Table 6.8, *Some Threats, Vulnerabilities, Scenarios, and Harm*. Note that many of the malicious attacks do not depend on the attacker using any technical means to gather information or to gain entrance to the information system.

The objective here is to identify as many scenarios as possible (utilizing known threats) within some agreed time period. A *brainstorming* approach is recommended where the initialization of the method consists of the tables of threats and vulnerabilities (for this organization) that have been identified. The analysts should formulate as many scenarios as possible and then decide which of these are not feasible or which are not relevant to this organization's mission or size. The remaining ones are the feasible threat scenarios for this organization. The potential harm (i.e., potential loss in dollars) that can be done for each threat scenario is dependent upon the vulnerabilities of the organization's information system and the values of the information system assets.

6.2.5 Risk Computations

Risk computations require that the security analysts develop a list of costs for each particular type of threat scenario, perhaps a worst-case cost for each threat. The probability of each threat occurring successfully is required in order to compute the risk associated with a particular threat.

Table 6.8 Some Threats, Vulnerabilities, Scenarios, and Harm

Threat	Vulnerability	Scenario	Potential Harm
Unauthorized physical entry	Facility has no fences, gates, or guards.	Unauthorized person enters the facility and disables a server with important applications and information stored in its hard disk.	Loss of critical information that is not backed up.
Malicious caller	Users have no knowledge of social engineering.	Caller calls and gets password that can be used to access the information system.	Malicious hacker gains access and erases or steals critical database.
Sniffer for malicious hacker	No encryption of sensitive info.	Malicious hacker sniffs networks and identifies user ID and password information.	Malicious hacker gains access and erases or steals critical database.

Threat	Vulnerability	Scenario	Potential Harm
Session attack	No encryption of or checksum of sensitive info.	Malicious hacker intercepts messages and either copies or changes the information.	Malicious hacker changes messages to cause poor decision.
Virus attack	No anti-virus capability and no knowledge of "not to download."	Malicious hacker sends virus as an e-mail attachment, which is downloaded by addressee.	Virus causes harm to database information.
Distributed Denial of Service attack	No anti-DDoS capability.	Malicious hacker sends a SYN flood attack that causes denial of service to authorized users.	Attack denies service to authorized users.
Worm attack	Software has holes and client has not taken appropriate proactive actions.	Malicious hacker sends worm code through identified hole in operating system.	Attack causes information system to crash.
Ping of death	Client has no buffer size check.	Malicious hacker sends ICMP ping to close down information system.	Information system shuts down.
Malicious hacker guesses password	Password rules do not include "no guessability."	Malicious hacker guesses password and accesses information system.	Attacker causes harm to database information.
Trojan horse attack	No check for Trojan horse in software.	Malicious hacker sends free game with Trojan horse that opens back door for hacker to enter.	Attacker gains access to system and causes harm to a database.
Dumpster diving	Sensitive information thrown in trash.	Attacker goes through trash and finds sensitive information on company's new product to be issued soon.	Attacker uses information to gain market edge.
Recipient denies having received message	No digital signature on messages.	Recipient denies having received an important message that he actually received.	Recipient does not respond properly to message that he actually received.
Improper authorization to authorized user	No certificates with distinguished name.	Authorized user gains access to information that is not need-to-know sensitive information.	User uses sensitive information for personal gain.
User gains	No role-based	Authorized user gets access to	User uses

Threat	Vulnerability	Scenario	Potential Harm
access to improper information	access control.	improper information.	improper information for personal gain.
Authorized (internal) user intentionally damages data	No intrusion detection or auditing.	Authorized user intentionally damages database information.	User causes harm to database information.
Authorized (internal) user unintentionally damages data	No intrusion detection or auditing.	Authorized user unintentionally damages database information.	User causes harm to database information.
Unauthorized user access to system	No intrusion detection or auditing.	Unauthorized user gains access to system and intentionally damages or steals database information.	Attacker causes harm to stored info or uses info to gain market advantage.
Disgruntled former employee gains access	No timely removal of names from access control list.	Disgruntled former employee gains access to system and damages database information.	Attacker causes harm to stored info or uses info to gain market advantage.
Sniffer	No sniffer detection and no encryption.	Attacker places sniffer on network link and collects plaintext information.	Attacker gets readable information and uses it to gain market advantage.
Brute force attack	Short key word encryption.	Attacker collects transmitted information (using sniffer) and decrypts is using brute force method.	Attacker gets decryptable information and uses it to gain market advantage.

Risk is a probabilistic cost of potential damage to the system, given the threats, vulnerabilities, and values of the system's assets. By applying probabilities to the various threats and identifying the potential damage costs in case of a successful attack, the risks can be determined. However, it is not a good idea to attempt to quantitatively identify exact damage costs since the result most likely will be a lot of meaningless guesswork. Moreover, the probability of a certain type of attack is also just guesswork. Some examples of risks, activities, and responsibilities are shown in Table 6.9, *Risks, Activities, and Responsibilities.*

However, one can decide somewhat more accurately, in a qualitative way, what the probabilities or costs might be. So, it is best to categorize the probabilities and potential damages using a three- or five-level qualitative assessment. For example, probabilities can be assessed as high, medium, and low or very high, high, medium, low, and very low. Similarly, the potential cost of an attack (or loss) can be assessed as similarly. With the categorizations of the probabilities and losses, a multiplication table must be devised for risk, where:

Risk (for the ith threat, dollars) = Probability of Successful Attack Using the ith Threat *
Potential Loss from that Attack ($).

Table 6.9 Risks, Activities, and Responsibilities

Risk Parameter	Activity	Responsibility
Threats	Identify	Security Analyst, Users
Vulnerabilities	Identify	Security Analysts, Users
System Assets/Values	Identify	Owners, Users
Threat Scenarios	Create	Security Analysts
Countermeasures	Identify	Security Analysts
Selection of Countermeasures (General and Specific)	Identify	Security Analysts, Owners/Users
Migration Plan	Identify	Security Analysts
Countermeasures Implementation	Implement	Vendors/Organizational Technicians
Remaining Risks	Identify	Security Analysts

6.2.5.1   Determine Attack Probabilities

Attack probabilities, for this system, can be determined by assessing each identified threat and/or threat scenario. A security analyst working closely with the owner or named representative of the organization can probably best perform this action. A table for a sample situation is shown below in Table 6.10, *Attack Scenario Probabilities Versus Threats*.

Table 6.10 Attack Scenario Probabilities versus Threats

Threat	Attack (Scenario)	Attack Probability
Physical		
Unauthorized physical entry	An unauthorized person is successful in getting into the secure area. He then pulls a server loose from its electrical connections causing the destruction of database information stored inside.	0.1
Unauthorized entry into sensitive server location	An unauthorized disgruntled employee is successful in getting access to a sensitive server. He then pulls a server loose from its electrical connections causing the destruction of database information stored inside.	0.2
Power loss	The local power company suffers an outage of power to the information system facility and the power to the computer system is lost for two hours.	0.01
Behavioral		
Social engineering	An employee is called by a malicious attacker who asks for his password under some reasonable pretense, and the employee gives the caller the password.	0.05

Threat	Attack (Scenario)	Attack Probability
Lack of managerial training	A manager is confronted by an unusual security situation and does not know what action to take.	0.01
Technical		
Someone scans the system	A potential malicious hacker scans the network using a scanning tool.	0.05
Hacker intrusion	An unauthorized person is able to access the system and gains access to a critical database and clobbers the information.	0.05
Sniffer	A malicious person is able to attach a sniffer to the network to collect information such as passwords and data messages.	0.05
Denial-of-service attack	A malicious person commits a denial-of-service attack and blocks service for two hours.	0.01
Non-repudiation	An authorized user denies having received an e-mail message that was sent to him.	0.01
Man-in-the-middle attack	A malicious hacker achieves a man-in-the-middle situation and is able to pretend he is the receiver of information for some sender and is able to receive sensitive information.	0.01

6.2.5.2   Determine Potential Attack Costs

Potential attack costs can be assessed by examining each threat and/or threat scenario versus the gamut of organizational information system assets (and their values), such as that shown in Table 6.11, *Attack Scenarios Versus Potential Costs*. The potential cost (or loss) from a particular attack can be given as High, Medium, or Low, rather than specific numbers.

6.2.5.3   Determine Risks

Use the material presented in Section 6.2, *Risk Assessment*.

Table 6.11 Attack Scenarios versus Potential Costs

Attack (Scenario)	Assets Attacked and Potential Damage ($)
Physical	
An unauthorized person is successful in getting into the secure area. He then pulls a server loose from its electrical connections causing the destruction of database information stored inside.	Medium

Attack (Scenario)	Assets Attacked and Potential Damage ($)
An unauthorized disgruntled employee is successful in getting access to a sensitive server. He then pulls a server loose from its electrical connections causing the destruction of database information stored inside.	Medium
The local power company suffers an outage of power to the information system facility and the power to the computer system is lost for two hours.	Medium
Behavioral	
An employee is called by a malicious attacker who asks for his password under some reasonable pretense, and the employee gives the caller the password.	Medium - High
A manager is confronted by an unusual security situation and does not know what action to take.	Low
Technical	
A potential malicious hacker scans the network using a scanning tool.	High
An unauthorized person is able to access the system and gains access to a critical database and clobbers the information.	High
A malicious person is able to attach a sniffer to the network to collect information such as passwords and data messages.	High
A malicious person commits a denial-of-service attack and blocks service for two hours.	High
An authorized user denies having received an e-mail message that was sent to him.	Low-Medium
A malicious hacker achieves a man-in-the-middle situation and is able to pretend he is the receiver of information for some sender and is able to receive sensitive information.	Medium

6.2.6   Security Countermeasures

Security countermeasures are the *methods* (e.g., badge checking, education, and training) or *devices* (e.g., fences, gates, badges, firewalls, or anti-virus software) that can be used to minimize the negative effects of existing threats to the system, see Table 6.12, *Threats and Countermeasures*. The potential loss that can occur from a threat is the computed risk value.

Table 6.12 Threats and Countermeasures

Threats	Countermeasures
Physical	
Unauthorized physical intrusion	Fences, gates, guards, patrols, cyber locks, and user badges to gain entrance to the organizational facility. Server rooms with cyber locks.

Threats	Countermeasures
Unauthorized access to secure subsystems	Placing servers in secure with cyber locked doors where only authorized p4ersons know the combination.
Unauthorized data removal from the premises by internal users	Random examination of all material leaving the premises.
Behavioral	
Lack of Training	Periodic presentation of Training Courses designed for Managers, System Administrators, Security Administrators, Technical Personnel, System Users, etc.
Lack of Awareness	Periodic presentation of Awareness Education Courses designed to convey proper facts to organizational personnel.
Lack of Standard Operating Practices	Prepare and distribute (hardcopy and electronic) organizational standard operating procedures that are required knowledge for managers, system administrators, technicians, and other personnel. For example, "System administrators should ensure prompt removal from access control lists the identification of any user who no longer needs access to the system."
Technical	
Catastrophic Failures	
Any Massive Failure	Automated backup of sensitive and critical information at remote sites using servers, disks, or tape for storage. Development and practice of using a contingency plan in case of a massive failure. Storage of replacement equipment (clients, routers, servers, cabling, hubs, bridges, etc.) for quick restoration of system operations.
Non-Catastrophic Failures	
Power Failures	Use of Uninterruptible Power Supply and Power Generators.
Minor Water Damage	Use of Water damage control subsystems.
Minor Fire Damage	Use of Fire control subsystems.
Climate Control	Use of Air Conditioning Systems and Heating Systems.
Availability situations (e.g., an outage due to physical damage to a server)	Use of dual subsystems and clustered servers (e.g., firewalls, routers, bridges, and servers such as Web-servers, FTP-servers, and mail-servers) with load balancing and automatic failover software to avoid a single-point-of-failure situation, including the DMZ, if desired. Use of identification and authentication systems. Use of authorization systems. Use of Fences, gates, guards, patrols, cyber locks, and user badges to gain entrance to the organizational facility.

Charles L. Smith, Sr.

Threats	Countermeasures
	Server rooms with cyber locks.
Malicious Activities	
Malicious intruders	Firewalls (can include software for checking user IDs and passwords, smart cards and biometrics, intrusion detection, anti-virus software, anti-DDoS software, and VPN encryption) Routers
Interception of e-mail by unauthorized entity	Encryption of e-mail Private leased lines
Malicious hacker steals User IDs and Passwords	Encrypt the information in access control lists (i.e., Users IDs and Passwords) Limit access to ACLs to specific individuals
External Hacker/Cracker[60] Attack	Firewalls with rules, virtual private network, anti-virus, anti-denial-of-service, and intrusion detection software. Install software patches prescribed by COTS vendors. Role-based access control for system access. Encryption of stored and transmitted information with data integrity protection. Deploy User ID and Password software for entity identification and authentication. Strong authentication (e.g., biometrics or smart cards). Auditing and intrusion detection system software. Scanning devices to identify and repair system entry points. Shutting off all protocols not required. Use of one-time passwords.
Internal User Attack	Auditing software and intrusion detection system platforms and/or software.
Malicious E-Mail	Anti-virus software. Do not open any e-mail attachment from a source that you do not know or if the e-mail message seems to be suspicious.
Social Engineering	Never give out any sensitive information over the phone to persons whom you cannot unequivocally identify.
Man-in-the-middle Attack	Encryption of transmitted messages.
Brute Force Attack	Encryption of transmitted messages with large keys (e.g., 128-bits or more).
Dictionary Attack	Do not use dictionary words and make passwords with a length greater than 8-characters, with at least three numerical values and one a special character.
War Dialing	No countermeasure for this (that I know of).

[60] A malicious hacker is sometimes referred to as a "cracker" but often a cracker refers to an entity that attempts to electronically enter a system using guesses at User IDs and Passwords.

Threats	Countermeasures
Dumpster Attack	Ensure that all trash with sensitive information is shredded.
Replay Attack	Place a time stamp on all transmitted information. Then check for a reasonable time for the message and if not, reject the message.
Misrouting Attack	Encryption of transmitted messages. Place a routing description on all transmitted messages.
Malicious Subsystem Configuration Change	Use an out-of-band subsystem for inputting router and firewall rules.
Sniffing	Encryption of transmitted messages. Physical inspection to detect the presence of sniffers. Use of twisted pair hub technology. Scanning to determine if the network is in a promiscuous mode. Include policy statements forbidding organizational use of sniffers.
Spoofing	Configure routers to reject any inbound packets that claim to originate from a host within the internal network. Do not extend trust relationships across public networks.
Message alterations	Hashing functions to provide data integrity.
Repudiation of messages	Digital signatures to offer non-repudiation.
Malicious changes to traffic	Certificates, Secure Sockets Layer (SSL), Secure Shell (SSH), Transport Layer Security (TLS), and Public Key Infrastructure (PKI) to provide encryption for traffic confidentiality.
Unauthorized access	Provide role-based access control (RBAC) for ensuring only authorized users have access.
Unapproved access to sensitive information	Provide RBAC with privileges to ensure that appropriate pedigrees are applied and for separation of duties.
Non-Malicious Activities	
Accidents	Auditing software and intrusion detection system platforms and/or software. Clustered servers with load balancing software.
Slips	Auditing software and intrusion detection system platforms and/or software. Clustered servers with load balancing software.
Mistakes	Auditing software and intrusion detection system platforms and/or software. Clustered servers with load balancing software.
Improper Use of the Internet	Auditing software and intrusion detection system platforms and/or software. Clustered servers with load balancing software.

For each type of threat there is a countermeasure that can be employed to minimize its effects. Some countermeasures can minimize the effects of more than one type of threat (e.g., firewalls). Similarly, some threats require more than one type of countermeasure to be minimized (e.g., unauthorized physical entry). Since there are literally thousands of specific threats, the primary objective here is to identify families or types of threats and then list the countermeasures that one could deploy to minimize the potential damage. The reader, in many cases, if given an identified threat (that is not listed in the table below), and wishes to identify a set of countermeasures that could be deployed to minimize the identified threat, should be able to make use of Table 6.12 for that purpose.

Technical countermeasures include the following products:

1) Routers,
2) Firewalls,
3) Uninterruptible power supplies,
4) Backup heating, ventilating, and air conditioning systems,
5) Identification and authentication methods (including biometrics),
6) RADIUS and TACACS systems for authenticating remote users,
7) Intrusion detection systems (IDSs),
8) Encryption methods (symmetric, asymmetric (e.g., public key infrastructure), or both),
9) Hashing functions,
10) Digital signatures,
11) Anti-virus software,
12) Anti-DDoS software,
13) Certificates,
14) Smart cards,
15) Virtual private network (VPN) encryption,
16) Role-based access control (RBAC) and use of roles for access control and authorization,
17) Secure sockets layer (SSL) protocol,
18) Secure shell (SSH) program, and
19) Transport layer security (TLS).

The owner or named representative must decide whether each of the countermeasures selected to minimize the effects of the relevant threats is cost-effective.

6.2.6.1   Countermeasures for the Entire System

It is much less costly to implement the *countermeasures for an entire system* rather than have each application or subsystem consider its relevant countermeasures separately. The reason is that by considering the countermeasures separately for each application or subsystem causes a lot of redundancy in the security architecture, and hence is more costly than a security architecture that addresses all of the applications and subsystems as a unit thereby meeting the security requirements of each application or subsystem.

It is often the case that applications or subsystems are considered independently and this is an unfortunate situation. It is more desirable to consider all of the security needs for the entire information system network, identify the countermeasures for the whole system, implement the selected countermeasures, and test and evaluate the installed countermeasures. This approach also simplifies the implementation of the countermeasures, as well as the security test and evaluation process, thereby ensuring an even greater cost savings for the organization. For example, when considered separately, each application will require an identification and authentication capability that can be employed whenever a user selects the application, whereas when all the subsystem applications are considered as a unit, the operating system can perform identification and authentication for all the applications and display an icon for each application for which the user has a legitimate access, such as by using a single sign-on capability.

Countermeasures to permit or limit access by authorized or unauthorized users perform the following security capabilities: 1) identification and authentication, 2) access control, and 3) authorization.

6.2.6.1.1 Identification and Authentication

Identification and authentication are the software processes used to identify users (i.e., obtain the user's appropriate identification (not necessarily their name but names are often used) and authentication (determine that the entity is who they say they are). There are several methods used for authentication, namely:

1) Users IDs and Passwords,
2) Tokens and Biometrics,
3) Remote Authentication,
4) Routers,
5) Firewalls, and
6) Kerberos.

Often, a primary infraction of the security rules is when a user, who no longer needs access to the system, is not removed from the access control list by the security administrator. Some capabilities of the various authentication mechanisms are shown in Table 6.13, *Advantages and Disadvantages of Selected Authentication Mechanisms*.

Digital certificates that link an entity's identify to a set of encryption keys required for developing digital signatures and encrypting information provide far better security than passwords do. But digital certificates are more expensive, so the tradeoff is how much better security (i.e., a smaller probability of being intruded) is worth (i.e., in dollars) to the organization, given its threats, mission, and operating environment.

When a separate password is provided for each application, in the case of application-based security, it is difficult for the users to refrain from pasting their passwords on the monitor, since many different passwords will be required for access to all of the relevant applications for some specific user. On the other hand, if a single password (note that this is not single sign-on) is used for all the applications, then should an intruder get one password, this intruder can access all of the applications in the system. This is not a good situation.

Table 6.13 Advantages and Disadvantages of Selected Authentication Mechanisms

Authentication Mechanism	Advantages	Disadvantages
Biometric ID	1. Uniquely identifies users 2. Difficult, it not impossible, to spoof or bypass 3. Some methods, such as handwriting analysis are very user-friendly 4. Several alternatives: fingerprint, voice, facial infrared, retinal infrared, and handwriting	1. Expensive 2. Technology still being developed, not completely accepted as yet
Callback	1. Quick implementation 2. Inexpensive 3. Can resolve phone billing issues in addition to security features 4. Identifies user's location providing one factor in a two-factor authentication scheme	1. Can be spoofed by call forwarding 2. Can be spoofed if the phone system can't initiate a hang-up at the right time 3. Difficult to administer with large callback lists 4. Not effective for users who travel to unknown locations 5. Authentication location only, not the user at hat location 6. Relies on phone company's computers being secure
Caller ID	1. Difficult to spoof 2. Inexpensive 3. Easy to implement 4. Identifies user's location providing one factor in a two-factor scheme	1. Not feasible if multiple phone company areas are involved 2. Difficult to administer with large remote user lists 3. Not effective for users who travel to unknown locations 4. Not available everywhere 5. Authenticates location only, not the user at hat location 6. Relies on phone company's computers being secure
Node Identification	1. Uniquely identifies a PC, providing the "what you have" portion of a two-factor authentication scheme 2. Unobtrusive to the user since it is performed automatically in the background 3. No hardware tokens to worry	1. A proprietary solution 2. Difficult to administer with large remote-user lists 3. Identifies machines, not users

Authentication Mechanism	Advantages	Disadvantages
	about	
One Time Password	1. Cost-effective 2. Can be implemented without special HW or SW 3. Difficult to spoof technically	1. Difficult to administer with large remote-user lists 2. Referring to printed lists may be unacceptable for users
PC Card	1. Protects integrity and confidentiality of stored data 2. Easy to use 3. Relatively small 4. Works with newer PCs	1. Requires PCMCIA slot 2. Limited number of products available 3. Requires every user to have a PC card 4. Relatively expensive for many users
Smart Card	1. Easy to use 2. Relatively small 3. Protects integrity and confidentiality of stored data	1. Requires a smart card reader 2. Requires every user to have a smart card 3. Relatively expensive for many users
Smart Disk	1. Protects integrity and confidentiality of stored data 2. Easy to Use 3. Relatively small 4. Works with existing hardware	1. Proprietary technology 2. Limited choice of security mechanisms 3. Requires every user to have a Smart Disk 4. Relatively expensive for many users
Token Device	1. Protects integrity and confidentiality of stored data 2. Easy to use 3. Relatively small 4. Does not require any additional hardware 5. Works with dumb terminals	1. Requires user participation so cannot be used unattended 2. Decreases system usability 3. Relatively expensive for many users 4. Requires every user to have one
Traditional Password	1. Cost effective 2. Easy to use 3. Easy to administer 4. Universally supported	1. Easily bypassed 2. May be observed and reused 3. May be stolen through social engineering

6.2.6.1.1.1   User Identifications and Passwords

User identification is requiring that an entity provide an identifier (such as their name or IP address) when logging onto the system. The user's identification (ID) may be the user's name or a name given to their client computer system (i.e., workstation or PC).

To authenticate the entity another process is used. Most often a password is used. A password is a set of characters for the system to determine that the user is who they say they are. However, to ensure that the entities are who they say they are, other forms of identification may be required.

6.2.6.1.1.2   Tokens and Biometrics

Tokens and biometrics are methods used to ensure that users (or entities) are who they say they are. Biometrics are methods based on some personal aspect of being human: fingerprints, retinal or pupil pattern, facial characteristics, facial infrared pattern, voiceprint, hand form, hand writing, or some other aspect of a person's physical or behavioral characteristics. Any of these capabilities add expenses to the security capabilities of the system since there must be a hardware device to see the object, software to evaluate the object, and a template that represents the digital object against which the software exercises its comparison. Additionally, the biometrics system will fail sometimes, either by falsely recognizing an unauthorized user or improperly blocking an authorized user.

A token such as a *smart card* and *personal identifier number (PIN)* can be used to authenticate a user. Smart cards depend on the person having a card (such as a credit card) that can be swiped in a card reader and then being able to input a PIN number that authenticates the person. This has the advantage that should the user lose their smart card, the finder cannot use the card unless they know the PIN number. This number is usually four digits long, so there is only a one in ten thousand chance that the finder will guess the right number. Even with three tries, the chances are very small that the number will be correctly guessed.

Biometric identifiers are based on the use of some sort of hardware device for reading and recording a physical attribute and then a software device for comparing the reading with a template stored in memory and associated with the person's certificate. Biometric identifiers can be based on fingerprints, eye pupils or retinas, facial characteristics, handwriting characteristics, voice characteristics, hand shape characteristics, or other human physical or behavioral traits. In the aftermath of the September 11, 2001, terrorist attacks on the U.S., biometric identifiers may be used in U.S. airports to employ facial-recognition software to compare digital photos of faces of passengers against digital facial-templates of known or suspected terrorists (Matthews 2001).

The benefit of a biometric identifier is that it is not something that can be lost or forgotten, nor can it be stolen, at least not normally. There are rather bizarre counterexamples. For example, someone's finger could be amputated by a malicious hacker (thereby giving another definition to the term) and placed in a fingerprint reader for verification, provided it is the appropriate finger. The computer could be programmed to ask the user who wishes to sign on to input a particular finger into the reader with a presentation such as:

> ### Please place the middle finger of your left hand in the fingerprint reader.

where the specific finger and hand selection are randomly chosen by the computer. If this randomness is based on specific individuals (who have already identified themselves) then the computer-selected hand or finger selection can take into consideration any limitations on hands or fingers of the individual. The computer would have to know whether or not the identified user has both hands and all fingers, or if an appendage (finger or hand) is missing. For a user who has no hands (and uses artificial hands), another method of authentication would have to be used, such as a voiceprint.

Strong authentication can be based on tokens or biometrics, since a user ID and a password can be discovered or stolen by someone other than the person they are meant to represent. However, it is very difficult if not impossible to discover or steal a biometric physical attribute. However, one of the issues with biometrics is that there may be type I and type II errors that are too large, namely too often incorrectly granting access to an unauthorized user (type I error) or incorrectly denying access to an authorized user (type II error).

Memory tokens can be smart cards and a PIN number. Smart tokens or cards require a concomitant reader and software (NIST 1995). One-time passwords are passwords that are used once and then are discarded (NIST 1995). A challenge-response protocol works by having the computer generate a challenge, such as a random string of numbers from the client to the server. The server then generates a response based on the challenge that is sent back to the client. This method is based on cryptography (NIST 1995).

6.2.6.1.1.3   Remote Authentication

Remote authentication is the software process of identifying users who sign on from remote sites usually through an Internet Service Provider system.

Remote Authentication Dial-In User Service (RADIUS) is a client/server protocol that enables remote authentication and access using a variety of authentication mechanisms such as token cards, smart cards, and biometrics.

RADIUS provides a solution for distributed security. This solution eliminates the need for special hardware and provides access to a variety of state-of-the-art security solutions. Based on a model of distributed security previously defined by the Internet Engineering Task Force (IETF), RADIUS provides an open and scalable client/server security system. It can be easily adapted to work with third-party security products or proprietary security systems. Any communications server or network hardware that supports the RADIUS client protocols can communicate with a RADIUS server. To provide distributed security, *the RADIUS server has been customized to work with the Kerberos Security System* for authenticating user names and passwords.

The Terminal Access Controller Access Control System (TACACS) is an authentication protocol, developed by the Defense Data Network (DDN) community, which provides remote access authentication and related services, such as event logging. User passwords are administered in a central database rather than in individual routers, providing an easily scalable network security solution.

The DDN is the U.S. military network composed of an unclassified military network (MILNET) and various secret and top-secret networks. DDN is operated and maintained by the Defense Information Systems Agency (DISA).

RADIUS allows remote access devices to share the same authentication database. RADIUS provides a central point of management for all remote network access by validating the user's credentials through a challenge and response process. The RADIUS capability is provided by a number of vendors (Brenton 1999). The TACACS is a similar system to RADIUS but is produced by a different vendor.

A Cisco router can operate in either a non-privilege or privilege mode. The *non-privilege mode* allows the user to check connectivity and to look at statistics, but not to make any type of configuration changes to the device. This helps to limit the damage that can be done by a hacker. From the non-privilege mode the user can enter the *privilege mode* which allows the user to change or even delete any configuration parameters. There are up to 16 different levels of privilege-level access. Cisco routers have the capability to use external authentication. TACACS provides a way to validate every user on an individual basis before a user can gain access to the router.

The RADIUS protocol uses a frame format based on the UDP/IP protocols whereas TACACS use the TCP protocol. RADIUS encrypts only the password with the rest of the packet in the clear whereas TACACS+ (a follow-on protocol to TACACS) encrypts the entire body of the packet. TACACS+ provides support for multiple protocols, whereas RADIUS does not support many protocols. Because of frequent middleware changes, it is recommended that any potential customer review the current capabilities of the RADIUS and TACACS+ protocols to ensure that the latest capabilities are considered when comparing the two.

6.2.6.1.1.4   Router Authentication Actions

Routers are computers that are programmed to relay network traffic to the next router or destination address on the basis of its source and destination addresses. Routers have a basic and speedy operating system that is capable of handling the rules used to relay messages. Routers can be used to route traffic and to block traffic, as does a firewall. In the case of unwanted traffic, the router can be programmed (with a rules base) to reject traffic from certain users. Routers are multiport devices that decide how to handle the contents of a frame, based on protocol and network information.

IP routing can be done by a specified route (i.e., source routing) or by a dynamic method where the next router is determined dynamically. Usually, routers assume that the *identified source address* is in fact the IP address where the message began and hence are subject to spoofing (NIST 1994). Routers are used to connect logical networks by stripping off the header that contains the destination address and providing a "nearest router address" for relaying the message. Routers use 1) static, 2) distance vector, or 3) link state entries to create a blueprint of the network known as a routing table (Brenton 1999):

1) *Static routing* is the most secure method of building routing tables but requires a high level of maintenance. Static routing defines a specific router to be the point leading to a specific network. It does not require other routers to exchange route information. Each static router is responsible for maintaining its own routing table, so if one router is compromised, the effects of the attack are not automatically passed on to other routers.

2) *Distance vector routing* is the oldest and most popular form of creating routing tables. The routing information protocol (RIP) is based on the distance vector. Distance routing is a dynamic routing option. Distance vector routers build their routing tables by examining tables being advertised by other routers and adding 1 to the hop values to create its own table. With distance routing, every router broadcasts its routing table once per minute. If the distance vector routing breaks down, for whatever reason, then the computation of its hop value using the "add 1" method does not work.

3) *Link state routing* is a dynamic routing method that is similar to distance vector routing but it uses only firsthand information when developing its routing tables. When the link state router powers up, it transmits a "Hello" message in an RIP packet. A link state routing protocol (LSP) frame is created and transmitted to its contiguous routers (from both ports) containing the following information (Brenton 1999):

a) The router's name or identification,

b) The networks it is attached to,

c) The hop count or cost of getting to each network, and

d) Any other routers on each network that responded to its "Hello" frame.

6.2.6.1.1.5   Firewall Authentication Actions

Firewalls are similar to routers, being a computer with a minimal operating system. Routers and firewalls are becoming more alike and soon may inhabit the same platform, if the speed of the platform is sufficient to handle the traffic and implement all of the router and firewall rules without appreciably slowing the traffic. The latest firewalls incorporate many functions within the platforms software, namely, firewall instructions (based on the security rules base), routing instructions, VPNs, IDSs, automatic failover and load balancing (for clustered platforms), auditing, and other functions that rely on performing similar activities.

Firewalls, like routers, operate on the basis of a set of rules. If you can afford the cost and response delay, it is wise to have more than one firewall in tandem (from different vendors) because the methods used to block and pass messages by the different vendors are complementary and serve to enhance one another. Thus, it is more likely that a malicious user who attempts to enter the system will be blocked if more than two or more firewalls (from different vendors) are used to pass/block the traffic.

Firewalls operate on the basis of one of two philosophies: 1) block everything except what is specifically listed to be allowed through, and 2) pass everything except what is specifically listed to be blocked. Modern firewalls have the following capabilities:

1) Access controlling,

2) User authenticating,

3) Network address translating,

4) Virtual private networking,

5) High availability,

6) Content security (such as anti-virus software and screening of Java objects),

7) Auditing and reporting,

8) Use-managing based on the LDAP protocol,

9) Intrusion detecting, and

10) Malicious activity detecting.

However, it is may be more effective for some of these tasks to be performed by the operating system or other servers rather than by the firewall. In some cases, a firewall may be an accumulation of several subsystems and their concomitant rules, such as a combination of gateways, hosts, routers, switches, and packet filters. Most often though, a firewall is a single computer platform with an operating system and an application that uses a set of security rules to either pass or block any message that enters the firewall communications link.

A firewall security policy is a set of rules (security rules base) defined in terms of other firewalls, services, users, and resources that govern the interactions among them. Once these rules have been specified, the code for the rules base can be developed and installed on the firewall (or other mechanisms as defined) to enforce the security policy.

In addition, a Firewall Management Module, can control and monitor one or more firewall modules. Normally, the Firewall Management Module operates independently of the System Management Module. In some cases, firewalls are installed at each perimeter of an organizational system to provide in-depth security (i.e., tiered security) and it is recommended that several different types of firewall (i.e., different vendors) be used since they operate differently and tend to complement one another to enhance security (Smith 1999). An example of a three-tiered system is shown in Figure 6.4, *An Illustration of a Three Level Perimeter Organizational System*.

The firewall vendors have exhaustive tests to perform security test and evaluations of their products, e.g., Check Point Firewall-1 uses the Common Criteria tests performed by an unbiased objective subcontractor, Computer Sciences Corporation (CSC 1999). Thus, it is not a prudent objective to perform unit security tests of firewall products. Once a firewall product is installed, the vendor can perform much of the ST&E testing required to ensure that the product is performing properly.

One of the new technologies introduced in the more advanced firewalls is *stateful inspection*, a *de facto* standard created by Check Point Software Technologies. It combines both communication-derived and application-derived state and context information that is stored and updated dynamically. Stateful inspection provides cumulative information that is used for evaluation of link traffic (i.e., packets) to determine if later communications should be passed or blocked. It provides full application-layer (i.e., OSI Model Layer 7) awareness without requiring a separate proxy for every service resulting in improved performance, scalability, and supportability for new applications (Check Point 2001).

Firewalls can also provide *anti-spoofing* and *anti-spamming* capabilities. eSafe Gateway, from Aladdin, examines HTTP, SMTP, and FTP traffic to block or clear undesirable content, providing advanced security features such as file type, anti-spoofing, macro removal, scanning and stripping of malicious JavaScripts, scanning and blocking of malicious Java/Active objects, blocking untrusted cookies, email anti-spam and anti-spoofing features, as well as auto-learn features for blocking offensive Web sites (Aladdin 1999). A source of firewall information is from *Web Trends* at the following URL site: http://www.webtrends.com/products/firewall/reporting.htm.

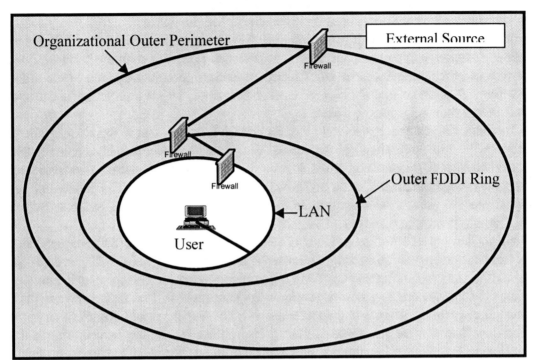

Figure 6.4 An Illustration of a Three Level Perimeter Organizational System

6.2.6.1.1.6   Kerberos Authentication

Kerberos is an authentication method designed to provide a single sign-on to a heterogeneous environment. It is used by Microsoft in its new Windows NT OS for authentication. It allows mutual authentication and encrypted communication between users and services. Kerberos relies on each user remembering and maintaining a unique password. Once a user has been authenticated by Kerberos, the user is issued a token that can be used thereafter for any Kerberos-aware authenticated service. The user is also issued a set of encryption keys for encrypting all data sessions (Brenton 1999). Unlike the RADIUS protocol, Kerberos also verifies that the user is logging onto the correct network, an important capability for wireless LAN users.

There are two types of credentials used in the Kerberos authentication model: tickets and authenticators. Both are based on private key encryption, but they are encrypted using different keys. A ticket is used to securely pass the identity of the person to whom the ticket was issued between the authentication server and the end server. A ticket also passes information that can be used to make sure that the person using the ticket is the same person to which it was issued. The authenticator contains the additional information which, when compared against the identity in the ticket proves that the client presenting the ticket is the same one to which the ticket was issued.

A ticket is good for a single server and a single client. It contains the name of the server, the name of the client, the Internet address of the client, a timestamp, a lifetime, and a random session key. This information is encrypted using the key of the server for which the ticket will be used. Once the ticket has been issued, it may be used multiple times by the named client to gain access to the named server, until the ticket expires. Because the ticket is encrypted in the key of the server, it is safe to allow the user to pass the ticket on to the server without having to worry about the user modifying the ticket.

Unlike the ticket, the authenticator can only be used once. A new one must be generated each time a client wants to use a service. This does not present a problem because the client is able to build the authenticator itself. An authenticator contains the name of the client, the workstation's IP address, and the current workstation time. The authenticator is encrypted in the session key that is part of the ticket.

A fundamental determination of a user's identity is the user's password. The initial exchange with the authentication server is designed to minimize the chance that the password will be compromised, while at the same time not allowing a user to properly authenticate herself or himself without knowledge of that password. The process of logging on appears to the user to be the same as logging on to a timesharing system. Behind the scenes, though, it is quite different.

The user is prompted for her or his user ID. Once it has been entered, a request is sent to the authentication server containing the user's name and the name of a special service known as the ticket-granting service. The authentication server checks that it knows about the client. If so, it generates a random session key that later will be used between the client and the ticket-granting server. It then creates a ticket for the ticket-granting server which contains the client's name, the name of the ticket-granting server, the current time, a lifetime for the ticket, the client's IP address, and the random session key just created. This is all encrypted in a key known only to the ticket-granting server and the authentication server.

The authentication server then sends the ticket, along with a copy of the random session key and some additional information, back to the client. This response is encrypted in the client's private key, known only to Kerberos and the client, which is derived from the user's password.

Once the client has received the response, the user is asked for her or his password. The password is converted to a Data Encryption Standard (DES) key and used to decrypt the response from the authentication server. The ticket and the session key, along with some of the other information, are stored for future use, and the user's password and DES key are erased from memory.

Once the exchange has been completed, the workstation possesses information that it can use to prove the identity of its user for the lifetime of the ticket-granting ticket. As long as the software on the workstation had not been previously tampered with, no information exists that will allow someone else to impersonate the user beyond the life of the ticket.

Another method for secure authentication is the Secure Shell (SSH), which is a powerful method for performing client authentication and safeguarding multiple service sessions between two systems. When two systems are using SSH to establish a connection, they validate one another by performing digital certificate exchanges. *Brute force and playback* attacks are ineffective against SSH because encryption keys are periodically changed. 3DES is used for message encryption (Brenton 1999).

To ensure confidentiality of e-mail traffic, Multimedia Internet Mail Extensions (MIME) allows the server to inform the Web browser as to what type of data it is about to receive

(Brenton 1999). Secure MIME (S/MIME) is a protocol for encrypting e-mail traffic. S/MIME is based on either RSA public key security or PGP security.

6.2.6.1.1.7 Security Scenario Example

In a simplified manner, the following scenario demonstrates how the security mechanisms described above function in an information system environment.

Anne dials up and establishes a connection with the information system network. Anne's computer and the information system present to each other their respective public keys. The public key of the information system has been signed by a certificate authority that is recognized and trusted by Anne's browser software. Likewise, Anne's public key was signed by an entity that the information system recognizes and trusts (perhaps even by the information system itself.)

Anne and the information system both can now be certain of the identity of the other (meaning that neither will accidentally reveal to the other any information, such as a password, that could be misused by an impostor). In addition, both can use their respective public keys to encrypt all data that they exchange. Their communications are confidential, meaning that nobody else can intercept this information or gain access to it.

Anne now decides to resume previous work with the Widget Application System (WAS). Calling up the WAS software causes the WAS to request from the information system authorization (for Anne) to offer its services to Anne. Through Anne's session information, the information system knows that Anne is a member of the "Able Contracting System" users. Members of this group are allowed to use the WAS application, according to the access control services used by the information system, so the WAS receives the "OK" that Anne may go ahead.

After Anne finally completes her work, the WAS application creates a performance record that becomes part of Anne's permanent profile (audited information). The WAS software digitally signs this performance record or audit trail, so that its authenticity can be verified at any time in the future. Any attempt to modify or delete this audit trail will fail. Anyone recognizing the identity of the signing entity (namely the WAS software, in this case, or the owner of the key that the WAS software uses) will recognize that Anne has properly obtained the performance record through legitimate use of the WAS software.

6.2.6.1.1.8 Linux Security

Linux is an open architecture operating system that provides both network and client control with superior security capabilities to the Windows operating systems. Security for the Linux OSs can be found at http://www.linuxsecurity.com/ or at http://www.infoworld.com/news/hnlinuxworldny.html.

In regard to OS security, the most sought-after account on your machine is the superuser account. This account has authority over the entire machine, which may also include authority over other machines on the network. A top priority for any system is keeping the superuser account secure (Wreski 2000). Only use the root account for very specific tasks that can be quickly run and running as a normal user most often. Running as root all the time is quite a bad idea and should be forbidden.

The Linux OS is an alternative to the Windows OSs that is worth investigating. The newest version, Red Hat 7.2, is described by O'Brien (O'Brien 2001).

6.2.6.1.2 Access Control

Access control is principally concerned with controlling access to information. Access rules, such as least privilege and data separation, are incorporated into the access control mechanism. It cannot be successfully implemented without proper authentication such as was established according to the principles presented in the previous section. In particular, the confidentiality and integrity (i.e., coherency) of secure records must be ensured. That is, private material may only be available to users who have been, in some other way, given access to the materials, such as by being a member of a particular group.

The focus of the information system to serve not merely as a single-user, but also as a collaborative environment comprised of groups (teams) and collections of groups (classes), raises additional issues of permitting access to private material based on a user's associations, rather than simply the user's individual identity.

Access control is a software mechanism that provides the type of access that any particular entity has to the organization's information system database assets or applications. There are three principal access control mechanisms, a) role-based, b) mandatory, and c) discretionary. For each alternative, access control (in an unclassified data environment) is provided through an access control list (ACL) that stipulates the privileges or permissions that a user has. The levels of privilege usually include the following:

1) *Read* - The user can access a file and display the contents for reading only.
2) *Update* - The user can access a file and input a different value in the file from the one already there. This privilege cannot be used to alter the contents of an ACL, unless the user is a security administrator (or similar official).
3) *Insert* - The user can access a file and input a new element and value into the file.
4) *Delete* - The user can access a file and delete an element, or all, of the file.

For the Oracle DBMS, there are other actions (such as Create a file, Grant a privilege, or Join two or more tables) that can be performed by special users, for example, system administrators, security administrators, and the like. Often, the organization will deploy software that talks to the databases so that the user can never possess the privilege of directly addressing a database. This approach offers protection against intentional or unintentional damage to the database information.

Access privileges are often provided to users on the basis of their membership at one of three levels, namely:

1) *Owner* - Refers to someone who owns a file or resource. Owners of information are trusted for their access to that information.
2) *Group* - Refers to users who share a common bond. This group of users may or may not be trusted.
3) *World or Public* - Refers to the access level that everyone has to a resource. Usually this group of users is not trusted.

6.2.6.1.2.1   Role-Based Access Control

A commonly employed metaphor for managing groups of people according to a classification or a relationship to a group or organization is reflected in the *Role-Based Access Control (RBAC)* model, which was developed by the NIST. In an RBAC environment, administrators think of individuals in terms of their job position or their role within an organization. This view is easily translated to the information system by viewing an individual as a "member of the Able Contracting System" or a "user of the Widget Application System."

Considering that RBAC is a natural representation of existing administrative policies and procedures, the adoption of RBAC in the information system lowers the cost and complexity of administering an information system's access control policy. It also reduces the chance of errors when compared to other access control mechanisms, in particular when such mechanisms directly associate users with specific permissions using manually prepared tables.

The RBAC model also supports non-trivial access control policies without compromising efficient implementation, which is an important consideration when large-scale information system environments need security policies enforced through such a role-based model for access control.

RBAC is not the only means by which information system applications and objects are protected from unauthorized access. An object could make authenticated requests to other information system objects to obtain additional authorization beyond what the RBAC model offers, and thus extend these basic services in ways that are dictated by the unique needs of a particular object.

6.2.6.1.2.2   Other Access Control Methods

There are two other primary access control methods, mandatory and discretionary access control. *Mandatory access control* is based on subjects (e.g., users) and objects (e.g., databases) having security labels and access is based on an appropriate matching of these labels. *Discretionary access control* is based on authorized users having the right to award authorization to others. The definitions of these access control methods (and other terms used in this book) are contained in the Glossary.

Both of these were developed by the Department of Defense (DoD) to provide access control for environments where the data is classified. They are not appropriate for commercial or civilian government environments. There are other access control methods but they are not significant and will not be discussed here.

6.2.6.1.2.3   A Further Explanation of Role-Based Access Control

The RBAC method has been implemented in the Oracle database management system, although it is different from the method as defined by the NIST. In addition to its own services, the RBAC model is flexible enough to allow an actual implementation to choose from, or implement, custom authorization systems that match the organization's needs, size, and other critical aspects of its environment. In this way, RBAC not only adapts itself to existing policy models, but may also be extended as the need arises.

For commercial and many government applications, it is important for the analyst to understand the basics of RBAC, which is a different approach than the other two principle methods discussed above which many organizations use who are not aware of the existence of alternative mechanisms.

Roles are generally defined as the different job functions that exist within a particular organization. *Access privileges* are the different rights to information system assets that an entity possesses. Fundamentally, RBAC consists of the following:

1) A list of each user's identification and the roles to which the user is a member (an access control list),
2) The roles for the organization, and
3) The access privileges for each role.

In general, for any organization, the roles and access privileges will change very slowly, whereas the list of users may change much more often as users leave the organization (or move within it) or enter the organization for various reasons. This means that for any organization, the roles and access privileges can be determined and then applied to a specific user as the user is added to the access control list (Smith 1997) or deleted from the ACL.

Using an RBAC approach, the administration of the system is greatly simplified, since the administrator need only add the name of a new organizational employee to the ACL or delete the departing user's name from the organization's ACL and then the information system can automatically remove this user's accesses and privileges from the appropriate ACLs and hence from the entire system. In contrast with other access control methods that require access control lists for many different applications and databases, the RBAC method requires a single principal ACL.

An example of what an ACL in an RBAC environment looks like is presented in Figure 6.5, *An ACL in an RBAC Environment*. An example of the privileges for each role is shown in Figure 6.6, *Roles versus Privileges*.

If a user's name or user ID (they may be the same) does not appear in the ACL list for a person attempting to log on, then that person will not be allowed to log on to the system.

If the name or ID is present, then to determine the access privileges a particular user has can be accomplished by using the ACL matrix to determine the roles to which the user is a member, and then assign the privileges that correspond to the roles as denoted in the Roles versus Privileges matrix.

```
                              Role Memberships
           Name        User ID   R₁  R₂  R₃  R₄  ...  Rₙ
         ⌠ Joe Blow     nnn900    1   0   0   1  ...  1  ⌉
         │ George Smith nnn899    0   1   1   0  ...  0  │
         │ ...                                           │
         │ Dan Fowler   nnn200    1   0   1   1  ...  0  │
         ⌡                                               ⌡
```

Figure 6.5 An ACL in an RBAC Environment

```
                    Access Privileges
         Roles     P₁  P₂  P₃  P₄  ...  Pₖ
          R₁     ⌠ 0   1   1   0  ...  1 ⌉
          R₂     │ 1   1   0   1  ...  0 │
          R₃     │ 0   1   1   0  ...  0 │
          R₄     │ 1   0   1   0  ...  1 │
          ...    │                       │
          Rₙ     ⌡ 0   1   0   1  ...  0 ⌡
```

Figure 6.6 Roles versus Privileges

A simple example of how RBAC would be implemented in an organization follows. Suppose the organization has been analyzed and the analyst finds that there are 50 organizational employees and 10 different roles for all of the activities performed in the organization. I then create an access control list that is a 50 x 10 matrix that looks like Figure 6.7, *An Access Control List*, which has privileges as shown in Figure 6.8, *Roles versus Privileges Matrix*. The analyst also creates a "Roles versus Privileges" matrix by identifying all of the access privileges that exist for each of the 10 organizational roles. Suppose that the analyst finds that there are 7 different types of access for the 10 roles, so the "Roles versus Privileges" matrix is 10 x 7, as shown in Figure 6.8.

Suppose that user number 10 (U_{10}) is a member of roles 3 and 7 (R_3, R_7), and that the privileges for R_3 is P_2 only and the privileges for R_7 are P_1 and P_6, then the access control process would conclude that this user (U_{10}) has privileges P_1, P_2, and P_6. Whatever these privileges are, U_{10} has them.

6.2.6.1.2.4   Security Considerations for RBAC

Users of network-based distributed applications require assurance that their interactions and data are reliable, private, and accessible only to properly authorized individuals. Such assurance is provided by applications that make use of the following four security mechanisms: 1) authentication, 2) access control, 3) data integrity, and 4) confidentiality.

Roles

Users	R_1	R_2	R_3	R_4	R_5	R_6	R_7	R_8	R_9	R_{10}
U_1	0	0	1	0	0	0	0	0	0	0
U_2	1	0	0	0	1	0	0	0	0	0
U_3	0	1	0	0	0	0	0	0	0	0
U_4	0	1	1	0	0	0	0	0	0	0
...										
U_{10}	0	0	1	0	0	0	1	0	0	0
...										
U_{50}	0	0	0	0	0	0	1	0	0	1

Figure 6.7 An Access Control List

Effective security depends on the use of the four mechanisms mentioned above. Information system security depends on the cryptographic algorithms used to implement these mechanisms. Since the effectiveness of cryptographic algorithms depends on computing speed, these algorithms will lose their effectiveness over time as computational speed increases. And computational speed will increase with the passing of time since computers get faster each year.

Access Privileges

Roles	P_1	P_2	P_3	P_4	P_5	P_6	P_7
R_1	0	0	1	0	0	0	0
R_2	1	0	0	0	1	0	0
R_3	0	1	0	0	0	0	0
R_4	0	0	1	1	0	0	0
...							
R_7	1	0	0	0	0	1	0
...							
R_{10}	0	0	0	0	0	0	1

Figure 6.8 Roles versus Privileges Matrix

6.2.6.2   Authorization

Authorization is actually a part of the access control process but is sometimes treated as a separate activity. Authorization is the process of deciding which applications, services, and privileges that a particular entity can rightfully use. In some cases, the entity's rights include read, write, delete, and create database privileges and are usually listed in the entity's privileges table.

6.2.6.2.1 Privileges

For RBAC, the entity's *privileges* appear in a table associated with the entity's roles, which were identified based on the ACL table with the user's identification versus approved roles. The presence of a privilege enables an entity to access an entire database or some portion of it and to read, update, insert, delete, create, or perform other actions depending based on the entity's granted privileges.

6.2.6.2.2 Pedigrees

The *pedigree* of an entity is the level of access privileges that are available to the entity. Often the pedigree is indicated by the entity's *distinguished name* as it appears in the entity's certificate. The distinguished name (a unique appellation) usually includes the entity's actual name, the entity's organization or company, and any additional information required to uniquely indicate to the computer what access privileges this entity has.

In an RBAC environment, the user's pedigree will likely be determined from the "Roles versus Privileges" table, which provides the various privileges that have been granted to the entity.

6.3 Interdependence of the Mechanisms

Each of the security mechanisms is dependent on other mechanisms, so all of the security mechanisms (resulting from the risk assessment) must be applied to ensure overall security for the computer system. For example, access to information cannot be properly controlled (access control) if the correct identity of a user has not been properly determined (authentication). Similarly, a user's alleged identity cannot be trusted if that user's identity was established in a way that could have allowed other users to forge his or her credentials.

It is clear that implementing less than all of the mechanisms of secure communications leaves open one or more system fallibilities that expose the information system in such a way that the effects are as if the information system had no security countermeasures at all.

6.3.1 Implementation Issues

Because the information system security are services separated from the actual implementation through the use of an interface specification, then the actual implementation of these security services may range from minimal to complex, as dictated by the needs of a particular installation.

6.3.2 Security in the Future

Many security mechanisms in an information system are dependent on some form of cryptography. Thus, some security mechanisms can be circumvented by decrypting information used to support these mechanisms. Given enough time, any encrypted message can be broken. However, this might require many millennia in the case of encryption with large key lengths and attempting decryption with today's computer technology.

Security does not derive its strength from the obscurity of its algorithms (in fact the popular security algorithms are quite well known and are published in various journals and books, even the advanced encryption standard, Rijndael, can be found on the NIST Web page). Instead the security of the mechanisms comes from the unreasonably large investment in computational time that is required to break any given encryption algorithm.

Faster computers and improved parallel computation significantly decrease the time required to decrypt any specific encryption algorithm. The 56-bit RC5 encryption that required a single desktop computer thousands of years to decrypt (in the late 1980s or early 1990s) was, in 1997, decrypted in less than a year with a massively parallel computational effort. Even this decryption does not mean that the method is not effective, for who would devote a yearlong effort and the computational power to anything other than something extremely important and the importance of a message is not known until it is decrypted?

This example still illustrates that no specific algorithm should be considered the basis of a security system, unless the effective lifetime of the mechanism is meant to be limited from the start. An essential design principle for a security mechanism is that the security interface be separate from the implementation of the security mechanism, allowing the implementation to be upgraded to keep pace with advances in anti-cryptographic systems. This approach will ensure that system security can be maintained at a sufficient level to satisfy the future requirements of authentication, access control, data integrity, and confidentiality.

6.4   Information System Security Rationale

An information system requires security for at least the following reasons:

First, personal or private information must be safe from unauthorized access. Information system security must also provide safeguards from possible falsification of such performance records and prevent forged records of unearned accomplishments from being added (e.g., adding false transactions to a customer's file).

Second, the information system must ensure that whoever uses its resources is actually authorized to do so, not merely to restrict access to sensitive material but also to provide a framework for performing services (e.g., ensuring that an authorized customer can easily access the system to perform business transactions).

Third, information exchanged between software components within the information system, or between different information systems and users, must be secured to ensure that the data cannot be compromised and thereby violate user privacy (e.g., by preventing the process of providing customer information to an unauthorized system).

Fourth, the information system and its information or applications must always be available for access by authorized entities or customers (e.g., implementing an anti-denial-of-service mechanism to ensure malicious code is prevented from causing unavailability of the system).

Fifth, sensitive information, whether stored or transmitted, must always be protected from modification, viewing, or damage by unauthorized entities (e.g., ensuring that VPN tunneling is used for transmitting sensitive information over an unprotected link).

The information system specifications should not dictate the use of any particular technology to implement any part of its security features. Instead, these specifications should include high-level interfaces to security mechanisms, preferably heterogeneous mechanisms that can be installed in an open architecture and are acquired for their positive attributes, including low cost. The separation of security features into interfaces versus implementation improves the portability and interoperability of information system components, but this separation also ensures that new security technologies can be integrated into the information system to keep pace with advances in anti-cryptographic capabilities or other advances in the capabilities and varieties of information system threats.

6.4.1 Authentication and Confidentiality

Computers on the Internet today may employ a concept referred to as cryptography in order to keep information from being viewed by others. This technology is also used to establish trust in the identity of two entities (individuals or applications) that exchange information over an insecure line (one whose data integrity cannot be known and which cannot be guaranteed to be free of the possibility of having the information intercepted).

Cryptographic technology plays an important role in computing today because it permits two users to trust each other's identity and then engage in private communications without either user ever having met the other. Traditionally, an exchange of secret keys was the only means of ensuring secure communications. Cryptography eliminates this need.

Communications between two users are encrypted and decrypted with a private key and the other party's public key, in the case of the PKI environment. Each user has ready access to the other's public key, but neither is ever privy to the other's private key.

As the computational overhead of public keys is significant, secure communications between two users usually involve the exchange (via public keys) of an alternate encryption method based on secret keys. The computationally less-expensive secret keys enable significantly faster exchange of confidential information.

It is important to recognize and understand that the public technology is by no means trusted in and of itself. A public key establishes no identity, unless at least one other trusted entity verifies that identity. Verifying identities is usually the role of so-called Certificate Authorities, whose authority is trusted by virtue of their established presence and widely-accepted trust, and whose signature on someone's public key guarantees the identity of the owner of that key.

Once both sides have assured themselves that the public keys that were exchanged actually validate the other's identity, a cryptographically secure (confidential) and trusted (authenticated) communication link has been established.

Another situation where public keys play an important role is in authenticating data. A digital signature provides a means of signing (for the purpose of authentication) arbitrary information, and when based on public keys allow unrelated parties to verify the authenticity of the signature. This is also closely related to ensuring the integrity of information because any change in the transmitted information is easily detected by checking data against the sender's digital signature.

6.4.2   Certificate Uses

A *digital certificate or digital ID* is the electronic equivalent of a passport or business license. It is a credential that is issued by a trusted authority that an entity can present electronically to prove her, his, or its right to use some network capability, such as "transmit or receive" a sensitive message. A user's digital certificate may be provided by the user directly, or by a known public key server, provided that the user has registered his or her public key with such a server. The public key supplied by a third party in this manner may require an additional verification step by which the public key server's identity is authenticated. Towards this end, the public key server would need to present an authenticating certificate that has been signed by a known and trusted certificate authority, such as by VeriSign.

Authentication and confidentiality services may be provided through the use of security mechanisms that support public key algorithms. Such systems include the Secure Socket Layer (SSL) protocol, the Secure Shell (SSH), the Common Object Request Broker Architecture (CORBA), and Kerberos.

Although public key certificates are in widespread use, an organization is likely to require investment in additional infrastructure to make consistent use of such certificates across multiple systems. While the technology to support this infrastructure is available, there are serious organizational and legal policy questions that must be decided by an organization before using the technology. Many organizations do possess the trust required to provide digital certificates and the concomitant security services.

6.4.3   Object Security

In security jargon, an object is any application or information that can be declared as sensitive but unclassified or nonsensitive unclassified. In order for information system objects to be securable, that is, to be able to take advantage of authentication, access control, data integrity services, and confidentiality assurance, these objects must implement a so-called sensitive interface. Through this interface, arbitrary objects can be queried for certain security-related properties, thus making these objects securable.

Methods (functions) that implement the security interface are:

1) Authorized Roles - Obtain information on the authorized roles that may access the object.

2) Digital Certificate - Obtain the object's digital certificate.

3) Invoke Secure - Invoke the object's purpose or function using security credentials.

6.4.4   Confidentiality

Confidentiality is the protection of sensitive information, whether stored or transmitted, from access by unauthorized entities or those who have no need-to-know authorization. This protection is provided by encrypting the information and by blocking access to unauthorized entities.

Encryption requires an encryption algorithm and at least one key (R. Smith 1997).

6.4.5 Availability

Availability is a measure of the capability of an information system to provide subsystem or information access at the time that it is requested by an authorized entity. Generally, availability can be computed as follows:

$$\text{Availability} = (\text{Total Uptime} - \text{Total Downtime})/\text{Total Uptime}$$

where Total Uptime (TU) is the time that the system should be up (and for most systems this is all the time) over some specified period and Total Downtime (TD) = total time that the system is down over some specified period.

However, since Availability $= (TU - TD)/TU = (TU/N_f - TD/N_f)/(TU/N_f)$, then,

$$\text{Availability} = (\text{MTTF} - \text{MTTR})/\text{MTTF}.$$

where Mean time to failure (MTTF) = Total Uptime (= total time)/N_f (= number of failures) and Mean time to restore = Total Downtime/N_f.

Availability can be adversely affected by random subsystem failures that normally occur for any system (and hence may not be considered a part of security, by some). These failures can be due to equipment malfunctions (such as firewall or cable outages) and to malfunctions of support systems (such as power, heating, or air conditioning equipment outages). All of this is considered a part of the security of the system. In addition, other events affecting availability are catastrophic failures, denial-of-service attacks, unauthorized electronic entry, and physical damage to the system.

Availability is dependent upon the following four capabilities:

1) *Protection from attack* - This is partly achieved by closing holes in operating systems, firmware, and network configurations. It is also achieved by protecting the system against Denial-Of-Service attacks. Anti-virus software also protects the system against hackers with malicious intent (e.g., viruses, worms, etc.).

2) *Protection from unauthorized use* - This is achieved by ensuring that no unauthorized entity (person, application, or server) gets access to an application or data such as by improperly using legitimate User IDs and Passwords.

3) *Resistance to failures* - There are two types of failures, catastrophic and non-catastrophic. Non-catastrophic failures, such as power outages, can be minimized through "loss of support" protection such as uninterruptible power supplies (UPSs) and redundant subsystems such as clustered servers with load balancing

and automatic fail-over software. Protection against catastrophic failures, such as earthquakes and floods, can be provided through backup and recovery capabilities where critical information (data and software) is stored in off-site backup systems (e.g., servers) and can be used to replace lost information in a timely manner.

4) *Resistance to undue latent response times* - Examples of these excessive times are: a) logging onto a system that takes an excessive amount of time to conclude that the user is authenticated; b) excessive time to connect following a single sign-on request for another application or database; and c) excessive time to respond to any action by the user. These types of availability problems can be addressed through the use of top quality programmers and programming techniques for the development of operating system, middleware, and application software (such as the WebLogic Sever (BEA), Web Sphere Server (IBM), or iPlanet Web Server (Sun)).

Thus, improved availability can be achieved through system adjustments that include encryption and authentication improvements, software patches, DoS detection and anti-DoS software, as well as the use of backup servers and associated software.

Many modern systems have an availability requirement of 0.99999 or greater and it is usually expressed as a sequence of 9s to the right of the decimal point, for example, 0.9999.

6.6   Countermeasures

Countermeasures are determined based on the desire by the organization owners to protect their system from the various identified threats and risks. These countermeasures consist of the following mechanisms and methods:

1) Identification and authentication - e.g., User IDs and passwords,
2) Confidentiality - e.g., PKI,
3) Data integrity - e.g., checksums and hashing functions,
4) Access control - e.g., RBAC,
5) Authorization - e.g., RBAC,
6) Availability - e.g., Clustered servers, anti-DoS software, and anti-Virus software,
7) Non-repudiation - e.g., Digital signatures and certificates,
8) Data separation, least privilege, and no object reuse.

Some of these have already been discussed and others will be discussed below.

6.6.1   Anti-Virus Software

Anti-virus software is software that has been developed to counter malicious code that comes into a network over the Internet or a public WAN as a malicious message or as a seemingly innocuous message but with a malicious attachment. Often, the anti-virus software is placed on the information system's firewall or it may be housed on the client PC systems. When working properly, this software will remove viruses and repair damaged files before the viruses can invade organizational networks. Remote administration, such as automatic updates (by the vendor) to the anti-virus software, makes it easy for an organization to benefit from virus protection.

The Norton Anti-virus software for firewalls, offers the following seven attributes:

1) *Flexibility and control* - The server configuration can be accomplished remotely through an HTML interface, with automatic updates as they are updated and programmed by the vendor.

2) *Comprehensive protection* - The software can scan HTML, FTP, SMTP, and other user-configurable firewall-supported protocols. It can scan Zip-files and MIME-encrypted files.

3) *No performance degradation* - The software minimizes the effects on network performance by scanning only suspicious traffic. The traffic load can be managed by system administrators.

4) *Maximum flexibility for administration* - The software supports remote configuration using an HTML interface, which enables administrators to quarantine, pass through, or repair infected files.

5) *Up-to-date definitions* - Because updates are automatic and scheduled, the software keeps virus definitions current without interrupting work efforts.

6) *Easy flexible administration* - The software minimizes the effects of the software on any of the applications and setup is simple.

7) *Auditing* - The software includes a detailed log of all activities, including statistical logs and customizable virus alerts.

Should a virus or worm enter the system through a firewall because it is encrypted, then it can do no harm (since it cannot be downloaded or executed) unless the receiver knows how to decrypt the message, which is usually not the case.

6.6.2   Anti-Denial-Of-Service Software

Anti-virus software and firewalls often provide *anti-denial-of-service* (anti-DoS) capabilities. This type of software is available from a variety of vendors. The primary objective of this type of software sometimes called anti-distributed denial of service (anti-DDoS) software is to detect, trace, and block DoS attacks before they reach their intended online targets.

There is a new anti-DoS technology based on the fingerprints of malicious software that allows the anti-DoS system to monitor and trace sharp spikes in Web traffic thereby enabling the system to block DoS attacks from an organization's operations center before the attacks can reach their targets (McDonald 2001).

Recently, Web security firms have reported advances in the war against DoS attacks, developing automated detection systems that can assist organizations in developing a rapid response to an attack before it cripples their operations.

Unfortunately, DoS attacks are relatively simple to perform but can have disastrous effects. These attacks can disable Web site routers by flooding them with false information requests. These attacks can last up to several hours, depending on how quickly they are detected and confronted. To discover the sources of the attacks, if handled manually, technicians must sort through thousands of lines of computer code. Fortunately, there are automated methodologies that can gather information about traffic patterns and detect anomalies, such as unusually heavy traffic, coming from a single source. By generating a *fingerprint* of the DoS attacking mechanisms, the monitors can trace the origin of the attack, which can then be blocked by operators at the organization's site.

One vendor, Arbor Networks, Inc., has such a package (McDonald 2001). In addition, the System Administration, Networking, and Security (SANS) Institute, a non-profit Internet technology research center, has been working on a similar solution that uses sensors on WAN backbones and large routers, as well as tracking information using a database of fingerprints.

6.6.3 Clustered Servers

The manufacturers of servers include among their products *clustered servers* that are two or more servers included in a group of servers so that the owner can use some or all of them. They come with software that manages the server load so that the loading can be balanced among the servers (using load-balancing software) and if one should fail, there is an automatic failover (using automatic failover software) so that the clustered system will maintain its uptime and a technician can replace the failed server at some later time.

6.6.4 Demilitarized Zone

The *Demilitarized Zone (DMZ)* is an area between the firewall or router and the internal network. It is used for placing servers (e.g., Web servers, mail servers, and FTP servers) that interface with the untrusted users or the public. External users are routed to the DMZ where they can access information that is external to the organization's users and information. Should any information on a server in the DMZ be modified or erased, the organization can upload the relevant information to the DMZ and replace the modified or erased information.

6.6.5 Private Networks

One method for ensuring that transmitted information remains private is the use of *private networks*. These networks are more expensive than the public networks, such as the Internet, but they offer privacy since the only users of the network are members of the organization, and hence only authorized users are allowed on the network (unless the attacker can somehow plug their computer system into the link). However, should an authorized user become malicious, then the method for catching the culprit is an auditing mechanism or intrusion detection system. An example of a private network is the use of T1 links.

6.6.6 Encryption

Encryption can be performed by a device or by software. The device or the software can be purchased from a vendor who can provide installation services. Encryption can be performed just before transmission (e.g., using router or firewall VPN or from the encryption device, such as a mobile PDA), at the gateway (e.g., firewall encryption), or at an authenticated device (e.g., PC or mobile system with authentication at the device) (Slaton 2001).

Cryptography can provide for *confidentiality, data integrity, non-repudiation* (i.e., digital signatures), and *advanced user authentication* (NIST 1995). Cryptography depends on an *algorithm* (i.e., a transformation function) and *one or more keys*. Mathematically, a plaintext message P that is transformed to an encrypted message C using an encryption algorithm E with a cryptographic key K can be represented as follows: $C = E_K(P)$.

The encryption of information is accomplished using an algorithm and one or more keys. A modern method for encrypting information uses a crypto algorithm, which specifies the mathematical transformation that is performed on data to both encrypt it and to decrypt it (R. Smith 1997). An encryption algorithm can be either a stream cipher (which encrypts a digital data stream a bit at a time) or a block cipher (which transforms data in fixed-size blocks, one block at a time). The algorithm is a procedure that transforms plaintext into a form that is unreadable (called ciphertext) so that the information can be read only by someone possessing the decryption algorithm and the appropriate key (or keys).

When a *block cipher* is applied to a data stream, then there must be a *cipher mode* (there are four of these) to define how the block mode algorithm should be used to process the data stream. The simplest *stream ciphers*, which are *Vernam ciphers* (i.e., the *exclusive-or logical operation* invented originally for the encryption of teletype messages), are rather simple to break, especially for a computer.

If a single key (secret key) is used, then the process is called symmetric encryption. Both encryption and decryption use the same key. If two keys are used (private and public keys), then the process is called asymmetric encryption. For encrypted transmissions, the receiver's public key is used to transform the message into ciphertext (by the sender) and the receiver's private key is used to decrypt the ciphertext message. The asymmetric method is used for encryption in the PKI.

Three examples of block ciphers are the:

1) Data Encryption Standard,
2) International Data Encryption Algorithm (IDEA), and
3) SKIPJACK.

The only surefire method for decrypting an encrypted message is called the "brute force" method. This method requires the encryption algorithm and then tries every possible bit structure for the key (assuming that the malicious cryptanalytic has sniffed a copy of the ciphertext message and knows the key word length and the encryption algorithm). For a 56-bit long key, this means that 2^{56} tries may be required. There are techniques for reducing this number.

6.6.6.1 Data Encryption Standard

The *Data Encryption Standard (DES)* is the encryption standard used by the U.S. government for protecting sensitive but unclassified data. For export, a 40-bit key is used and a 56-bit key is used for U.S. organizations. Triple DES (3DES) can also be used for U.S. organizations. DES is a symmetric key (i.e., single key) algorithm (Brenton 1999). 3DES can be implemented using either two or three different keys. The NIST follow-on standard to DES is called *Rijndael* and is briefly explained in the next section on the advanced encryption standard.

IP Security (IPSEC) is a public/private key encryption algorithm that offers a set of open standards. It is implemented at the Transport or IP layer. It is convenient to use and has been implemented by Cisco on its routers. IPSEC is an obvious VPN solution. At present, IPSEC is limited to a 40-bit key encryption (Brenton 1999).

IPSEC protects IP packets from snooping or modification. It is implemented between the Internet (level 3) and Transport levels (level 4) of the OSI model. An IPSEC-based VPN is composed of AH and ESP modes.

1) An Authentication Header (AH) provide integrity-checking information (cryptographic checksum) so that one can detect if the packet's contents were forged or modified while en route over the Internet.

2) The Encapsulating Security Payload (ESP) encrypts the data contents of the remainder of the packet so that the contents cannot be extracted while en route over the Internet. The encryption information is selected using the *security parameter index (SPI)*.

An AH can defeat spoofing and provide data integrity. ESP provides packet confidentiality. Newer versions of the Simple Network Management Protocol (SNMP) contain *time stamps* to detect and block replay attempts.

Hybrid cryptography offers the advantages of symmetric cryptography (such as speed, 1,000 to 10,000 times faster) while offering the advantages of asymmetric cryptography (such as ensured protection).

Key escrow is the storage of cryptographic keys with escrow agents who can divulge a user's public and private keys if the requestor provides sufficient rationale for obtaining the keys. This can be especially beneficial should a user lose their keys or die or leave an organization without notice.

6.6.6.2   The Advanced Encryption Standard

The *Advanced Encryption Standard (AES)* will be a new Federal Information Processing Standard (FIPS) that will specify a cryptographic algorithm for use by U.S. Government organizations to protect sensitive but unclassified information. NIST also anticipates that the AES will be widely used on a voluntary basis by commercial organizations, institutions, and individuals outside of the U.S. Government, and even outside of the U.S. in some cases.

The block cipher *Rijndael* (pronounced "Rain - Doll") is the winner of the NIST AES four-year long competition for a cipher to replace the DES, which has been the NIST standard for many years. Rijndael is a block cipher, designed by Ms. Joan Daemen and Mr. Vincent Rijmen as a candidate algorithm for the AES. This cipher is designed to use only simple whole-byte operations. It provides extra flexibility over that required of an AES candidate, in that both the key size and the block size may be chosen to be any of 128, 192, or 256 bits (NIST CSRC 2001).

The combination of security, performance, efficiency, ease of implementation, and flexibility make Rijndael an appropriate selection for the AES. Rijndael is a good performer in both hardware and software across a wide range of computing environments regardless of its use in feedback or non-feedback modes. Its key setup time is excellent and its key agility is good. Rijndael has low memory requirements, which make it well suited for restricted-space environments, in which it also demonstrates excellent performance. Rijndael's operations are among the easiest to defend against power and timing attacks.

It appears that some defense can be provided against such attacks without significantly affecting Rijndael's performance. It is designed with some flexibility in terms of block and key sizes, and can accommodate alterations in the number of rounds (i.e., steps in the algorithm's execution), although these features would require further study and are not being considered at this time. Rijndael's internal structure appears to have good potential to benefit from instruction-level parallelism (NIST AES 2001), which should make it much faster than a sequential implementation.

During an early stage of the AES competition process, a draft version of the requirements required each algorithm to have three versions, with both the key and block sizes equal to each of 128, 192, and 256 bits. This was later changed to make the three required versions have those three key sizes, but only a block size of 128 bits, which is more easily accommodated by many types of block cipher design.

Keys with a length of 128, 192, or 256 bits can be used to encrypt blocks with a length of 128, 192 or 256 bits (all nine combinations of key length and block length are possible). Both block length and key length can be extended very easily to multiples of 32 bits. Rijndael can be implemented very efficiently on a wide range of processors and in hardware.

The design of Rijndael was strongly influenced by the design of the block cipher *Square,* another cipher designed by Daemen and Rijmen. Square is a 128-bit block cipher and its original design concentrated on resistance to differential and linear cryptanalysis. The standard techniques of differential and linear cryptanalysis can be adapted to be used against Rijndael. A specific attack, originally developed for use against Square, called the "Square attack," can be used as well.

If one uses 256 blocks of chosen plaintext, where every byte but one is held constant, and that one is given all 256 possible values, then after one round of Rijndael, four bytes will go through all 256 possible values, and the rest of the bytes will remain constant. After a second round, sixteen bytes will each go through all 256 possible values, without a single duplicate, in the encipherment of the 256 blocks of chosen plaintext. (For a 128-bit block, this is every byte; for larger blocks, the rest of the bytes will remain constant.) This interesting property, although not trivial to exploit, can be used to impose certain conditions on the key when one additional round, before or after the two rounds involved, is present.

The possibility of this attack was first noted by the developers of Square and Rijndael themselves, and was noted in the paper that initially described Square. Differential cryptanalysis attacks on DES become quite a bit easier for variants of DES. Thus, if Rijndael had been designed a little differently, there could have been an important weakness in Rijndael. There may still be a slight flaw, but existing cryptanalytic results against Rijndael do not exhibit a pattern indicating this (NIST CSRC 2001).

6.6.6.3   Hashing Functions

Hashing functions are algorithms that can be used to encode a message so that the message is reduced or compressed to a constant size, such as 160-bits, and is referred to as a message digest or checksum. This message digest or checksum is a unique numerical value that is changed even if one bit in the message is altered. Inverse hashing functions are virtually impossible to discover so that one cannot undo a hashed message to discover the original message.

6.6.6.3.1   Message Digest

A *message digest* is an algorithm that assures data integrity by generating a unique, 128-bit cryptographic message digest value from the contents of a file. If as little as a single bit value in the file is modified, the Message Digest 5 (MD5) checksum for the file will change. Forgery of a file in a way that will cause MD5 to generate the same result as that for the original file is considered extremely difficult.

The Secure Hash Algorithm (SHA) is an algorithm that produces a 160-bit message digest from any message of less than 264 bits in length. For messages longer than 264 bits, the user can just mask off the first 264 bits and hash these to get a message digest. This is slightly slower than MD5 but is considered to be more secure against:

1) Brute-force collision (i.e., decrypting by using every combination of decryption algorithms and keys (Canavan 2001)) and

2) Inversion attacks (viz., using knowledge of how a particular text was created, for example a text with a watermark, and then reversing the process to remove it (i.e.,   the   watermark),   see http://www.brunel.ac.uk/depts/ee/COM/Home_Saeed_Vaseghi/ew.pdf   for information on uses of watermarks and types of attacks on watermarked texts and potential defenses against these).

6.6.6.3.2   Checksum

A *checksum or cyclic redundancy check (CRC)* is a mathematical verification of the data within a file. If files have a checksum attached, then it is far more difficult to replace the file with an alternative file containing a Trojan horse. For example, the modification of any stored or transmitted messages or information can be identified using hashing functions or cyclic redundancy checksums.

Setting user-level file permissions ensures that executable files do not become infected. If all applications are launched from servers (i.e., thin clients), then you can decrease the likelihood of virus infection by setting the minimum level of required permissions. CRC checksums can be used to detect replacement of Telnet or FTP applications, even if the replacement has the same size and timestamp.

6.6.6.4   Symmetric Encryption

Symmetric encryption means encoding with an algorithm and one key called a "private or secret or session key." The prime issue with symmetric encryption is that the sender must be assured that the receiver has a key, the same key, and that this key has been transmitted somehow to the receiver in a trusted manner. It is used in sessions where the secret key is transmitted using asymmetric encryption and then the message traffic is transmitted using symmetric encryption with the session key. This greatly speeds up the transmission time, by about a factor of 1000 or more. The first encryption method to use this hybrid technique was the PGP method.

6.6.6.5 Asymmetric Encryption

Asymmetric encryption is based on the Diffie-Hellman dual key methodology that requires an encryption algorithm and two keys, a "public key" and a "private key." The *public key* is transmitted to the designated user in a certificate and is available to any user who can access the designated user's client computer. The *private key* is usually generated by the designated user and is known only to the designated user. It is usually kept in a secret location and is encrypted so that if it is somehow stolen, the key cannot be understood and used to decrypt messages.

A complex and well-defined process has been created for dual key management, called PKI. However, the public key infrastructure (PKI) is not itself well defined and fixed. As a result the PKI may be different depending on where the infrastructure is generated or obtained.

6.6.6.5.1   Diffie-Hellman

The initial public key system for encrypting information was proposed by Drs. Whitfield Diffie and Martin Hellman (Oppliger 1998). The original proposal was that two parties use a session key and it was known as the *Diffie-Hellman* key exchange. The suggested protocol was based on the following mathematical relationship:

A and B agree on a large prime number p and a generator **g** (i.e., a special number). A picks a random number x_a and transmits the value $y_a = \mathbf{g}^{x_a}$ (modulo p) to B. Similarly, B picks a random number x_b and transmits the value $y_b = \mathbf{g}^{x_b}$ (modulo p) to A. A can compute $K_{ab} = y_b{}^{x_a} = g^{x_a x_b}$ (modulo p) and B can compute $L_{ab} = y_a{}^{x_b} = g^{x_a x_b}$ (modulo p). Since $K_{ab} = L_{ab}$ both A and B have a shared secret which can then become the session key. It is based on the following mathematical equation: $x^a \, x^b = x^{ab} = x^{ba} = x^b \, x^a$.

Unfortunately, there are two disadvantages to the Diffie-Hellman key exchange. It is subject to a man-in-the-middle attack and to a clogging attack. A *man-in-the-middle attack* is when C (the man in the middle) postures as B for A and A for B, so that A and B end up negotiating secret keys with C who can then passively or actively attack the communications between A and B. A *clogging attack* is an attack in which an attacker requests a high number of keys in order to overload the victim.

6.6.6.5.2 Public Key Infrastructure

The *public key infrastructure (PKI)* encompasses certificate management, registration functions, and public key enabled applications. PKI provides the security cornerstones for authentication, encryption, integrity, and non-repudiation. The principle components of PKI are:

1) Digital certificates,
2) Public and private keys,
3) Secure sockets layer, and a
4) Certificate Authority (CA).

Additional components include:

5) Secure storage of certificates and keys,
6) Management tools to request certificates, access wallets, and adminster users,
7) Directory service for centralized repository for users, and
8) Machine identification and authentication.

Its framework and services provide the following:

1) Generation, production, distribution, control, revocation, archive, and tracking of public key certificates;
2) Management of keys;
3) Support to applications providing confidentiality and authentication of network transactions;
4) Data integrity; and
5) Non-repudiation (Lyons-Burke 2000).

DOD certification for PKI compliance requires the following characteristics (Lyons-Burke 2000):

1) Standards-based,
2) Support mulitiple applications and products,
3) Provide secure interoperability,
4) Support digital signature and key exchange applications,
5) Support key/data recovery (in case the key is lost),
6) Commercial-based, and
7) Support the Federal Information Processing Standards.

With the continued growth of distributed systems, the problem of user authentication and user management is now acute. A DBMS (e.g., Oracle 8i) can address the needs for strong security, single sign-on, and centralized user-management by offering the following services:

1) Single sign-on to multiple databases or applications throughout the organization,
2) Single organizational user account,
3) Reduced total cost of ownership through single station administration (SSA),
4) Well-integrated, standards-based PKI, and
5) Stronger security through centralized authorization management and strong authentication.

Secure Sockets Layer (SSL) is an important ingredient of the PKI. The SSL protocol is an industry standard for securing network connections that provides authentication, data encryption, and data integrity. SSL is used to secure communications from any client or server to one or more servers or from a server to any client. You can use SSL alone or with other authentication methods, and you can configure SSL to require server authentication only, or both client and server authentication.

The *Simple Key Management for Internet Protocol (SKIP)* is similar to the SSL protocol because it operates at the session level. SKIP requires no prior communication to establish or exchange keys on a session-by-session basis. A shared secret is generated using the public key encryption method for IP packet-based encryption and authentication. While SKIP is efficient at encrypting data, which improves VPN performance, it relies on the long-term protection of this shared secret to maintain the integrity of each session.

A method for accessing certificates granted by an authority from any vendor is the Certificate Arbitrator Module (CAM) developed by Mitretek Corporation. Mitretek maintains the CAM Open Source Web site (http://207.123.140.7/) and allows downloads of the software and evaluates the source code before it is submitted to the user (Frank 2001). Many organizations that have adopted the PKI technology will need this product that was developed for the government, in particular, it was developed by the General Services Administration (GSA) Federal Technology Service (FTS).

6.6.6.5.3 Key Management

Key management is the set of procedures and protocols, both manual and automated, that are used throughout the entire lifecycle of the keys. This lifecycle includes the generation, distribution, storage, entry, use, destruction, and archiving of the cryptographic keys (NIST 1992).

It is important to archive keys since long after the key has been declared to be invalid, it may still be needed to decrypt information that has been stored as encrypted information in a database. Without the key, it may be impossible to transform the encoded information into plaintext.

6.6.6.5.4 Digital Certificates for PKI

A *digital certificate* (using the X.509 format), sometimes called a Digital ID, can provide a very strong method for authenticating a user's identity. It can also be used for storing the user's public key and for performing single sign-on and advanced access control such as fine-grained access. It can also contain the user's *distinguished name (DN)*, a string of characters uniquely identifying the user in significant detail. Certificates are usually stored on a server, often designated as a digital certificate server (Brenton 1999).

Whenever a certificate has been revoked, for any reason whatever (e.g., it has expired), it should appear in a *Certificate Revocation List (CRL)*. A CRL is a list of user certificates for which the begin-and-end dates that appear on the certificates are no longer valid, or which for any other reason, the certificates have been revoked.

Digital certificate servers provide a central point of management for multiple public keys. Also known as certificate authorities (CAs), they can provide verification of digital signatures. With a certificate, the entity can encrypt information, or it can be encrypted using a virtual private network capability. X.509 certificates contain:

1) Certificate owner's distinguished name,
2) DN of the certificate's issuer,
3) Certificate owner's public key,
4) Issuer's signature,
5) Dates of certificate validity, and
6) Serial number.

In some systems, even though they may be certificate based, an entity can be authenticated to use some of the available services without a certificate. However, their authorizations are severely limited.

Certificates provide electronic certification that the entity's name, identifying information, and public key are valid and actually belong to the entity. The certificate usually contains information about the certificate authority that issued it as well as a serial number, expiration date, and information about the rights, uses, and privileges associated with the certificate.

A CA is a trusted third party that certifies that other entities are who they say they are. When a certificate authority generates a certificate, the authority verifies the user's identity and that the user is not on the certificate revocation list. The certificate authority signs the certificate using its private key. The certificate authority has its own certificate and public key that it publishes for servers and clients to use to verify signatures that have been made by the certificate authority. Certificate authorities can be external or internal to a company.

The X.509 v3 certificate format is shown in the Figure 6.9, *The X.509 v3 Certificate Format*, below:

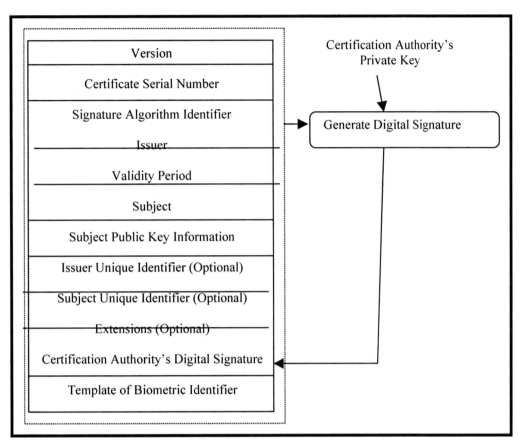

Figure 6.9 The X.509 v3 Certificate Format

1) The Version field contains the version number of the encoded certificate. The current legal values are 1,2, and 3.

2) The Certificate Serial Number field is an integer assigned by the certification authority. Each certificate issued by a given CA must have a unique serial number.

3) The Signature Algorithm Identifier field identifies the algorithm, such as RSA or DSA, used by a CA to digitally sign the certificate.

4) The Issuer Name field identifies the certification authority who has signed and issued the certificate.

5) The Validity Period field is the time interval during which the certificate remains valid, and the CA is obliged to maintain information about the status of the certificate. The validity field consists of a start date, the date on which the certificate becomes valid, and an end date, the date after which the certificate ceases to be valid. Note that a CA may elect to maintain status information for a longer period. VeriSign, for example, maintains revocation information about a certificate even after the validity period of the certificate ends.

6) The Subject Name field identifies the entity whose public key is certified in the subject public key information field. This field is often called the subject's distinguished name. The subject name must be unique for each subject entity certified by a given certification authority. A CA, however, may issue more than one certificate with the same subject name for the same subject entity.

7) The Subject Public Key Information field contains the public key material (public key and parameters) and the identifier of the algorithm with which the key is used.

8) The Issuer Unique Identifier field is an optional field to allow the reuse of issuer names over time.

9) The Subject Unique Identifier field is an optional field to allow the reuse of subject names over time.

10) The Extensions field is another optional field.

11) If some biometrics attribute is required, a template of the biometric identifier should be stored in the certificate (e.g., a right index fingerprint from the user).

6.6.6.6   Pretty Good Privacy

The *Pretty Good Privacy (PGP)* algorithm is a public key encryption program. The method was created by Phil Zimmerman and has become a *de facto* standard for e-mail on the Internet. There are PGP programs for most any platform (e.g., Windows, UNIX, MS-DOS, OS/2, Macintosh, Amiga, and Atari) that a user might have. Unfortunately, many people believe that PGP is illegal, however that is untrue because PGP is legal to use, provided that you choose the right version and don't download the program from a site in the United States if you are in another country.

The latest versions of PGP are version 6.0.2i, which is for Windows 95/98/NT and the Macintosh OS, and version 5.0i, which is for other platforms. For PGP international (PGPi) documentation you can go to Web URL http://www.pgpi.org/doc/.

PGP is a hybrid cryptosystem that combines features of both conventional and public key cryptography. The process for the PGP program is that it compresses the encrypted message (symmetric encryption) and then creates a one-time session key that is used to encrypt the compressed message and the session key, which are transmitted to the receiver. The recipient decrypts the encrypted session key with their private key, decrypts the message with the session key, and then decompresses the compressed message to get the original plaintext message, see Figure 6.10, *PGP Process*.

Hybrid encryption is about 1000 times faster than asymmetric encryption (i.e., public key encryption) (Network Associates 1999). PGP program information is available in many different languages (Zimmerman 2001).

PGP can also be used for generating digital signatures that can be used to ensure that a particular sender did indeed send the message and for non-repudiation. The benefit of a digital signature is that although an actual signature can be counterfeited, a digital signature is impossible to counterfeit. PGP can also be used to generate a hash function. The hash function can be used to create a digest of the plaintext message so that the digest, with the user's private key, can be used to create a kind of signature. In this case, the plaintext message plus the message digest and the private key are transmitted to the receiver. The benefit of the hashed message process is that it is faster than the use of the digital signature method (Network Associates 1999). However, both should have the same security capability of ensuring that the sender (i.e., entity) is the person (i.e., object) whom they say they are.

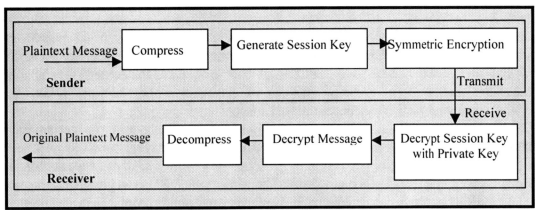

Figure 6.10 PGP Process

6.7   Virtual Private Networks

A *virtual private network (VPN)* session is an authenticated and encrypted communications channel across some form of public network (e.g., the Internet). VPNs between two networks (e.g., via two firewalls) require that each do the following:

1) Each site must set up a VPN-capable device on the network perimeter (e.g., at a firewall, router, or a device dedicated to VPN activity).
2) Each site must know the IP subnet addresses used by the other site.
3) Both sites must agree on a method of authentication and exchange digital certificates if required.
4) Both sites must agree on a method of encryption and exchange encryption keys as required.

A VPN is a network that extends remote access to users over a shared infrastructure. VPNs maintain the same security and management policies as a private network. They are the most cost-effective method for establishing a point-to-point connection between remote users and an enterprise customer's network (such as the Internet). There are three main types of VPNs:

1) Access VPNs that provide access to an enterprise customer's intranet or extranet over a shared infrastructure. Access VPNs use analog, dial, ISDN, DSL, mobile IP, and cable technologies to securely connect mobile users, telcommuters, and branch offices.
2) Intranet VPNs that connect link enterprise customer headquarters, remote offices, and branch offices to an internal network over a shared infrastructure using dedicated connections. Intranet VPNs differ from extranet VPNs by allowing VPN access only to the enterprise customer's employees.

237

3) *Extranet VPNs* that link outside customers, suppliers, partners, or communities of interest to an enterprise customer's network over a shared infrastructure using dedicated connections. Extranet VPNs differ from intranet VPNs by allowing users outside the enterprise to use the VPN.

Access VPNs use layer 2 tunneling technologies to create a virtual point-to-point connection between users and the enterprise customer network. A user anywhere in the world has the same connectivity as he or she would at the enterprise customer's headquarters. Access VPN architectures can be based on either network access server (NAS)-initiated access VPN or client-initiated VPN.

VPNs are private data networks that make use of the public telecommunication infrastructure, maintaining privacy through the use of a tunneling protocol and security procedures. A virtual private network can be contrasted with a system of owned or leased lines that can only be used by one company. The idea of the VPN is to give the company the same capabilities at much lower cost by using the shared public infrastructure rather than a private one. Phone companies have provided secure shared resources for voice messages. A virtual private network makes it possible to have the same secure sharing of public resources for data. Companies today are looking at using a virtual private network for both extranet and wide-area intranet.

Internet Protocol Security (IPSEC) is a public/private key encryption algorithm that is a Diffie-Hellman exchange to perform authentication and establish session keys. Use of IPSEC is transparent to the end user.

Firewall-1 can create VPNs through the support of SKIP and DES protocols. The Simple Key Management for Internet Protocol (SKIP) is similar to the SSL protocol because it operates at the session level. SKIP requires no prior communication connection to establish or exchange keys on a session-by-session basis. A shared secret is generated using the public key encryption method for IP packet-based encryption and authentication. While SKIP is efficient at encrypting data, which improves VPN performance, it relies on the long-term protection of this shared secret to maintain the integrity of each session.

Typically, VPN is service independent, so that information exchanged between any two hosts (e.g., Web, FTP, or SMTP) is transmitted along the same encrypted channel. Only when two networks have performed some preplanning can they use VPNs. A VPN only protects communications sessions between two encryption domains. Multiple VPNs require that the user define multiple encryption domains. There are only two specific applications for which VPNs are used:

1) Replacement for dial-in modem pools, and
2) Replacement for dedicated WAN links.

When selecting a particular VPN product, look at the following features:

1) Strong authentication,
2) Adequate encryption, and
3) Adherence to standards.

VPN product options fall into three categories:

1) Firewall-based VPN (best solution security-wise),
2) Router-based VPN (poorest solution security-wise), and
3) Dedicated software or hardware (next to best solution security-wise).

One way to think of a VPN is as *a hole in the firewall*. Someone with a VPN is allowed to tunnel through the firewall into the network. If a VPN user has an always-on Internet connection like cable or Digital Subscriber Line (DSL), it complicates things. In fact, a message that has been encrypted will be passed through the firewall because firewalls do not decrypt and hence are not aware of the contents of an encrypted message; it simply operates on the message header (source and destination addresses and port numbers, and possibly a flag).

The VPN server (or gateway or switch) is the point at which all VPN clients and other devices communicate to establish the VPN tunnel. Most VPN gateways are similar to firewalls because their access control lists must be configured to pass or block traffic to the destinations protected by the gateway (Thurman 2001). This is the reason why many firewall vendors offer VPN services on their firewalls, e.g., Check Point's Firewall-1. An out-of-band process is recommended for VPN server administration.

Authentication can be done using an LDAP server. If a hacker can identify the client's system configuration and somehow obtain a valid user ID and password (assuming that these are sufficient for system access), then no matter how much security capability resides in the rest of the system, the system can be compromised. The best way to deal with this issue is to create detailed user-friendly configuration instructions that are balanced with a policy for addressing security issues that might arise from improper configurations. By making the VPN architecture modular, the customer can easily apply the appropriate security best practices while balancing other administrative and support-related issues (Thurman 2001).

Two prominent Internet researchers, Steve Bellovin and Randy Bush, AT&T Labs, are among a growing number of experts raising red flags about Multi-Protocol Label Switching (MPLS), a next-generation traffic engineering technology backed by network industry leaders such as Cisco, Juniper Networks, and AT&T itself. They say MPLS creates serious network management challenges for Internet backbone providers. Even more serious are their warnings about potential security and privacy problems for organizations that deploy MPLS-based VPNs. Randy Bush says that MPLS VPNs are a great way to sell routers, but they greatly complicate the core of the Internet (Marsan 2001).

With MPLS VPNs, there's a potential for a network administrator doing the provisioning wrong and losing the privacy of the communication since MPLS VPNs do not automatically encrypt data.

The controversial MPLS VPNs are in use by companies such as IBM Canada and Canadian Life Assurance that want to outsource the management of their VPNs. But MPLS may be unnecessary because carriers can run frame relay or ATM traffic directly over an Internet backbone. Some users prefer VPNs using IPSEC, because if a communication is sent to the wrong person, that person can't read it, and IPSEC causes less stress on the Internet's backbone routers because customers handle provisioning.

Some analysts predict that MPLS VPNs running over the Internet will fail to gain widespread use but MPLS VPNs running on separate dedicated IP networks, such as that offered by AT&T, can be made more secure and might succeed.

Cisco says MPLS VPNs based on RFC 2547 are more scalable and just as secure as VPNs using frame relay or ATM and the amount of configuration involved with RFC 2547 VPNs is less than that of IPSEC VPNs, but that this burden is carried by ISPs, not customers. Moreover, MPLS-based VPNs are significantly less expensive to deploy than IPSEC VPNs (Marsan 2001).

Because the MPLS labeling hides the real IP address and other aspects of the packet stream, it provides data protection at least as secure as other Layer 2 technologies, including frame relay and ATM. MPLS-based isolation of packet streams can be viewed as the WAN equivalent to virtual LANs, the segregation of traffic over LANs that's enabled by IEEE 802.1p and 802.1q tags (Mier 2001).

MPLS without encryption doesn't provide the same level of security as IPSEC-based VPNs using 3DES encryption. However, 3DES's added processing also has a measurable effect on the latency and throughput of traffic that is sent through VPN tunnels. MPLS will not put IPSEC-based VPNs out of business but there seems to be no technical reason why an organization could not also apply IPSEC-based security, including encryption, to traffic that is being handled by the service provider via MPLS-labeled VPNs. That combination might well provide the best overall security that can be achieved for Internet-based data transmission today (Mier 2001).

Giacalone believes that MPLS VPNs have as much security as current WAN offerings (Giacalone 2001). An example of a network using MPLS routing is shown in Figure 6.11, *Using MPLS for Steering IP Traffic* (Duffy 2001). In this figure, the label switch routers represent MPLS switches and routers. MPLS-based networks may replace ATM and frame relay networks in the future. But ATM and frame relay networks provide four things that MPLS cannot, namely: high reliability, guaranteed performance, low latency, and security (Duffy 2001).

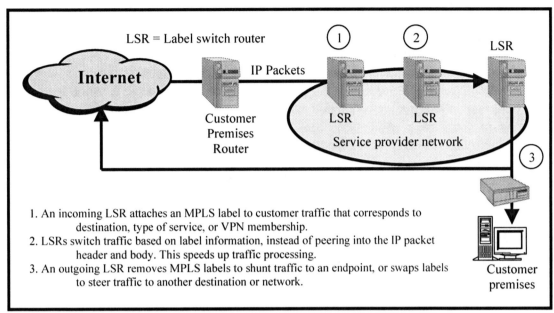

1. An incoming LSR attaches an MPLS label to customer traffic that corresponds to destination, type of service, or VPN membership.
2. LSRs switch traffic based on label information, instead of peering into the IP packet header and body. This speeds up traffic processing.
3. An outgoing LSR removes MPLS labels to shunt traffic to an endpoint, or swaps labels to steer traffic to another destination or network.

Figure 6.11 Using MPLS for Steering IP Traffic

At one time administrators did not have to worry about Address Resolution Protocol (ARP) cache poisoning and redirection because such assaults required attackers to gain direct access to the network segment that hosted the target systems. This network segment was protected by physical boundaries. Because of the increase in wireless network installations on the rise, access to ARP redirection and connection hijacking has been offered to anyone with an off-the-shelf wireless network card. This means that a remote attacker who can get within range of an access point could potentially poison client ARP caches into believing that the attacker is the default gateway, resulting in all traffic being routed through the attacker.

The solution is to treat the wireless network as a hostile network. Require all connections to authenticate and communicate via VPN in order to enter into the network properly. If your network has an Internet firewall, then ensure that all of the wireless connection points are routed through the firewall (Wysopal 2001).

Although the new software system (Next Generation Feature Pack 1 (NG FP1)) has a licensing scheme that is overly complex, the latest firewall/VPN offering from Check Point decreases the many time-consuming configuration hassles while increasing performance (Sturdevant 2001). Despite its simplification of the chores associated with network security, this task is still complex. One advantage of this new software over other competing products (such as Cisco's PIX firewall and VPN) is the One-Click policy tools that enable a system administrator to define VPN links and firewall policies.

The One-Click policy also eliminates configuration errors that can be introduced when setting up rules manually. In addition, sites with a large number of remote users can benefit from NG FP1's load balancing improvements.

6.8   Data Integrity

Data integrity refers to the state of data not being modified. Data integrity is the protection that ensures that information (stored or transmitted) is not changed from its original form. It is similar to error detection since data integrity does not provide for error or change correction, except by a request for a retransmission if a message is found to contain a change (whether intentional or unintentional). The manner in which the message change is detected is to transmit a checksum along with the message. The checksum cannot be modified and if the message is modified, then the checksum computed from the modified message is different from the checksum transmitted with the message and thus the receiver must request a retransmission.

Data integrity is defined as preventing unauthorized modifications of both stored and transmitted data. If a modification is detected during transmission, then that transmitted message is thrown out and a retransmission of that message is then requested. If a stored file has been modified, then that file must be replaced with an appropriate copy. Modifications may include changes, insertions, deletions, duplications, sequence changes, and replays, or even transmission errors (e.g., a bit error). Data integrity can be ensured through the use of data encryption for sensitive data (detection and protection from viewing) and through checksums for plaintext data (detection only, retransmission request required).

The Transport Control Protocol (TCP) provided by the Internet assures end-to-end reliability. It translates and manages message communications through the network and ensures data integrity and deals with packet sequencing.

Data integrity can also be provided using cryptography, but it can be provided without encryption as described above using checksums.

The use of a digital signature ensures both data integrity and non-repudiation, since the digital signature message would not decrypt properly if it had been modified during transmission.

For IPSEC, an Authentication Header (AH) can defeat spoofing and provide data integrity.

6.9   Accountability

Accountability is the process of determining that a malicious event has occurred and that some particular entity is responsible. The process of recording all events, malicious or not, is called auditing. Auditing coupled with identification and authentication ensures accountability.

Accountability requires at least two things to be known: 1) the identity of the entity (user, application, or server) and 2) the identity of the action taken that caused some network event to take place. The identified event should be in the audit trail, and the specific entity performing the action should be identified by the Identification and Authentication system. The audit trail is the tracking of an event that contains enough information for the analyst to be able to understand what actually happened. Ordinarily the audit trail should contain the following items:

1) The operation performed (i.e., the event),
2) The entity performing the operation,
3) The date and time of the operation,
4) Success or failure of the operation,
5) The object on which the operation was performed, and
6) The result of the operation.

6.9.1 Auditing

Auditing is the recording of events and who did it so that an evaluator can later reconstruct the event and identify the instigator. The audit trail must be such that no unauthorized user can erase or modify the audit trail. There are two forms of auditing: 1) host-based audit and 2) network traffic monitoring.

A *host-based (client computer) audit trail* should include the

1) Date,
2) Time,
3) Event name (e.g., attempt to log on),
4) Success or failure of the event, and
5) Any other pertinent information that is required to reconstruct the activity.

In addition, the audit trail should include the entity responsible for performing the event and this depends on having a link to the identification and authentication of the entity responsible for performing the action causing the event.

At least the following activities should be audited: logon attempts, logoff, DBMS actions, authorizations, attempts to bypass auditing, and attempts to modify or delete audit logs.

Network traffic monitoring or firewall auditing is a significant component of firewalls. Firewalls provide an ideal place to log all of the activity going on between networks. Most firewalls log the time, type of service, and source and destination ports of incoming packets but some firewalls allow the selection of certain events to be logged. For intrusion detection, firewalls can provide for automatic notification of the administrator via pager or e-mail given certain unauthorized access attempts. In some cases, the firewall may attempt to trace future attempts at access to gain more information about the intruder (NSA 1998).

Some firewalls offer anti-denial-of-service capabilities by either guarding against certain packets or shutting down entirely.

Despite advances in firewall technologies, current technologies fall short as far as the reporting and alerting associated with the log files are concerned. Some firewall vendors are partnering with third party vendors to include intrusion detection and audit trail analysis tools with their firewall offerings. It should be kept in mind, however, that as firewalls continue to incorporate more capabilities, the complexity of the product increases, thus increasing the number of avenues for attack.

6.9.2   Intrusion Detection System

An *intrusion detection system (IDS)* is an auditing mechanism that not only records events but also identifies them as malicious and only deals with these. The determination that an event is malicious can be done offline by going through the audit trails or in real-time (online) by examining all events as they occur and deciding that some of these are malicious.

There are two styles of intrusion detection: 1) pattern-based and 2) anomaly-based (Stillerman 1999). *Pattern-based* systems are explicitly programmed to detect certain known kinds of attack. Anti-virus detection programs are an example of pattern-based intrusion detection. Commercial IDSs for networks can recognize known intrusions. Although pattern-based systems tend to have a low rate of false alarms, they do have some limitations. For example, they cannot detect attacks that they have not seen before (or have a specific pattern which has been included in its pattern file); their complexity increases as the number of known attacks grows (because they are constantly having to check for all of the known malicious codes); and it is difficult to keep them updated as the catalog of attacks grows.

Anomaly-based systems address these problems by attempting to characterize normal operations and detecting any deviation from normal. The challenge in such systems is to define "normal" in a way that minimizes the false alarm rate and maximizes the detection efficiency.

Some vendors deploy their IDSs with "honeypots." What may look like an entire network can actually be a software system designed to monitor and report any activity by would-be hackers attempting to discover holes in a simulated device (Canavan 2001). The honeypot system is designed to lure a malicious hacker into a simulated safe network or server. When the hacker enters the simulated network, a trap is sprung and a security administrator is notified. The objective is for the security administrator to trace and identify the hacker. A similar system to the honeypot, but somewhat more ambitious, is the "honeynet" which attempts to trick attackers into wasting time and revealing their hand by attacking a fake that looks like an entire organization's information system that includes workstations, routers, and Web, e-mail, and domain name servers.

IDSs are vulnerable to direct attack. Whereas a firewall is often a perimeter guard and will typically fail open[61], the IDS is designed to run unobtrusively with a single collision domain and usually fails closed. One option the security analyst has when implementing the IDS is to place it on the network without an IP address. The IDS can be installed in-band but without a valid IP address (Brenton 1999).

Along with logging (recording suspicious activities) and alerting (notifying system administrators of suspicious activities), the IDS has two other countermeasures that can be employed:

1) Session disruption, and
2) Filter rule manipulation.

---

[61] *Fail open* means that the device is unable to pass traffic, whereas *fail closed* means that the device will continue to pass traffic.

Session disruption is when the IDS resets or closes each end of an attack session. This doesn't prevent the attacker from launching more attacks but it does prevent the attacker for causing further harm during the current session. Some IDS engines have the capability to modify the filter rules of a router or firewall in order to prevent continued attacks. This can block the attacking system from transmitting additional traffic to the target host. The IDS can add a new filter rule to the firewall that blocks all inbound traffic from the system IP address. Unfortunately, *filter rule manipulation* is not always totally effective. So the programmer should take care when implementing this option (Brenton 1999).

IDSs are either host-based or network-based. A *host-based IDS* is a real-time intrusion monitoring system that can detect unauthorized activity and security breaches and responds automatically. If it detects a threat, it will respond with an alarm to the security administrator or it may react according to pre-established security policies to avoid the loss, theft, or modification of information.

Security administrators can create, update, and deploy policies and securely collect and store audit logs for incident analysis, while maintaining the availability and integrity of the information system. As a complement to firewalls and other access controls, a host-based IDS can enable the development of security policies that prevent malicious hackers or authorized users from abusing the system. The IDS can provide complete control over systems with policy-based management.

While firewalls and VPNs offer perimeter and access controls—internal, remote and even authenticated users can attempt probing, misuse, or malicious acts. Internet access comprises most attack entries. A third of corporate Intranets are penetrated by external users from outside the corporation. Security strategies should provide countermeasures for the possibility of internal or external network attacks.

A *network-based IDS* can complement existing security countermeasures and strengthen any company's Web-based activities by offering dynamic detection that transparently examines network traffic. Some IDSs can identify, log, and terminate unauthorized use, misuse, and abuse of the system, its applications, or information from internal saboteurs or external hackers.

These systems can prevent intruders from exploiting hundreds of known codes and new security flaws in real time, but an attack definition wizard can enable network administrators to protect exposed corporate applications and stop even the most sophisticated assaults. Software wizards will perform many routine tasks and silent installation and remote tune-up capabilities make it easy to deploy and maintain the system (Symantec Intruder Alert 2001).

These systems can prevent intruders from exploiting hundreds of known codes and new security flaws in real time, but an attack definition wizard can enable network administrators to protect exposed corporate applications and stop even the most sophisticated assaults.

6.10 Non-Repudiation

Non-repudiation is the situation where the sender cannot deny having transmitted a particular message and the receiver cannot deny having received the message. Non-repudiation is based on digital signatures and can also be used in legal situations.

6.10.1 Digital Signature

The ISO X.800 defines a *digital signature* as a relatively low-level cryptographic mechanism. However, the term "digital signature" usually means an extension of that mechanism to enable use of, and/or reliance upon, certificate validation during signature verification. A digital signature (in a public key environment) is the encryption, with the user's private key, of a hashed value of the original message. This signature can only be decrypted with the user's public key. A digital signature that can be correctly decrypted provides strong authentication, data integrity, and non-repudiation (for the sender).

The digital signature is created when a public key algorithm is used to sign a message and assures that the document is authentic, has not been forged by another entity, has not been altered, and cannot be repudiated by the sender.

All users in a PKI system have both a public key and a private key. Consider a general situation. Bill and the receivers of his message can access Bill's public key (at the location of his PKI certificate). Bill has a message that he wants for his friends on the network to read and to know that it is from him. Bill will use his digital signature. The process for creating a digital signature is as follows:

> Bill writes the message and encrypts the message using his Private Key.
> Bill sends the message over the Internet to his friends.
> His friends receive the message and decrypt it using Bill's Public Key.

Any message decrypted using Bill's public key could only have been created using Bill's private key, which means it came from Bill.

The use of a digital signature ensures both data integrity and non-repudiation, since the message would not decrypt properly if it had been modified during transmission, and Bill cannot deny that the message came from him. If Bill wants to send his message so that it is unreadable by anyone other than the person to whom he is sending it, say Ann, then the process is as follows (Bernstein et al. 1996):

> Bill writes the message and encrypts it using His Private Key (signs it).
> Bill then encrypts the resulting message with Ann's Public Key (encrypts it again).
> Bill sends the now-twice-encrypted message over the Internet to Ann.
> Ann receives the message.
> Ann decrypts the message, first with Her Private Key and then with Bill's Public Key. Ann is reversing the steps used by Bill prior to sending the message.
> Ann now can read the message and be sure that it is both secret and that it is from Bill. Also, she can be assured that the message has not been modified.

In a more general sense, a digital signature is a cryptographic checksum computed as a function of a message and a user's private key, as shown in Figure 6.12, *General Digital Signature Methodology* below. It usually is computed using the original message, a hash of this message (a fixed length digest, say 160-bits), and an encryption of the hashed digest using the originator's private key. It is evaluated using the original message as transmitted, a hash of this message, and a decryption of the signature (using the originator's public key). The decrypted message is then compared with the original message as hashed, using the same hash function. The digital signature process, illustrated below, provides for message integrity, user authentication, and nonrepudiation with proof of origin. The secure hash algorithm is assumed to be the SHA-1 encryption algorithm developed by the National Security Agency (NSA).

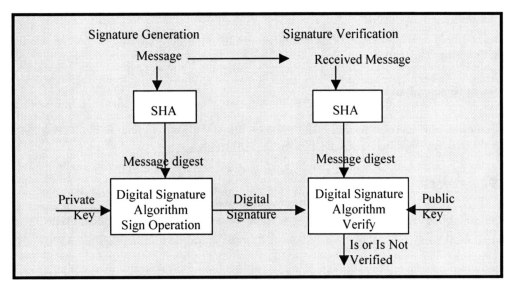

Figure 6.12 General Digital Signature Methodology

6.10.2 Dual Encryption

Dual encryption is when both the sender's private key and the receiver's public key are used to encrypt the transmitted message (in a PKI environment). When this is done, the non-repudiation goes both ways, that is, the sender cannot deny having transmitted the message and the receiver cannot deny having received the message, because both the sender's public key and the receiver's private key are required to decrypt the transmitted message. Only a message from *this sender* to *this receiver* can be decrypted in this way.

Mathematically, the situation is as follows:

$$T = f_{Sprivatekey} (f_{Rpublickey} (P))$$

where T is the transmitted message, P is the plaintext message, f is the encryption algorithm, $f_{Rpublickey}$ is the transformation function using the receiver's public key, and $f_{Sprivatekey}$ is the transformation function using the sender's private key.

This message can only be decrypted using the following process:

$$P = d_{Rprivatekey} (d_{Spublickey} (T))$$

where d is the decryption algorithm, $d_{Rprivatekey}$ is the transformation using the receiver's private key, and $d_{Spublickey}$ is the transformation using the sender's public key.

The transformations must be done in the order indicated.

6.11 No Object Reuse

No object reuse is a requirement that no information left in the computer by its present user will remain in the computer for a later user. Many COTS software packages, such as DBMSs, now provide this service by clearing all previous temporary information prior to a later user logging on.

6.12 Data Separation

Data separation, sometimes called "separation of duties," is the denial of access for any user to information that may cause a conflict of interest.

6.13 Documentation

Documentation is a requirement that all relevant security documentation will be provided to ensure that the various phases of a security process are properly completed.

6.14 Countermeasures

The primary objective of a risk assessment is to determine an appropriate set of cost-effective *countermeasures* that will minimize the system's vulnerabilities against the relevant threats to the system. By comparing the performance of these countermeasures with the vulnerabilities, the analyst can determine the effectiveness of the new set of countermeasures against relevant threats in conceivable attack scenarios.

The alternative countermeasures discussed in Chapter 8 can be assessed to determine which ones are cost-effective and then purchased, installed, tested, evaluated, and appropriately modified, and operated by trained technicians, programmers, and system administrators.

6.15 Remaining Risks

Once the decision has been made to purchase a set of countermeasures, then one or more risks (that are not cost-effective to counter) will be left to control without any countermeasure to protect the system. These are the *remaining risks* and are the result of some countermeasures being more costly than the risk associated with a particular threat, or the result of a countermeasure that does not fully protect the system against one or more threats.

These remaining risks should be quantified and explained so that the owners of the information system can understand what the extent of the protection is and what the remaining risks are that need to be accepted.

6.16   Recapitulation

After identifying a list of relevant threats, determining the system's vulnerabilities, and creating a list of threat scenarios, compute the probabilities of successful attack for each of the scenarios and the likely loss (in dollars) so that the various risks can be determined. Tables 6.15, *Scenarios, Descriptions, Probabilities, and Losses*, and 6.16, *Scenarios, Countermeasures, and Costs*, will be useful in performing the analysis steps.

6.16.1 Sample Case

A simple example is chosen to illustrate the security analysis methodology. The example is an organization that does most of its business through the Web getting orders via e-mail messages. The business manufactures Widgets, which are sent to its customers using the regular post office mail. Customer orders are received in three forms: 1) over the Internet, 2) via post office mail, or 3) over the telephone. All orders are placed in the organization's information system.

6.16.1.1   Example Security Policy

The security policy for this system is the basic policy depicted in Appendix D.1, *Security Policy: An Example*.

6.16.1.2   Example Security Requirements

The security requirements for this system are the ones shown in Appendix J, *Minimum Security Requirements*. These requirements should be traceable back to the policy statements. For example, the security requirement for "confidentiality" is traceable back to the policy statement "Sensitive information shall be protected from unauthorized viewing or access during storage or transmission."

6.16.2 Identified Threats

The identified threats, based on an analysis of the gamut of known threats versus the definition of the information system for the example organization, are listed below:

1) Unauthorized person enters the facility housing the information system,
2) Unauthorized physical access by employees to information system critical devices,
3) Loss of power from local electricity supply company,
4) Hackers enter the system (electronically) and steal or damage database information,
5) Denial-of-service (DoS) and distributed DoS (DDoS) attacks,
6) Internal users intentionally or unintentionally access unneeded information, access conflicting information, damage information, and record critical information.

7) Internal users remove critical information from the organization's premises on floppy disks.

8) Hacker calls in and requests password information from an internal user,

9) Attacker obtains sensitive information by going through organization's trash,

10) Attacker obtains plaintext transmitted sensitive information by sniffing the link from the organization's information system to the ISP line,

11) The information system is totally destroyed by a catastrophic event,

12) Customer denies placing an order via the Internet for a Widget (when she did place the order), and

13) E-mail with a virus attachment.

Without a doubt, one can theorize many more threats than those listed above but the objective here is to present an example, not perform an exhaustive survey.

6.16.3 Identified Vulnerabilities

I will assume that the organization has no implemented countermeasures at present, so all of the methods for minimizing the identified risks are needed. That is, the organization's information system is vulnerable to all of the identified threats. Usually, the analyst would scan the network to identify the system vulnerabilities. Since the system has no installed or implemented countermeasures (our assumption), scanning the system and questioning administrators are not required.

6.16.4 Developed Threat Scenarios

The threat scenarios can be developed from the listed threats in section 6.16.2 and the reasonable losses can be determined from the theorized scenarios, see Table 6.15, *Scenarios, Descriptions, Probabilities, and Losses*. One can determine the minimum, maximum, or reasonable values of any expected losses. I am using the third option, namely, reasonable values of expected losses. The numbers used for probabilities, losses, and costs (in this example) are sample values used to illustrate the situation, and should not be considered as necessarily reflecting realistic values.

The threat scenarios, with the estimated probabilities of attack and estimated reasonable losses, are listed below:

1) An unauthorized person enters the facility and proceeds to steal a server (which he detaches from the local area network), then using a dolly (that happens to be located nearby) he removes the server to a van and drives away. The server has sensitive information stored in it worth $500,000, much more than the value of the server (say $4,000) which can be ignored here. The probability that this attack is committed and successful is estimated to be 0.01.

Table 6.15 Scenarios, Descriptions, Probabilities, and Losses

Scenario	Description	Probability of Successful Attack	Reasonable Loss ($)
1	Unauthorized person entering the premises and committing malicious act.	0.01	$500,000
2	Unauthorized internal person physically accessing and committing malicious act.	0.1	$700,000
3	Loss of power	0.01	$100,000
4	Hacker enters the system and steals or damages sensitive information.	0.2	$150,000
5	Attacker launches DDoS attack.	0.01	$250,000
6	Internal user intentionally erases sensitive database.	0.3	$50,000
7	Internal user accesses database that can cause a conflict of interest.	0.1	$250,000
8	Internal user unintentionally accesses unneeded sensitive database.	0.8	$20,000
9	Internal user copies sensitive information, writes it onto a floppy disk, and removes the disk in his briefcase. He then sells the information to a competitor.	0.5	$50,000
10	Hacker calls in and requests user ID and password information, which he gets. He then is able to enter the system and access sensitive information, which he sells to a competitor.	0.1	$100,000
11	Attacker is able to obtain sensitive information by sifting through the organization's trash. He then sells the information to a competitor.	0.01	$10,000
12	Sensitive information is transmitted in plaintext and the information is sniffed by an attacker who sells the information to a competitor.	0.25	$50,000
13	A tornado hits the facility and totally destroys the facility.	0.001	$25,000,000
14	A customer places a large order, then subsequently denies having placed the order.	0.9	$5,000
15	An e-mail arrives with a virus attachment which is opened by an internal users thereby activating the virus which then contaminates the system, causing the administrator to have to work many hours to repair the damage.	0.01	$200,000

2) An unauthorized employee enters the system facility and damages a server with sensitive information that is lost. The probability of this attack is estimated to be 0.1 and the potential loss is $700,000.

3) The local power company experiences a blackout in the area of the organization's information system facility. The probability of this occurrence is estimated to be 0.01 and the potential loss due to the downtime of the system is $100,000.

4) A hacker enters the system (electronically) using a discovered user ID and password and accesses a sensitive database containing information on a new version of the Widget, which the attacker copies and sells to a competitor. The probability of this attack is estimated to be 0.2 and the potential losses amount to $150,000.

5) An attacker launches a DDoS attack that disables the Web servers for four hours. The probability of this attack is estimated to be 0.01 and the potential losses amount to $250,000 in lost profits.

6) An internal user intentionally causes damage to sensitive database information. The probability of this attack is estimated to be 0.3 and the potential losses amount to $50,000.

7) An internal user intentionally accesses unneeded (for him) database information. The probability of this event is 0.1 and the potential losses amount to $250,000.

8) An internal user unintentionally causes damage to sensitive database information. The probability of this event is 0.8 and the potential losses amount to $20,000.

9) An internal user stores critical information on a floppy disk and removes the disk from the organization's premises, then sells the disk to a competitor. The probability of this attack is estimated to be 0.5 and the potential loss amounts to $50,000.

10) A hacker calls in and requests password information from an internal user. The password information is used to gain access to the system and sensitive database information is stolen. The probability of this attack is estimated to be 0.1 and the potential losses amount to $100,000.

11) An attacker obtains sensitive information by going through the organization's trash. This information is sold to a competitor. The probability of this attack is estimated to be 0.01 and the potential losses amount to $10,000.

12) An attacker obtains plaintext transmitted sensitive information by sniffing the link from the organization's information system to the ISP line. This information is sold to a competitor. The probability of this attack is estimated to be 0.25 and the potential losses amount to $50,000.

13) The information system is totally destroyed by a catastrophic event. The probability of this attack is estimated to be 0.001 and the potential losses amount to $25,000,000.

14) A customer denies placing an order via the Internet for a Widget (when she did place the order). She is able to cancel her order because the organization does not have any proof that she actually placed the order. The probability of this event is estimated to be 0.1 and the potential losses amount to $5,000. However, since this event is likely to occur at least 10 times per year, the probability of at least one occurrence of the event is 0.65 (which is $1 - (1 - 0.1)^{10}$).

15) E-mail with a virus attachment is received. Several internal employees who received the message open the virus-infected attachment, which then clogs their PC. A programmer than has to go to each user's PC and remove the virus and return the

system to its original state. The probability of this attack is estimated to be 0.01 and the potential losses amount to $200,000.

In this analysis, I did not consider the monetary value that many organizations place on their concern for the protection of sensitive company files, or files of private information on customers such as credit card numbers, and other such information, because it is difficult to ascertain the monetary value of this information. However, a given organization's management will likely have some idea of the value (to them) of this sensitive information. Also, just the fact that this information has been compromised is something that many companies do not wish to discuss nor to enter into assessing its loss or compromise.

6.16.5 Computed Risks

Using the formula,

Risk (i) = Probability of successful attack (for the ith threat scenario) * likely loss from the attack (as shown in Table 6.15 above),

the risk values for the fifteen scenarios are found to be:

1)	$5,000.00
2)	$70,000.00
3)	$1,000.00
4)	$30,000.00
5)	$2,500.00
6)	$15,000.00
7)	$25,000.00
8)	$16,000.00
9)	$25,000.00
10)	$10,000.00
11)	$100.00
12)	$12,500.00
13)	$25,000.00
14)	$4,500.00
15)	$2,000.00

6.16.6 Selection of the Countermeasures

The total costs associated with each countermeasure are compared with the risks associated with the identified countermeasures. Below is a list of the identified countermeasures:

1) Fences, gates, badges, guards, patrols, inspections
2) Locked rooms with cyber locks for servers housing sensitive information
3) Uninterruptible power supply system

4) a. User IDs and passwords for identification and authentication, b. authorization for database access, and c. firewall with rules for blocking unauthorized access
5) Anti-DDoS software
6) Auditing capability and IDS
7) An organizational policy that states that no information shall be copied or removed from the premises, guards randomly inspect employees as they leave the premises, and other rules for ensuring security
8) An organizational policy that states that employees who damage data are subject to immediate removal
9) Employee training and education against social engineering attacks, et al.
10) Encryption of all sensitive information prior to transmission using SSL, certificates, PKI, and VPN (based on IPSEC standards)
11) An organizational policy rule that states that all sensitive information must be shredded
12) Encryption of all sensitive information prior to transmission using SSL, certificates, PKI, and VPN (based on IPSEC standards)
13) Implementation of a backup and recovery system, creation of a contingency plan, and acquisition of remote site facilities with an appropriate LAN system
14) Implementation of digital signatures
15) Training of employees to ensure that they do not open e-mail attachments from unrecognized sources

Because of the interdependencies of the various countermeasures and the fact that many different threats are countered by the same mechanisms, it is not a straightforward task to determine the return on investment for the implemented countermeasures, except when the costs of all of the countermeasures (CM Cost) can be aggregated, namely,

$$CM\ Cost = \Sigma_i\ C(i))$$

is compared with the total risk (TR),

$$TR = \Sigma_i\ R(i).$$

The ROI for all of the countermeasures is then,

$$ROI = TR - CM\ Cost.$$

If this ROI > 0, then the set of all considered countermeasures is cost-effective. And, of course, the greater the ROI value, the more cost-effective is this set. A ratio of ROI to CM Cost of 2 or more is highly desirable, that is, it is advantageous to the organization when,

$$ROI/(CM\ Cost) > 2.$$

In many cases, the same set of countermeasures serves to counter more than one threat. And similarly, a single countermeasure may be a part of the countering of more than one threat. For example, items 10 and 12 (encryption) in Table 6.16, *Scenarios,*

Countermeasures, Costs, and ROI, are the same. Note that the ROIs for each countermeasure are also listed in this table. The ROIs will be discussed in the next section.

Table 6.16 Scenarios, Countermeasures, Costs, and ROI

Scenario	Countermeasure	Total Cost ($)	Risk ($)	ROI
1	Fences, gates, badges, guards, patrols, inspections	$4,000	$5,000	$1,000
2	Sensitive servers in locked rooms with cyber locks	$2,000	$70,000	$68,000
3	Uninterruptible power supply system	$2,000	$1,000	-$1,000
4	User ID and passwords for identification and authentication, authorization for database access, and firewall with rules for blocking unauthorized access	$5,000	$30,000	$25,000
5	Anti-DDoS software	$500	$2,500	$2,000
6	Auditing capability and IDS	$2,000	$15,000	$13,000
7	An organizational policy that states that no information shall be copied or removed from the premises, guards randomly inspect employees as they leave the premises	$100	$25,000	$24,900
8	An organizational policy that states that employees who damage data are subject to removal	$100	$16,000	$15,900
9	Employee training and education against social engineering, et al.	$3,000	$25,000	$22,000
10	All sensitive information is encrypted prior to transmission, using SSL, certificates, PKI, VPN	$6,000	$10,000	$4,000
11	Organizational policy states that all sensitive information must be shredded	$100	$100	$0
12	All sensitive information is encrypted prior to transmission, using SSL, certificates, PKI, VPN	$5,000	$12,500	$7,500
13	Backup and recovery, contingency plan, remote site facilities	$12,000	$25,000	$13,000
14	Digital signatures	$2,000	$4,500	$2,500
15	Training of employees to ensure that they do not open e-mail attachments from unrecognized sources	$2,000	$2,000	$0

6.16.7 Return on Investment

The ROI to counter the threat in the ith threat scenario is as follows:

ROI (i) = Costs (i) - Risk (i)

The ROIs for the various countermeasures are listed above in Table 6.16. If the ROI is negative or zero (i.e., ROI $\leq$ 0), then the owner may not wish to purchase the countermeasure, because the mathematics implies that the countermeasure is not cost-effective. The larger the ROI is relative to the costs associated with the countermeasure, then the more cost-effective is the countermeasure where the ith ratio is computed as follows,

Ratio (i) = ROI (i)/Costs (i)

where ROI (i) is the ROI for the ith countermeasure and Costs (i) are the costs associated with acquiring, installing, testing, and operating the ith countermeasure.

Thus, the sixth countermeasure (viz., the IDS) is very cost-effective since Ratio (6) = 6.5.

However, even for low probability threats with small or negative ROIs, if the costs associated with acquiring, installing, testing, and operating the countermeasure are low, the owner still may wish to implement the countermeasure.

In many, if not most, cases, spending on security is usually reactive, that is, some type of attack on an information system has occurred (or management has heard about an attack) and the organization is responding to the attack with a security upgrade. In cases where the organization has been attacked, the organization has some idea of the costs associated with this specific attack and has decided that to stay in business, they need some protection.

Convincing owners that it is wise to provide proactive spending on security is a very difficult task and an ROI approach is one method for supporting such a proposal. However, a comparative analysis might be desirable also. For example, a security analyst (for organization A) might be able to point to a similar organization B that had its information system attacked and the attack cost organization B the amount of $2.5M and for a mere fraction of this (say 1/50), organization A can purchase protection against similar attacks. This sort of argument might be more convincing than the ROI approach.

6.16.8 Migration Plan for Upgrades

The organization, in an effort to limit the risks (i.e., security, technical, schedule, and performance) for upgrading the legacy system to the target system decides to migrate to the target system with two interim upgrades. The target system consists of the following improvements:

1) Install a firewall and an intrusion detection system at the perimeter of the organization's information system network (i.e., between the information system and the wide area network).

2) Install a PKI capability implemented with certificates, SSL, digital signatures, VPN, and PGP encryption.

3) Install two servers at a remote site, activate the replication capabilities of the Oracle database management system (DBMS) to replicate the online critical information twice daily (12 noon and 6 PM) to a remote site, and create a contingency plan for ensuring that critical information is properly replicated and can be quickly accessed for replacing lost information in case of a catastrophe.

To minimize risks, the organization decides to upgrade to the target system (legacy with firewall/IDS, the PKI capability, and remote site storage) in three steps, performing ST&E following each upgrade. In this case, the firewall and IDS are installed to provide the initial interim system. The firewall and IDS should be tested and evaluated and modified if needed.

Next, the PKI capability is installed, tested, and evaluated. Once this upgraded system (i.e., the secondary interim system) has been certified as being in operational condition, the third upgrade (to achieve the target system capability) is implemented.

This last upgrade consists of installing:

1) Two storage servers at a remote site and

2) Activating replication software for placing critical information on the remote storage servers at a frequency determined by the organization's users.

Again, following this third upgrade, the system with its contingency capability is tested and evaluated, and then certified following any required modifications.

If all of the upgrades are installed in one major upgrade, it is much more difficult to identify any malfunctions, to ensure that the system is working properly, and to properly test and evaluate the system.

6.16.9   Considerations of Remaining Risks

All of the identified risks were countered in this example. But if a risk had been considered to be small enough to remain, then this risk should be addressed with rationale for not countering with some sort of security mechanism or method. The argument should state that the cost of the countermeasure is more (perhaps much more) than the risk.

6.16.10   Identified Countermeasures

Taking the risks (and threats) in the order that they were listed, it is apparent that each countermeasure will protect against several threats and in some cases, several countermeasures may be needed to address a single threat.

Digital signatures require the implementation of an encryption methodology and this can be accomplished with the implementation of a public key infrastructure (PKI). The customer will wish to perform a tradeoff of the cost and benefits of an encryption methodology (such as PGP) versus PKI. For an organization with a small budget for security protection, PGP may be the best alternative for meeting confidentiality, data integrity, digital signature, and authentication requirements. The internal version of PGP (PGPi) can be found at the following URL address: http://www.pgpi.org/.

REFERENCES AND BIBLIOGRAPHY

6.1   NSA (1998). "Network Security Framework," Network Security Group, Technical Directors, Security Solutions Framework, Release 1.1, National Security Agency, 3 December 1998.

6.2   Oppliger, Rolf (1998). "Internet and Intranet Security," Artech House, 1998.

6.3   Dawes, R. (1979). "The Robust Beauty Of Improper Linear Models In Decision Making," American Psychologist, 34, pp. 571-582, 1979.

6.4   Smith, Sr., Charles L. (1998). "Computer-Supported Decision Making: Meeting the Demands of Modern Organizations," Ablex Publishing Corporation, Greenwich, Connecticut and London, England, 1998.

6.5   DITSCAP (1999). "Department Of Defense Information Technology Security Certification And Accreditation Process (DITSCAP)," Application Document, DOD Manual, 5200.40-M, 21 April 1999.

6.6   Bernstein, Terry et al. (1996). "Internet Security for Business," John Wiley & Sons, pp. 150-151, 1996.

6.7   Brenton, Chris (1999). "Mastering Network Security," SYBEX Network Press, 1999.

6.8   Smith, Richard E. (1997). "Internet Cryptography," Addison Wesley, Reading, MA, 1997.

6.9   Krause, Micki and Harold F. Tipton (1999). "Handbook of Information Security Management," Auerbach, Washington, DC, 1999.

6.10   Canavan, John E. (2001). "Fundamentals of Network Security," Artech House, Boston, 2001.

6.11   Russell, Deborah and G. T. Gangemi, Sr. (1992). "Computer Security Basics," O'Reilly, Cambridge, MA, 1992.

6.12   Stillerman, Matthew, et al. (1999). "Intrusion Detection for Distributed Applications," Odyssey Research Associates (ORA), Ithaca, NY, 1999.

6.13   NIST (1995). "An Introduction to Computer Security: The NIST Handbook." NIST Computer Policy, SP 800-12, October 1995.

6.14   Smith, Sr., Charles L. (1999). "A Process for the Development of a Security Architecture for an Enterprise Information Technology System," Command and Control Research and Technology Symposium, U.S. Naval War College, Newport, RI, 29 June - 1 July 1999.

6.15   Swanson, Marianne (1998). "Guide for Developing Security Plans for Information Technology Systems," NIST Special Publication 800-18, Federal Computer Security Program Managers' Forum Working Group, December 1998.

6.16   NIST (1994). "Keeping Your Site Comfortably Secure: An Introduction to Internet Firewalls," NIST SP 800-10, December 1994.

6.17   Smith, Sr., Charles L. (1997). "A Survey to Determine Federal Agency Needs for a Role-Based Access Control Security Product," *International Symposium on Software Engineering Standards '97*, IEEE Computer Society, Walnut Creek, CA, June 1997.

6.18   CSC (1999). "Evaluation Technical Report for Check Point Software Technologies, LTD, Firewall -1," Version 4.0, CSC, Hanover, MD, October 1999.

6.19 Curtin, Matt (1997). "Introduction to Network Security," Kent Information Services, Inc, March 1997. This document can be found at: (http://www.interhack.net/pubs/network-security.html).

6.20 Stallings, William (2001). "SSL: Foundation for Web Security," Cisco Web Page, http://www.cisco.com/warp/public/759/ipj_1-1/ipj_1-1_SSL2.html, 2001.

6.21 Check Point (2001). "Firewall-1, A Complete Solution for Securing the Internet," Check Point Software Technologies, 2001. This document can be found at: http://www.checkpoint.com/products/downloads/FW1-4.1_Brochure.pdf

6.22 Aladdin (1999). "eSafe Gateway," Aladdin, 1999. This document can be found at: http://www.checkpoint.com/opsec/partners/aladdin.html

6.23 Cisco Cable (2001). "Cisco Cable Web Page," Cisco, 2001. This information can be found at: http://www.cisco.com/cable/products/.

6.24 Zimmerman, Phil (2001). "Why Do You Need PGP?" PGP Web Page, http://www.pgpi.org/doc/whypgp/en/).

6.25 Network Associates (1999), "An Introduction to Cryptography," Network Associates, ftp://ftp.pgpi.org/pub/pgp/6.5/docs/english/IntroToCrypto.pdf, 1999.

6.26 NIST (2001). "NIST Subject List," http://www.nist.gov/public_affairs/siteindex.htm.

6.27 Cisco Switch (2001). "Switch Clustering: A Breakthrough for Scalable and Affordable LANs," http://www.cisco.com/warp/public/cc/pd/si/casi/ca3500xl/prodlit/swclu_ov.htm, 2001.

6.28 Lyons-Burke, Kathy (2000). "Federal Agency Use of Public Key Technology for Digital Signatures and Authentication," NIST SP 800-25, Federal Public Key Infrastructure Steering Committee, October 2000.

6.29 NIST Handbook (1995). "NIST Computer Policy, An Introduction to Computer Security: The NIST Handbook," SP 800-12, October 1995.

6.30 NIST (1992). "Key Management Using ANSI X9.17," FIPS 171, NIST, 27 April 1992.

6.31 iDefense (2001). "Code Red FAQ v1.0," iDefense, http://www.idefense.com/pages/ialertexcl/coderedfaq.htm, 2001.

6.32 Thurman, Mathias (2001). "VPN Security Review Moves to the Front Burner," COMPUTERWORLD, p. 54, 10 September 2001.

6.33 Matthews, William (2001). "IT May Alter Air Travel," Federal Computer Week, p. 16, September 17, 2001.

6.34 McDonald, Tim (2001). "Companies Race To Solve Denial-of-Service Riddle," NewsFactor Network, http://www.newsfactor.com/perl/story/7282.html, February 6, 2001.

6.35 Symantec NetProwler (2001). "NetProwler," Symantec, www.symantec.com, 2001.

6.36 Symantec Intruder Alert (2001). "Intruder Alert," Symantec, www.symantec.com, 2001.

6.37 Slayton, Joyce (2001). "Security: How Much Is Too Much?" Business Daily, pp. 22-30, October 2001.

6.38 NIST AES (2001). "Advanced Encryption Standard (AES): Questions and Answers," NIST, http://csrc.nist.gov/encryption/aes/aesfact.html, 2001.

6.39 NIST CSRC (2001). "AES Algorithm (Rijndael) Information," NIST, http://csrc.nist.gov/encryption/aes/rijndael/, 2001.

6.40   Chen, Anne and Matt Hicks (2001). "How to Stay Afloat," eWeek, pp. 49-56, October 8, 2001.

6.41   Marsan, Carolyn Duffy (2001). "Experts call MPLS bad for 'Net," Network World, http://www.nwfusion.com/news/2001/0806mpls.html, September 6, 2001.

6.42   Mier, Edward (2001). "Tester's Choice: MPLS takes on Security Role," Network World, http://www.nwfusion.com/research/2001/0521feat2.html, May 21, 2001.

6.43   Giacalone, Spencer (2001). "MPLS VPNs Improve WAN Connectivity," Network World, p. 43, October 15, 2001.

6.44   Wysopal, Chris (2001). "New Class of Wireless Attacks," http://archives.neohapsis.com/archives/vulnwatch/2001-q4/0008.html, October 16, 2001.

6.45   Frank, Diane (2001). "GSA Leaps PKI Hurdle," Federal Computer Week, p. 29, October 15, 2001.

6.46   Wreski, Dave (2000). "Root Security," http://www.linuxsecurity.com/tips/tip-12.html, July 30, 2000.

6.47   O'Brien, Bill (2001). "Red Hat 7.2: Pain-free Linux," Enterprise, http://www.zdnet.com/products/stories/reviews/0,4161,2816836,00.html, October 9, 2001.

6.48   Duffy, Jim (2001). "MPLS Facing Slow Adoption, Despite Flurry of Market Hype," Network World, pp. 1, 14, November 5, 2001.

6.49   Sturdevant, Cameron (2001). "Security Made Simpler," eWeek, pp. 39-42, December 31, 2001.

PART III. FINDING A SOLUTION

This section contains the process for finding a solution to the defined issue of "what protection do I need for my particular information system?" This question or hypothesis is framed by the various scenarios that were generated to characterize the gamut of threats confronting this particular information system.

CHAPTER SEVEN

COUNTERMEASURES

7.1   Identification of Countermeasures

Risk is a probabilistic cost (for a specific threat (T_i)) or potential loss resulting from an attack and can be computed with the following formula:

R (T_i) = Risk (for a specific threat) =
Probability of a successful attack by this specific threat * Potential cost (i.e., loss) caused by the attack

A recapitulation of the risk assessment process is as follows:

1) *Identify the system assets and estimate their value to the* organization (probably based on high-level organization management interviews),
2) *Define the security requirements* (either as deduced from the organization's security policy or a relevant organization document that contains the overall organizational security requirements),
3) Identify the relevant threats $(T_i,)$ to the organization (based on an analysis),
4) *Identify the organization's vulnerabilities* $(V_k,\ k = 1,\ 2,\ ...,\ m)$ (either from a vulnerability analysis or from an offline scanning of the systems to be installed later as online systems),
5) For each relevant threat $(T_i,\ i = 1,\ 2,\ ...,\ n)$, determine the organizational assets likely to be affected (based on a threat versus organizational asset analysis),
6) For each relevant threat $(T_i,)$, determine the probability of attack $(P_A\ (Ti))$ (launched against a particular vulnerability) and using the total value of the affected assets determine the risk, namely, R $(T_i) = P_A\ (T_i) * V\ (A_j)$, where $V\ (A_j)$ is the total value (i.e., cost) of the organizational assets potentially affected by the specific threat (determine the various risks from the risk computations for various identified potential scenarios),
7) For each identified vulnerability and risk, identify the cost-effective countermeasures that can reduce the vulnerability to a tolerable level (based on an analysis of the risks and the total costs (acquisition, installation, test and evaluation, and operation and maintenance) of the various relevant countermeasures),
8) Perform a risk assessment analysis of the organizational network with the implemented countermeasures as suggested by the risk assessment analysis, (based on an analysis of the risks, cost-effective countermeasures, the assumed performance of these countermeasures, and remaining risks that are not countered because of cost considerations),
9) *Write a Risk Assessment Report* (based on the risk analysis and results produced above).

The process described above has some mathematics in it, but, alas, risk assessment is a mathematical issue.

Some treat risk in a quasi-mathematical way that non-mathematicians may prefer (as already stated earlier) using the equation below to compute risk (for some given threat and vulnerability),

$$R = L * (C+I+A)$$

where the L represents the Likelihood of the attack, C represents Confidentiality, I represents (data) Integrity, and A represents Availability. It is not clear what the L-value or C-value or I-value or A-value actually mean, The likelihood values are also shown and appear to be numbers from 1 to 3, but should be probabilities that must be in the interval [0,1]. The risk rating can have values in the interval of integers [0, 27].

For each threat, the effects on the confidentiality of data has four gradations, "0" meaning "no effect" to "3" meaning "an extremely negative effect." Similarly, the other risk factors, "A" and "I" are evaluated. The likelihood of the attack "L" can also be rated with "0" meaning "not likely at all" and "3" meaning "highly likely."

As mentioned earlier, it is not clear what these values mean except perhaps that when combined to compute the risk value using the equation above, "a larger value reflects a greater risk than does a smaller value." There appears to be no instructions for how the analyst following the template directions is supposed to arrive at values for C, I, and A, nor for the value of L.

Once the risk values have been determined, then the alternative countermeasures to minimize these risks can be selected from the following list:

1) Firewalls (perhaps with virtual private network (VPN) encryption/decryption),
2) Routers,
3) Improved security cabling,
4) Uninterruptible power supplies (UPSs),
5) Backups for the heating, ventilation, and air conditioning (HVAC),
6) Backup and recovery devices and procedures,
7) Anti-virus software,
8) Anti-denial-of-service software (anti-DoS),
9) Encryption capabilities (symmetric, asymmetric, and hybrid) for protection of sensitive information (either during storage or transmission) and for non-repudiation using digital signatures,
10) Hashing functions or checksums for data integrity,
11) Strong authentication,
12) Intrusion detection systems (IDSs),
13) Auditing capabilities,
14) Badges,
15) Guards,
16) Fences,
17) Cyber locks,

18) Gates,

19) Inspections,

20) Training and education of personnel,

21) Dual servers,

22) Load balancing and automatic failover software, and

23) Other countermeasures relevant to organizational security needs.

The process of selecting the appropriate countermeasure to minimize a particular risk requires an understanding of the capabilities of the various countermeasures, their performance capabilities, and their costs. They are sort of magically mixed and "Presto!" out comes the appropriate cost-effective countermeasure for the specific risk.

A process of determining directly the risk values (qualitatively) is shown in Table 7.1, *Risk Values*. The security analyst, in this case, would use the qualitative risk values to determine the recommended countermeasures.

Table 7.1 Risk Values

| Threat Difficulty | Vulnerability Severity | | | | |
	Negligible	Minimum	Major	Hazardous	Catastrophic
Low	Low	Moderate	High	Severe	Severe
Low-Medium	Low	Low	Moderate	High	Severe
Medium	Low	Low	Low	Moderate	High
Medium-High	Low	Low	Low	Low	Moderate
High	Low	Low	Low	Low	Low

In this case, *Threat Difficulty* means the difficulty for the attacker, so "Low" means that "the threat is highly **likely**" and "High" means that "the threat is highly **unlikely**." *Threat Severity* indicates the effects on the organization from a successful attack on the organization. "Negligible" means that the vulnerability that a successful attack would have little if any effects on the organization, whereas "Catastrophic" means that a successful attack would totally disable the system. The qualitative values for the risks are certainly subjective. Much of the table entries are "Low" which indicates a risk that is not very great.

These empirical processes for computing risk are not recommended. They are presented here only to show what some organizations have resorted to in order to avoid the mathematical notions originally intended for the definition of risk (which they must consider too difficult to understand and use). The process presented earlier based on the mathematical definition of risks using the concepts of threats, vulnerabilities, and threat scenarios is the process that should be used.

7.2   Discussion of Countermeasures

In this section, I will discuss several countermeasures that can be used to provide certain security services.

7.2.1 Secure Sockets Layer and the Transport Layer Security Protocols

A *socket* is a method for establishing a communications link between a client program and a server program. The *Secure Sockets Layer (SSL)* protocol is a general-purpose solution, placed between the underlying TCP transmission protocol and the application, for implementing security as a protocol. The SSL is an industry standard protocol for securing network connections that provides authentication, data encryption, and data integrity. SSL contributes to the public key infrastructure (PKI). SSL is used to secure communications from any client or server to one or more servers or from a server to any client. You can use SSL alone or with other authentication methods, and you can configure SSL to require server authenction only, or you can require both client and server authentication.

The SSL protocol, along with the digital certificates, provide support for the following functions (VeriSign SSL 2001):

1) *Mutual Authentication* - Identities of both the server and the client can be verified so that both parties know exactly who is on the other end of the transaction.
2) *Confidentiality* - All traffic between the server and the client is encrypted based on a session key that is unique.
3) *Data Integrity* - The contents of all traffic between the server and the client are protected from being altered while en route.

If a hacker attempts to use an invalid falsified certificate, the user will receive a warning message over the Internet stating that a secure connection with this site (i.e., the hacker) cannot be verified. Both of the primary Internet browsers, the Netscape Navigator and the Microsoft Internet Explorer, incorporate the SSL protocol.

A subsequent Internet standard that follows from the SSL protocol is the Transport Layer Security (TLS) protocol. SSL has become a *de facto* standard for protecting **http** traffic and becomes **https** (for secure **http**) when traffic is encrypted. The current draft version of TLS is very similar to SSL Version 3 (SSLv3) (Stallings 2001).

TLS is an Internet Engineering Task Force (IETF) standardization initiative that has the goal of producing an Internet standard version of SSL. The charter for the TLS working group states:

"The TLS working group is a focused effort on providing security features at the transport layer, rather than general purpose security and key management mechanisms. The standard track protocol specification will provide methods for implementing privacy, authentication, and integrity above the transport layer."

This means that TLS can be used to provide security services to any application that uses TCP or the User Datagram Protocol (UDP). Both Microsoft and Netscape have agreed to include the SSL capability in their next generation browsers; however, SSL can be used independently of any Web applications.

TLS uses slightly different cryptographic algorithms for such things as the message authentication code function generation of secret keys and includes more alert codes. For full generality, SSL (or TLS) can be provided as part of the underlying protocol suite and therefore it will be transparent to applications. Or they can be embedded in specific packages.

Whenever SSL is used in a Web environment, the hypertext protocol used is denoted by **https**. By using the **https** protocol, the system is requiring the use of the SSL protocol for encrypting the Web traffic. Netscape Communicator, version 4, displays this with a closed padlock in a lower status window. Microsoft's Internet Explorer indicates it with a padlock in a lower information window.

The current work on TLS has a goal of producing an initial version as an Internet Standard. This first version of TLS can be viewed as essentially an SSLv3.1, and is very close to SSLv3 (Stallings 2001).

Three higher-layer protocols, Handshake, Change CipherSpec, and the Alert, are defined as part of SSL. These protocols are used to manage all exchanges of SSL information. Two important concepts for the SSL protocol are (Stallings 2001):

1) *Connection:* For SSL, logical client-to/from-server connections are peer-to-peer relationships. The connections are transient and are associated with one session.

2) *Session:* Sessions are created by the Handshake Protocol as associations between clients and servers and define a set of cryptographic security parameters, which can be shared among multiple connections. The Session Connection avoids having to negotiate new security parameters for each connection.

There may be multiple secure connections between any pair of parties. Several states are associated with each session. When a session is established, there is a current operating state for both read and write (i.e., receive and send). In addition, during the Handshake Protocol, pending read and write states are created. After the conclusion of a successful Handshake Protocol, the pending states become the current states. A session state is defined by the following parameters (Stallings 2001):

1) An arbitrary byte sequence, *Session Identifier*, chosen by the server to identify an active or resumable session state.

2) A *Peer Certificate* using the X509.v3 standard format. This element of the state may be null.

3) A *Compression Algorithm* used to compress data prior to encryption.

4) The *CipherSpec* for specifying the bulk data encryption algorithm (such as DES) and a hash algorithm (such as MD5 or SHA-1). The CipherSpec also defines cryptographic attributes such as the hash size.

5) A 48-byte *Master Secret* shared between the client and server.

6) A *Resumable Flag* indicating whether the session can be used to initiate new connections.

A connection state is defined by the following parameters:

1) A *byte sequence* that is chosen by the server and client for each connection.

2) A *first secret key* used in message authentication code operations on data sent by the server.

3) A *second secret key* used in message authentication code operations on data sent by the client.

4) A *first conventional encryption key* for data encrypted by the server and decrypted by the client.

5) A *second conventional encryption key* for data encrypted by the client and decrypted by the server.

6) When a block cipher in Cipher Block Chaining (CBC) mode is used, an *initialization vector* is maintained for each key.

7) Each party maintains separate *sequence numbers* for transmitted and received messages for each connection.

Another protocol, the SSL Record Protocol, provides two services for SSL connections: *confidentiality*, by encrypting application data; and *data integrity*, by using a message authentication code. The Record Protocol is called a base protocol. It can be utilized by some of the upper-layer protocols of SSL, one being the handshake protocol, which is used to exchange the encryption and authentication keys. It is vital for security reasons that this key exchange not be visible to anyone who may be watching this session.

The Change CipherSpec Protocol is one of the three SSL-specific protocols that use the SSL Record Protocol, and it is the simplest. It consists of a single message containing a single byte with the value 1. Its one and only purpose is to cause the pending state to be copied into the current state thereby updating the CipherSuite so that it can be used on this connection. This signal is used as a coordination signal. The client must send it to the server and the server must send it to the client. After each side has received it, all of the following messages are sent using the agreed-upon ciphers and keys.

Other protocols developed for TCP/IP security are the IP Security Option (IPSO), the Secure Data Network System (SDNS), and the Network Layer Security Protocol (NLSP). These three protocols are briefly explained as (R. Smith 1997):

1) *IPSO* - This protocol is an evolving series of protocols that insert sensitivity labels into IP packets. These labels operate in association with special trusted routers and other network components that prevent sensitive information from reaching untrusted entities.

2) *SDNS* - This is a family of protocols created to provide encrypted messaging at several protocol layers. It has been a commercial failure probably because of a lack of technical support and because there was little interest at the time the SDNS was first produced and marketed (1980s).

3) *NLSP* - This protocol evolved from earlier protocols but suffered from the fact that potential customers felt that there were too many protocol options.

7.2.1.1   The Alert Protocol

The Alert Protocol is used to convey SSL-related alerts to the peer entity. As with other applications that use SSL, alert messages are compressed and encrypted, as specified by the current state.

Each message in this protocol consists of two bytes. The first byte takes the value "warning" or "fatal" to convey the severity of the message. If the level is fatal, SSL immediately terminates the connection. Other connections on the same session may continue, but no new connections on this session may be established. The second byte contains a code that indicates the specific alert. An example of a fatal message is **illegal_parameter** (a field in a handshake message was out of range or inconsistent with other fields). An example of a warning message is **close_notify** (notifies the recipient that the sender will not send any more messages on this connection; each party is required to send a **close_notify** alert before closing the write side of a connection).

7.2.1.2   Handshake Protocol

The most complex part of SSL is the Handshake Protocol. This protocol allows the server and client to authenticate each other and to negotiate an encryption and message authentication code algorithm and cryptographic keys to be used to protect data sent in an SSL record. The Handshake Protocol is used before any application data is transmitted. The Handshake Protocol consists of a series of messages exchanged by the client and the server (Stallings 2001).

Figure 7.1, *Activities for the Handshake Protocol*, shows the initial exchange needed to establish a logical connection between the client and the server. The exchange can be viewed as having four phases.

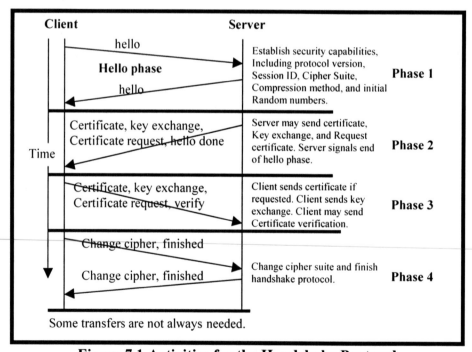

Figure 7.1 Activities for the Handshake Protocol

270

The process begins by initiating a logical connection and to establish the security capabilities that will be associated with it, which is *Phase 1*. The exchange is initiated by the client, which sends a **client_hello** message with the following parameters:

1) *Version:* The highest SSL version understood by the client.

2) *Random:* A client-generated random structure, consisting of a 32-bit timestamp and 28 bytes generated by a secure random number generator. These values serve as nonces and are used during key exchange to prevent replay attacks.

3) *Session ID:* A variable-length session identifier. A nonzero value indicates that the client wishes to update the parameters of an existing connection or create a new connection on this session. A zero value indicates that the client wishes to establish a new connection on a new session.

4) *CipherSuite:* A list that contains the combinations of cryptographic algorithms supported by the client, in decreasing order of preference. Each element of the list (each CipherSuite) defines both a key exchange algorithm and a CipherSpec; these are discussed later.

5) *Compression Method:* A list of the compression methods that the client supports.

After sending the **client_hello** message, the client waits for the **server_hello** message, which contains the same parameters as the **client_hello** message. For the **server_hello** message, the following conventions apply. The Version field contains the lower of the version suggested by the client and the highest version supported by the server. The Random field is generated by the server and is independent of the client's Random field. If the SessionID field of the client was nonzero, the same value is used by the server; otherwise the server's SessionID field contains the value for a new session. The CipherSuite field contains the single CipherSuite selected by the server from those proposed by the client. The Compression field contains the compression method selected by the server from those proposed by the client.

The first element of the CipherSuite parameter is the key exchange method (that is, the means by which the cryptographic keys for conventional encryption and message authentication code are exchanged). The following key exchange methods are supported:

Diffie-Hellman—
    Fixed: This technique requires a key exchange in which the server's certificate contains the Diffie-Hellman public parameters signed by the certificate authority. That is, the public-key certificate contains the public-key parameters. The client provides its public key parameters either in a certificate, if client authentication is required, or in a key exchange message. This method results in a fixed secret key between two peers, based on the Diffie-Hellman calculation using the fixed public keys.
    Ephemeral: This technique is used to create ephemeral (temporary, one-time) secret keys. In this case, the public keys are exchanged, and signed using the sender's private RSA or Digital Signature Standard key. The receiver can use the corresponding public key to verify the signature. Certificates are used to

authenticate the public keys. This option appears to be the most secure of the three Diffie-Hellman options because it results in a temporary, authenticated key.

Anonymous: This technique requires a base algorithm, with no authentication. That is, each side sends its public parameters to the other, with no authentication. This approach is vulnerable to man-in-the-middle attacks, in which the attacker conducts anonymous Diffie-Hellman exchanges with both parties.

RSA - The secret key is encrypted with the receiver's RSA public key. A public-key certificate for the receiver's key must be made available.

The ten steps for a handshake process are listed as follows (Stallings 2001):

1) The client sends the server the client's SSL version number, cipher settings, randomly generated data, and other information the server needs to communicate with the client using SSL.

2) The server sends the client the server's SSL version number, cipher settings, randomly generated data, and other information the client needs to communicate with the server over SSL. The server also sends its own certificate and, if the client is requesting a server resource that requires client authentication, requests the client's certificate.

3) The client uses some of the information sent by the server to authenticate the server (see Server Authentication for details). If the server cannot be authenticated, the user is warned of the problem and informed that an encrypted and authenticated connection cannot be established. If the server can be successfully authenticated, the client goes on to Step 4.

4) Using all data generated in the handshake so far, the client (with the cooperation of the server, depending on the cipher being used) creates the premaster secret for the session, encrypts it with the server's public key (obtained from the server's certificate, sent in Step 2), and sends the encrypted premaster secret to the server.

5) If the server has requested client authentication (an optional step in the handshake), the client also signs another piece of data that is unique to this handshake and known by both the client and server. In this case the client sends both the signed data and the client's own certificate to the server along with the encrypted premaster secret.

6) If the server has requested client authentication, the server attempts to authenticate the client. If the client cannot be authenticated, the session is terminated. If the client can be successfully authenticated, the server uses its private key to decrypt the premaster secret, then performs a series of steps (which the client also performs, starting from the same premaster secret) to generate the master secret.

7) Both the client and the server use the master secret to generate the session keys, which are symmetric keys used to encrypt and decrypt information exchanged during the SSL session and to verify its integrity—that is, to detect any changes in the data between the time it was sent and the time it is received over the SSL connection.

8) The client sends a message to the server informing it that future messages from the client will be encrypted with the session key. It then sends a separate (encrypted) message indicating that the client portion of the handshake is finished.

9) The server sends a message to the client informing it that future messages from the server will be encrypted with the session key. It then sends a separate (encrypted) message indicating that the server portion of the handshake is finished.

10) The SSL handshake is now complete, and the SSL session has begun. The client and the server use the session keys to encrypt and decrypt the data they send to each other and to validate its integrity.

7.2.1.3 The CipherSpec Protocol

Following the definition of a key exchange method is the CipherSpec, which indicates the encryption and hash algorithms and other related parameters.

The server begins *Phase 2* by sending its certificate, if it needs to be authenticated; the message contains one or a chain of X.509 certificates. The certificate message is required for any agreed-on key exchange method except anonymous Diffie-Hellman. Note that if fixed Diffie-Hellman is used, then this certificate message functions as the server's key exchange message because it contains the server's public Diffie-Hellman parameters.

Next, a **server_key_exchange** message may be sent, if it is required. It is not required in two instances:

1) The server has sent a certificate with fixed Diffie-Hellman parameters; or
2) An RSA key exchange is to be used.

Next, a non-anonymous server (server not using anonymous Diffie-Hellman) can request a certificate from the client. The **certificate_request** message includes two parameters: 1) **certificate_type** and 2) **certificate_ authorities**. The certificate type indicates the type of public-key algorithm. The second parameter in the **certificate_request** message is a list of the distinguished names of acceptable certificate authorities (Stallings 2001).

The final message in Phase 2, and one that is always required is the **server_done** message, which is sent by the server to indicate the end of the server hello and associated messages. After sending this message, the server waits for a client response. This message has no parameters.

Upon receipt of the **server_done** message, the client should verify that the server provided a valid certificate, if required, and check that the server hello parameters are acceptable. If all is satisfactory, the client sends one or more messages back to the server in *Phase 3*. If the server has requested a certificate, the client begins this phase by sending a certificate message. If no suitable certificate is available, the client sends a **no_certificate** alert instead.

Next is the **client_key_exchange** message, which must be sent in this phase. The content of the message depends on the type of key exchange.

Finally, in this phase, the client may send a **certificate_verify** message to provide explicit verification of a client certificate. This message is only sent following any client certificate that has signing capability (that is, all certificates except those containing fixed Diffie-Hellman parameters).

Phase 4 completes the setting up of a secure connection. The client sends a **change_cipher_spec** message and copies the pending CipherSpec into the current

CipherSpec. Note that this message is not considered part of the Handshake Protocol but is sent using the Change CipherSpec Protocol. The client then immediately sends the finished message under the new algorithms, keys, and secrets. The finished message verifies that the key exchange and authentication processes were successful.

In response to these two messages, the server sends its own **change_cipher_spec** message, transfers the pending to the current CipherSpec, and sends its finished message. At this point the handshake is complete and the client and server may begin to exchange application layer data.

After the records have been transferred, the TCP session is closed. However, since there is no direct link between TCP and SSL, the SSL state may be maintained. For further communications between the client and the server, many of the negotiated parameters are retained. This may occur if, in the case of Web traffic, the user clicks on another link that also specifies secure HTTP (HTTPs) on the same server. If the clients or servers wish to resume the transfer of records, they don't have to again negotiate encryption algorithms or totally new keys.

The SSL specifications suggest that the state information be cached for no longer than one day (24 hours). If no sessions are resumed within that time, all information is deleted and any new sessions have to go through the handshake again. The specifications also recommend that neither the client nor the server have to retain this information, and shouldn't if either of them suspects that the encryption keys have been compromised. If either the client or the server does not agree to resume the session, for any reason, then both will have to go through the full handshake protocol process.

7.2.2   The Transport Layer Security Protocol

The *Transport Layer Security (TLS)* protocol was describe earlier in Section 7.2.1z

7.2.3   Secure Communications

If a user attempts to transmit sensitive information to a site that is not secure, the browser (Netscape Navigator or Microsoft Internet Explorer) will show a warning (see Figure 7.2, *Security Warning* (VeriSign Server 2001). If the site has a valid digital certificate and an SSL connection, then this warning message is not shown and the user can assume that the transmission has confidentiality, data integrity, identification, and mutual authentication (of clients and servers).

This capability to provide a secure link for users no matter where they are located on the Internet provides the protection that users trading over the Web wish to have. Conducting business online requires that transactions are secure so that any sensitive information, such as a credit card number, remains private.

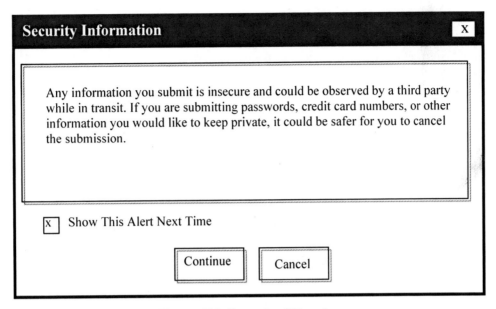

Figure 7.2 Security Warning

If a site has a certificate issued by an untrusted certification authority, then the browser will display a warning such as shown in Figure 7.3, *Certificate Warning*. If a site is falsifying its claim to a certificate, then a warning message similar to the one shown in Figure 7.4, *False Certificate*, will appear (VeriSign Server 2001).

7.2.4 Secure Shell

The *Secure Shell (SSH)* is a program that can be used to securely logon to a remote machine, to execute commands on that machine, and to move files from one machine to another (Oppliger 1998). It is meant to act as a complete replacement of the Berkeley tools such as **rlogin, rsh, rcp,** and **rdist**. In many cases it can replace telnet. The Web home page for the SSH program is http://www.ssh.fi/.

www.widget.com is a site that uses encryption to protect transmitted information. However, Netscape does not recognize the authority who signed its Certificate.

Figure 7.3 Certificate Warning

A secure connection with this site cannot be verified. Would you still like to proceed?

The certificate you are viewing does not match the name of the site you are trying to view.

| Yes | No | View Certificate | More Info |

Figure 7.4 False Certificate

There are currently several products developed under the SSH Communications Security name for security software, namely:

1) *SSH Secure Shell Software* is a *de-facto* standard for encrypted terminal connections on the Internet.

2) *SSH Complete VPN* is a totally private networking solution for making the Internet Secure.

3) *SSH Sentinel* is a revolutionary new IPSEC client that takes full advantage of certificate-based authentication technologies and the emerging global PKI.

4) *SSH Certifier* offers a cost efficient way to issue and manage certificates from a central location.

5) *SSH IPSEC Express* is a full, portable implementation of the IPSEC protocol intended for original equipment manufacturers (OEMs) and system integrators who want to include IPSEC functionality into their own products.

6) *SSH NAT (Network Address Translation) Traversal* toolkit is a revolutionary solution for developers creating secure IPSEC applications. It enables true end-to-end VPN security in the presence of NAT devices.

7) *SSH Internet Key Exchange (IKE),* also known as ISAKMP/Oakley, is a toolkit for adding key management functionality to an IPSEC product. Delivered as an easy-to-use toolkit, the SSH IKE can be used as a part of the SSH IPSEC Express solutions, or as a separate key management solution for third party IPSEC implementations.

8) *SSH Certificate Toolkit* offers a complete set of tools required for using X.509 certificates in security applications. The library includes all the necessary components, such as X.509 certificate handling, Certification Revocation List (CRL) processing, a Lightweight Directory Access Protocol (LDAP) client, and the required cryptography.

The SSH protocol can protect against IP spoofing, IP source routing, Domain Name Service (DNS) spoofing, and interception of plaintext passwords and other data by intermediate hosts. Several vendors offer SSH products (Canavan 2001).

7.2.5 Remote Procedure Protocols

The *Remote Access Dial-In User Service (RADIUS)* allows remote access devices to share the same authentication database. RADIUS provides a central point of management for all remote network access by validating the user's credentials through a challenge and response process. The RADIUS capability is provided by a number of vendors (Brenton 1999). The *Terminal Access Control Access Control System (TACACS)* is a similar system to RADIUS but is produced by a different vendor. Both of these products offer a user the capability to identify and authenticate remote clients through software and servers housed at their sites. There are many alternative vendors for these products.

The *Point-to-Point Tunneling Protocol (PPTP)* is sometimes used for authentication but has been shown to be quite easy to crack and hence is of little use as a secure protocol. It is sometimes called the kindergarten crypto. If this is the security protocol your system is using, then you might wish to change to another protocol.

The SSH is a powerful method for performing client authentication and safeguarding multiple service sessions between two systems. When two systems that are using SSH establish a connection they validate one another by performing digital certificate exchanges. Brute force and playback attacks are ineffective against SSH because encryption keys are periodically changed. For the SSH protocol, 3DES is used for message encryption (Brenton 1999).

7.2.6 Internet Protocol Security

The *Internet Protocol Security (IPSEC)* is a public/private key encryption algorithm (i.e., asymmetric encryption) that is a Diffie-Hellman exchange to perform authentication and establish session keys at the IP level. IPSEC is a set of general-purpose protocols for protecting messages transmitted over the TCP/IP communications links (i.e., the Internet) (R. Smith 1997). Use of IPSEC is transparent to the end user. IPSEC is an obvious choice for a VPN. All data that is not needed for routing the message is encrypted by IPSEC.

IPSEC protects IP packets from snooping or modification. It is implemented between the Internet (level 3) and Transport levels (level 4) of the OSI model. An IPSEC-based VPN is composed of two modes.

1) An *Authentication Header (AH)* provides integrity-checking information (cryptographic checksum) so that one can detect if the packet's contents were forged or modified while en route over the Internet.

2) The *Encapsulating Security Payload (ESP)* encrypts the data contents of the remainder of the packet so that the contents cannot be extracted while en route over the Internet. The encryption information is selected using the *security parameter index (SPI)*.

277

An AH can defeat spoofing and provides data integrity. ESP provides packet confidentiality. Together they provide both data integrity and confidentiality. Newer versions of the Simple Network Management Protocol (SNMP) contain *time stamps* to detect and block replay attempts.

IPSEC contains a key management process that offers four standard approaches (R. Smith 1997):

1) *Manual keying* - This is a manual method of configuring security associations.

2) *Simple Key Interchange Protocol (SKIP)* - This is a program for negotiating and exchanging session keys between IPSEC hosts.

3) *Internet Security Association and Key Management Protocol (ISAKMP)* - This is a broad, general-purpose protocol intended for managing security associations and for handling key exchanges. It uses the Diffie-Hellman algorithm for key exchange.

4) *Photuris* - This alternative to ISAKMP also establishes security associations and exchanges keys using the Diffie-Hellman algorithm.

7.2.7 Open Systems Interconnection Model

The *Open Systems Interconnection (OSI) Model* is shown in Table 7.2, *Open Systems Interconnection Model*, below. The ISO[62] X.800 is the standard for the OSI Security Architecture, which outlines measures that can be used to secure data in a communicating open system by providing appropriate security service in each layer. X.800 defines a digital signature as a relatively low-level cryptographic mechanism. However, the term "digital signature" usually means an extension of that mechanism to enable use and/or reliance upon certificate validation during signature verification.

The *OSI model* represents a description of several protocols that provide a well-defined set of communications processes to simplify the transmission of messages over a network. A *protocol* is a set of rules that describe how information is exchanged over a communications network. A protocol provides both the message format and the sequences of the messages that are transmitted from a sender to a receiver and establishes how errors should be handled. A protocol suite is a collection of protocols and the OSI is an example of a protocol suite.

Table 7.2 Open Systems Interconnection Model (after Brenton 1999)

Layer	Data Types & Functions	Equipment	Explanation (Form of Communication)
Application (7)	Application protocols and programs (SW)*	Client PCs	Manages program requests that require access to services provided by a remote system. (Messages)

---

[62] ISO is the International Organization for Standards.

Layer	Data Types & Functions	Equipment	Explanation (Form of Communication)
Presentation (6)	Translation (SW)	Client PCs	Translates data formats of sender to data format of receiver. Also performs encryption. Provides data compression, translation, and encryption. (Messages)
Session (5)	Connection (SW)	Gateways	Connection negotiation, establishes and maintains connection, and synchronizes dialog. (Messages)
Transport (4)	Network protocols (SW) *Transport Control Protocol (TCP)*	Routers	Assures end-to-end reliability. Translates and manages message communications through network. Ensures data integrity and deals with packet sequencing. (Segments)
Network (3)	Network routing (SW) *Internet Protocol (IP)*	Routers	Defines network segmentation and network address scheme. Connectivity over multiple network segments. Cornerstone on which all upper layers are based. (Packets)
Data-Link (2)	Network interface cards (HW/SW)*	Bridges and Switches	Creates packet headers and checksum trailers. Packages datagrams into frames. Detects errors. Regulates data flow. Maps hardware addresses. (Frames)
Physical (1)	Cable and connectors (HW)*	Repeaters	Defines physical and electrical specifications for transmission. Defines connector types and pin-outs, voltage, and current. (Bits)

*SW means that the OSI model is software implemented. HW/SW means that the OSI model is either hardware or software implemented and HW means that the OSI model is hardware implemented.

7.2.8 Simple Network Management Protocol

The *Simple Network Management Protocol (SNMP)* is used to monitor and control network devices. The controlling station is called the "SNMP management station" and the controlled device is referred to as an "SNMP agent." The agent can be a manageable hub, router, or server. Both static and dynamic data are used when reporting to the management station. *Static information* is data stored within the device in order to identify it uniquely and *dynamic information* is data that pertains to the current state of the device.

The SNMP can be used to collect statistics as well as make configuration changes to a router. SNMP has poor authentication and has a major flaw because it transmits in the clear; hence it is an insecure and dangerous protocol. It has no place on a security device. You should avoid using SNMP on your routers, if possible. Unsecured SNMP access can be used to monitor and even change settings on the router remotely.

7.2.9   Lightweight Directory Access Protocol

The *Lightweight Directory Access Protocol (LDAP)* was developed at the University of Michigan as a simpler and user-friendlier version of the IEEE protocol standard for directory access (X.500). It has been quite successful and has become a *de facto* standard.

The LDAP represents the emerging solution for security information for providing an Internet-ready, lightweight implementation of the ISO's X.500 standard for directory services. A major feature of LDAP is that it requires a minimal amount of networking software on the client side, making it particularly attractive for Internet-based thin client applications. LDAP databases are hierarchical.

The LDAP Data Interchange Format (LDIF), as defined by the Internet Engineering Steering Group (IESG), is a widely used file format that describes directory information or modification operations that can be performed on a directory. LDIF is completely independent of the storage format used within any specific directory implementation and is therefore implementation-neutral. Typically, it is used to export directory information from, and import data to, LDAP servers.

7.2.10 LAN Alternatives

There are three products for providing a *local area network (LAN),* namely Ethernets, Token rings, and Fiber Distributed Data Interface (FDDI). By far, the most popular is the Ethernet.

7.2.10.1   Ethernet

Ethernet is the most commonly used LAN protocol (Canavan 2001). With this protocol, any device on a network segment can monitor communications with any other device on the same network segment. Thus, it is often recommended that an organization should segment its information system networks. The segmentation process can separate a large network into several smaller networks. One way to accomplish this is to group associated users together on a hub[63] or similar network device. The hub can act as a channel for traffic traveling from one device to another so that when a packet arrives at one port it is then copied to each of the other ports meaning that all segments of the LAN can view all packets on the LAN.

Segmentation provides both performance and security advantages. It prevents packets from traversing the entire network. The performance advantage is due to the fact that the packets stay within a segment and do not traverse the entire network. This means that traffic on the entire network is reduced and the physical distance that a packet must travel is reduced. The security advantage emanates from the fact that if someone should place a sniffer on the network, it can only sniff the traffic on the particular segment to which it is attached. Therefore, an attacker would need multiple sniffers to find what the entire network is doing.

---

[63] A hub is a network device with multiple ports into which other network devices can be plugged.

An alternative to standard hubs is Ethernet switches, called *switching hubs*. These hubs can be used for a switched Ethernet. A switched Ethernet can provide virtual dedicated connections between devices. This dedicated connection restricts who can see the traffic. It also improves the network throughput because packets are only forwarded to the required port and not to all ports. The standard Ethernet provides a 10Mbps speed, the fast Ethernet provides 100Mbps, and the Gigabit Ethernet provides 1Gbps bandwidth. Switched Ethernets are somewhat more expensive than the traditional Ethernet hub.

Ethernet has become the topology of choice for most networking environments. When used with twisted-pair cabling, the topology can be very resistant to failure due to cabling problems on any single element. However, a single system is capable of gobbling up all of the available bandwidth.

Ethernet's communication rules are called carrier sense multiple access and collision detection (CSMA/CD):

1) *Carrier sense* means that all Ethernet stations are required to listen to the wire at all times (even when transmitting). Multiple access means that more than two stations can be connected to the same network and all stations are allowed to transmit whenever the network is free.

2) *Multiple access* means that more than two stations can be connected to the same network and that all stations are allowed to transmit whenever the network is free.

3) *Collision detection* resolves the issue of a simultaneous transmission by two or more stations.

It is possible to configure a system to read all information that it receives (*promiscous mode*). This capability is the biggest security flaw with the Ethernet, namley that all systems on an Ethernet are operating in promiscuous mode.

7.2.10.2   Token Rings

The *Token Ring* was designed to be fault tolerant and is a wonderful topology when all systems operate as intended. Since Token Ring requires that each system successively pass a token to the next, a single network interface card set to the wrong speed can bring down the entire ring. Token Ring switches are less popular and far more expensive than their Ethernet counterparts. In an Ethernet network, if a single system were set to the wrong speed, only that one system would be affected. Also, Token Rings require that all media access control numbers be unique.

7.2.10.3   Fiber Distributed Data Interface

The *Fiber Distributed Data Interface (FDDI)* is a commercial Local Area Network (LAN) that is compatible with the ISO 9314 standard. The FDDI, or fiber optical cabling, can provide communications support for up to 1000 nodes with a maximum distance between nodes of two kilometers and a maximum network length of 200 kilometers. The FDDI medium is optical fiber with a signaling rate of 100Mbps. The FDDI network uses an orderly ring structure of two counter-rotating rings so it is fault-tolerant. If either ring is broken, the remaining ring can be used instead. If both rings should break, then the two rings are joined to form a single ring bypassing the break.

An important security benefit for fiber optic cabling is that it is not susceptible to electromagnetic interference (EMI), which can be monitored to ascertain the information traveling down the link. Wire cabling is susceptible to EMI.

The FDDI ring uses a timed token for bandwidth allocation to each station as they join the ring. It allows for variable frame sizes as large as 20,000 bytes although much smaller frame sizes are more typical. Because of the large frame size potential and the timed token, the FDDI ring can achieve very high efficiency, nearly the highest theoretical bandwidth.

An important feature is that it has a distributed architecture. All FDDI algorithms are distributed meaning that control of the rings is not centralized. If any component should fail, other components can reorganize and continue to function, including fault recovery, token initialization, and configuration control.

The FDDI is also a ring topology but a second ring has been added to rectify many of the problems found in Token Rings. FDDI is considered to be a dying technology since no effort has been made to increase speeds beyond 100Mbps. FDDI can be run in a full duplex mode, which allows both rings to be active at all times. However, the redundancy of the second ring is not in force in this case. A FDDI network can have either a ring[64] or star[65] topology.

7.2.11 Virtual LANs and Switches

Virtual LANs (VLANs) are based on switches. VLANs provide for alternative configurations of LAN elements (i.e., clients and servers) using switches so that there can be alternative configurations of these elements that compose different local area networks, even though they may not be directly connected. A sample VLAN is shown in Figure 7.5, *Example of a VLAN*.

---

[64] A *ring topology* is one in which each host is connected to contiguous hosts (two) around the periphery of a ring and if one host should go down, the entire ring goes down.
[65] A *star topology* is one in which each peripheral host is connected to a center host and must go through the center host in order to communicate with any other peripheral host.

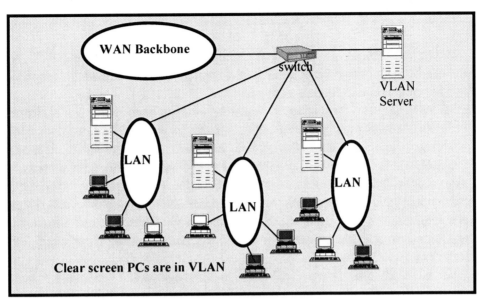

Figure 7.5 Example of a VLAN

The basic networking devices are:

1) *Repeaters* - These devices are simple two-port signal amplifiers used in a bus topology[66] to extend the maximum distance that can be spanned on a cable run.

2) *Hubs* - These devices are boxes of varying sizes that have multiple female connectors designed to accept one twisted-pair cable outfitted with a male connector for connecting a workstation or other device to the hub.

3) *Bridges* - These devices are small boxes with two network connectors that attach to two separate portions of the network. It is nearly identical to a repeater but it looks at frames of data.

4) *Switches* - These devices are the marriage of hub and bridge technology. A switch functions as though it has a little miniature bridge built into each port. A switch will keep track of the media access control (MAC) addresses attached to each of its ports and route traffic destined for a certain address only to the port to which it is attached.

5) *Routers* - These are multiport devices that decide how to handle the contents of a frame, based on protocol and network information.

Switch clustering provides a capability for system administrators to quickly expand and upgrade their networks across multiple wiring closets and various LAN media without having to add resources or replace existing switching equipment (Cisco Switch 2001).

Without switch clustering, the administration and configuration of LANs required someone to manually configure, monitor, and upgrade each LAN switch, one at a time. In

---

[66] A *bus topology* is the linear LAN used by Ethernet networks.

dynamic environments where high availability is critical to organizational success, switch clustering enables system administrators to quickly provide scalability and network management.

Switch clustering, although not a specific enhancement of security capability (except in the area of availability), does enable security and system administrators to devote more time to security rather than more mundane activities.

Enterprise switches are sophisticated multi-service devices designed to form the core backbones of large WANs thus complementing the duties of high-end multi-protocol routers that are commonly in use today. Enterprise switches are capable of supporting LAN switching, packet WAN interfaces such as frame relay, ATM-connected servers, and ATM routers. Enterprise switches can also serve as the single point of integration for all of the various services and technologies employed in modern enterprise backbones. By merging all of these services onto a common platform and ATM transport infrastructure, modern network planners can achieve greater manageability and eliminate the need for complex overlay networks (Alles 2000).

Enterprise switches are often deployed in homogeneous[67] networks because this maximizes the benefits of their service integration capabilities (Alles 2000). In heterogeneous networks (such as campus and workgroup settings) support for standard devices is much more critical for enterprise switches.

7.2.12   Cabling

Cabling can be provided through copper wire or fiber optics. Most information system network owners are interested in obtaining best-of-class, end-to-end cable networking solutions. These solutions should:

1) Provide links to deliver and receive information,
2) Conduct communications, and
3) Provide entertainment through data, voice, and video (Cisco Cable 2001).

Cable networking solutions should offer scalability, security, flexibility, and manageability, backed up by vendor programs, services, and support capabilities designed to assist users with the transition to emerging technology cable networks that can support the needs of modern information systems.

Emerging new technologies and the dynamics of the regulatory environment give cable companies new opportunities to offer data and voice-over-cable services. Along with these new markets comes the challenge of managing the cable infrastructure that delivers these services. The vendors of cabling products need tools for deploying, maintaining, troubleshooting, and monitoring the equipment in the cable network.

Some cable manager systems provide fault tolerance, performance, and configuration management for the implemented cabling. The emergence of new technologies and changes in the regulatory environment give cable companies new opportunities to offer data and voice-over-cable services. Along with these new markets comes the challenge of managing

---

[67] *Homogeneous* networks use the same vendor for their network devices. *Heterogeneous* networks use different vendors for their network devices.

the cable infrastructure that delivers those services. Cable providers need tools to deploy, maintain, troubleshoot, and monitor equipment in the cable network.

A cable manager system should provide a graphical, client/server application for monitoring faults, performance, and configuration of the cabling products. The cable manager system should be a part of the end-to-end solution for managing cable and multi-service networks.

Some vendors implement a security feature in their *twisted pair hub technology* that prevents packets from traveling everywhere within a shared media network. The approach can control the threat of aunauthorized physical sniffers and packet capture programs. While twisted-pair cabling has become very popular due to its low cost, twisted-pair cabling is also extremely insecure due to its *Electromagnetic Interference* (EMI) radiation.

EMI is produced by circuits that use an alternating signal, like analog or digital communications (referred to as an alternating current or AC circuit). EMI is not produced in DC circuits. *Fiber Optic Cabling* consists of a cylindrical glass thread center core 62.5 microns in diameter wrapped in cladding that protects the central core and reflects the light back into the glass conductor and is encapsulated in a jacket of tough KEVLAR fiber and sheathed in PVC or Plenum with a total diameter of 125 microns and is often referred to as 62.5/125 cabling. Fiber optic cabling does not produce EMI and hence is much more secure.

The light in a fiber optic cable is produced by a *light emitting diode (LED)* and another diode also receives the signal. Light transmission can take two forms:

1) *Single-mode transmissions* consist of an LED that produces a single frequency of light that is faster and can travel long distances but is expensive and installation can be difficult.
2) *Multimode transmissions* consist of multiple light frequencies and are less expensive than single mode but are subject to light dispersion (the tendency of light rays to spread out as they travel).

Twisted-pair and fiber optic cabling are *bounded media*, whereas the atmosphere is an *unbounded medium*. Unbounded media have a host of security problems. *Light transmissions* through the atmosphere use lasers to transmit and receive network signals but because of precise alignment and a focused signal, these transmissions are relatively secure. However, *radio waves* broadcast using microwave signals that can be more easily intercepted.

Fixed frequency signals, such as radio station signals, use a single frequency as a carrier wave and can easily be monitored once an attacker knows the carrier frequency. *Spread spectrum signals* use multiple frequencies for transmission and are more difficult to monitor. Most transmissions rely on encryption to scramble the signal so that it cannot be monitored by outside parties. *Terrestrial transmissions* are completely land-based radio signals and are usually line-of-sight systems and can only be received locally. *Space-based transmissions* are signals that originate from a land-based system but are then bounced off one or more satellites that orbit the earth above the atmosphere. The greatest benefit of these systems is the range. But the larger the range, the more likely the signal will be monitored.

7.2.13   Portable or Mobile Systems

One way to avoid the issue of attempting to decide which wired network to acquire is to go to a scalable wireless network based on *portable or mobile systems*. The portable or mobile personal digital assistants (PDAs) should have a security capability such as that provided by the Pointset 4.0 package. This package has been tested and approved to meet the FPS 140-1 or 140-2 requirements. All mobile systems should meet the FPS 140-1 requirements. Wireless LANs are alternatives to wired LANs.

7.2.14   Wireless local area networks

Wireless local area networks (WLANs) are flexible data communications systems implemented as an extension to, or as an alternative for, a wired LAN. Using radio frequency (RF) technology, WLANs transmit and receive data over the air, minimizing the need for wired connections. Thus, WLANs combine data connectivity with user mobility (Proxim 2001).

WLANs offer the following advantages over traditional wired networks:

1) *Mobility* - WLANs can provide LAN users with access to real-time information anywhere in their organization while supporting productivity and service opportunities not possible with wired networks.
2) *Installation Speed and Simplicity* - Installing a WLAN can be fast and easy and can eliminate the need to pull cable through walls and ceilings.
3) *Installation Flexibility* - WLAN technology allows the network to go where wire cannot go.
4) *Reduced Cost-of-Ownership* - Even though the initial investment required for WLAN hardware can be higher than the cost of wired LAN hardware, overall installation expenses and life-cycle costs can be significantly lower.
5) *Scalability* - WLANs can be configured in a variety of topologies to meet the needs of specific applications and installations. Configurations are easily changed and range from peer-to-peer networks suitable for a small number of users to full infrastructure networks supporting thousands of users so that they (users) may roam over a broad area.

Manufacturers of WLANs have a range of technologies to choose from when designing a WLAN solution, namely narrowband, spread spectrum, frequency-hoping spread spectrum, direct-sequence spread spectrum, and infrared. Each of these has certain advantages and limitations (Proxim 2001).

A *narrowband radio system* transmits and receives user information on a specific radio frequency at a nominal rate of 1.6 Mbps. Narrowband radio keeps the radio signal frequency as narrow as possible just to pass the information. Undesirable cross talk between communications channels is avoided by carefully coordinating different users on different channel frequencies. A drawback of narrowband technology is that the end-user must obtain an FCC license for each site where it is employed.

Most WLAN systems use *spread-spectrum technology*, a wideband radio frequency technique developed by the military for use in reliable, secure, mission-critical communications systems. Spread-spectrum is designed to trade off bandwidth efficiency for reliability, integrity, and security. In other words, more bandwidth is consumed than in the case of narrowband transmission, but the tradeoff produces a signal that is, in effect, louder and thus easier to detect, provided that the receiver knows the parameters of the spread-spectrum signal being broadcast. If a receiver is not tuned to the right frequency, a spread-spectrum signal looks like background noise. There are two types of spread spectrum radio: frequency hopping and direct sequence.

Frequency-hopping spread-spectrum (FHSS) uses a narrowband carrier that changes frequency in a pattern known to both transmitter and receiver. Properly synchronized, the net effect is to maintain a single logical channel. To an unintended receiver, FHSS appears to be short-duration impulse noise.

Direct-sequence spread-spectrum (DSSS) generates a redundant bit pattern for each bit to be transmitted. This bit pattern is called a *chip* (or chipping code). The longer the chip, the greater the probability that the original data can be recovered (and, of course, the more bandwidth required). Even if one or more bits in the chip are damaged during transmission, statistical techniques embedded in the radio can recover the original data without the need for retransmission. To an unintended receiver, DSSS appears as low-power wideband noise and is rejected (ignored) by most narrowband receivers.

Another technology is *infrared*. Infrared (IR) systems use very high frequencies, just below visible light, to carry data. Like light, IR cannot penetrate opaque objects. It is either directed (line-of-sight) or diffuse technology. Inexpensive directed systems provide very limited range (three feet or less) and typically are used for personal area networks but occasionally are used in specific WLAN applications. High-performance directed IR is impractical for mobile users and is therefore used only to implement fixed sub-networks. Diffuse (or reflective) IR WLAN systems do not require line-of-sight, but cells are limited to individual rooms (Proxim 2001).

Because wireless technology began as a military application, security has been a design criterion for wireless devices. Security provisions are typically built into WLANs, making them more secure than most wired LANs. It is very difficult for unintended receivers (eavesdroppers) to listen in on WLAN traffic. Complex encryption techniques make it impossible for all but the most sophisticated to gain unauthorized access to network traffic. In general, individual nodes must be security-enabled before they are allowed to participate in network traffic.

WLANs from different vendors may not be interoperable because: 1) they may be based on different technologies, 2) they may be using different frequency bands, or 3) the implementations may be different.

7.2.15   Wired Equivalent Privacy

The 802.11 standard describes the communication that occurs in wireless LANs. A part of the 802.11 standard is the *Wired Equivalent Privacy (WEP)* algorithm, which is used to protect wireless communications from eavesdropping. A secondary function of WEP is to prevent unauthorized access to a wireless network. Although the latter function is not an explicit goal in the 802.11 standard, it is frequently considered to be a feature of WEP. Several flaws in the WEP algorithm, which seriously undermine the security claims of the system, have been discovered. Namely, WEP is susceptible to (Borisov et al. 2001):

1) Passive attacks to decrypt traffic based on a statistical analysis.
2) Active attacks that inject new traffic from unauthorized mobile stations based on known plaintext.
3) Active attacks to decrypt traffic based on tricking the access point.
4) Dictionary-building attacks that, after analysis of about a day's worth of traffic, allows real-time automated decryption of all traffic.

Analysis suggests that all of these attacks are practical to mount using only inexpensive off-the-shelf equipment. It is recommended that anyone using an 802.11 wireless network not rely on WEP for security, and employ other security measures to protect their wireless network, such as VPNs.

7.2.16 New Mechanisms for Countering DDoS Attacks

Practical solutions for stopping DDoS attacks are new and infrequent. Often the best that a security administrator can achieve is to minimize the down time from such an attack. However, a device that will ensure a minimum downtime is the "Gap Appliance." This new countermeasure consists of a single device that comprises three subsystems: 1) an external CPU, 2) an internal CPU, and 3) a selective switching mechanism (Carmeli 2001).
The device works as follows:

1) The external CPU downloads traffic from the Internet (via the router and firewall). This device physically terminates the entire traffic flow.
2) The traffic is then routed to the internal CPU via a selective switched mechanism.
3) Using a set of rules for passing and blocking traffic, the internal CPU either passes traffic, if it is determined to be non-malicious, or blocks traffic, if it is determined to be malicious.

In case of a DoS attack, the gap appliance essentially sacrifices itself, on the untrusted side (the external CPU). Although severed from the outside world for a few minutes, the servers never go down and the system is able to recover from the DoS attack by simply rebooting the gap appliance, which should take only a few minutes. If after rebooting, the DoS attack is continuing, then put in a set of rules (in the internal CPU) to block all traffic from the source or sources of the attack. The gap appliance is available from Spearhead Security Technologies (http://www.spearhead.com/home.html).
A software mechanism for stopping DDoS attacks is the FloodGuard package from Reactive Network Solutions. This product offers protection against attacks that use spoofing, against routers, Domain Name Service machines, and other sensitive network devices. FloodGuard is composed of both detectors and actuators that can identify unusual traffic patterns (even encrypted malicious messages) and block them if they appear to be malicious, such as SYN flood DDoS attacks. Unlike other anti-DDoS solutions, FloodGuard does not rely on a database of known attack signatures (Fisher 2001).

Many attacks can be handled by modern routers, firewalls, and anti-virus and anti-DDoS products, such as the following attacks (the definitions of these attacks were obtained by searching the Internet with the www.google.com search engine):

IP spoofing - This is a technique used to gain unauthorized access to computers, where the intruder sends messages to a computer with an IP address indicating that the message is coming from a trusted host. To engage in IP spoofing, a hacker must first use a variety of techniques to find an IP address of a trusted host and then modify the packet headers so that it appears that the packets are coming from that host. Newer routers and firewall arrangements can offer protection against IP spoofing. It should be noted that IP spoofing is not the actual attack but a step in the attack. The attack is actually exploitation of a trust-relationship.

Land attack - This attack consists of sending a packet to a machine with the source host/port the same as the destination host/port. This attack causes a lot of systems to crash. Vulnerable systems include Windows 95 and Windows NT 4.0.

Ping of death - Many computer systems will crash if they are sent IP packets if their packet size exceeds the maximum length (65,535 bytes) allowed by the software.

IP with zero length key - Data integrity and authentication for IP datagrams can be provided by the IPSEC Authentication Header. A message digest for the transmitted message is produced using the MD5 algorithm using the AH key. When combined with the AH key, authentication data is produced. This value is placed in the Authentication Data field of the AH. This value is also the basis for the data integrity service offered by the AH protocol. The AH key is used as a shared secret between two communicating parties. The AH key (shared secret) is hashed with the transmitted data ensuring that an intervening party cannot duplicate the authentication data. Keys either need to be chosen at random or generated using a cryptographically strong pseudo-random number generator based on a random seed. A *key length of zero* is prohibited and implementations must prevent key lengths of zero from being used with this transform, since authentications with a zero-length key are ineffective.

Smurf attack - Denial-of-service attacks involving forged ICMP echo request packets (commonly known as "ping packets") sent to IP broadcast addresses can result in large amounts of ICMP echo reply packets being sent from an intermediary site to a victim. These attacks can cause network congestion or outages and are referred to as "smurf attacks" because the name of one of the exploit programmers who executed this attack is called "smurf."

User datagram protocol (UDP) port loopback - A loopback interface is one that bypasses unnecessary communications functions when the information is sent to an addressed entity within the same system. UDP is an unreliable packet-based protocol that does not provide a connection-oriented protocol. Instead each packet of data has to be individually addressed and the user is responsible for handling lost packets (corrupted packets are detected by the IP layer and discarded). This is useful where a machine must talk to multiple machines and where it does not want the overhead of a connection-oriented protocol like TCP. The loopback interface is present on all machines and always has address 127.0.0.1.

Snork attack - This is a denial-of-service attack where an attacker with minimal resources can cause a remote Windows NT system to consume 100 percent of its CPU Usage for an indefinite period of time. Also, the snork attack allows the attacker to utilize a large amount of bandwidth on a remote NT network by inducing vulnerable systems to engage in a continuous bounce of packets between all combinations of systems. This type of attack will affect most systems that use the Windows NT OS.

TCP null scan - Attackers can use the TCP Null Scan to identify listening TCP ports. This scan uses a series of strangely configured TCP packets, which contain a sequence number of 0 and no flags. This type of scan can get through some firewalls and boundary routers that contain rules for filtering incoming TCP packets with standard flag settings. If the target device's TCP port is closed, the target device sends a TCP RST packet in reply. If the target device's TCP port is open, the target discards the TCP NULL scan, sending no reply.

TCP SYN flooding - Another denial-of-service attack is the TCP SYN Flooding attack, which consists of a tool that only implements one portion of the Sequence Number Guessing attack, with a completely different focus. TCP SYN Flooding causes servers to quit responding to requests to open new connections with clients causing a denial of service. The TCP SYN Flooding attack takes advantage of the way the TCP protocol establishes a new connection. Each time a client, such as a browser, attempts to open a connection with a server some information is stored on the server. Because the stored information takes up memory and operating system resources, only a limited number of in-progress connections are allowed, typically less than ten. When the server receives an acknowledgement from the client, the server considers the connection open, and the queue resources are released to accept another connection. The attacking software generates spoofed packets that appear to be valid connections to the server. These spoofed packets enter the queue, but the connection is never completed thereby leaving these new connections in the queue until they time out. Only a few of these packets need to be sent to the server, making this attack simple to carry out even using a slow, dial-up (like PPP or SLIP) connection from the attacker's computer. The system under attack quits responding to new connections until some time after the attack stops. The newer routers and firewalls can counter this type of attack.

Sequence number guessing - Sequence number guessing occurs when an intruder/client attempts to guess the SYN/ACK sequence number of a packet and impersonates a trusted host. This type of attack can be defeated using TCP Wrappers.

TCP wrappers - TCP Wrappers act much like a soldier at a checkpoint, verifying a host's clearance prior to allowing entry by using the client/server relationship necessary for most TCP/IP applications. It inserts itself into the middle of the relationship and acts as the server until the client/host is authenticated and then uses its access control feature to authenticate hosts. It does all of this with no overhead to the system. TCP Wrappers is located in one of the files at the Web URL ftp.porcupine.org/pub/security/.

7.3   List of Countermeasure Alternatives

A list of the available countermeasures was contained in Table 6.12, *Threats and Countermeasures,* in Chapter 6. The process by which they are identified is explained below.

7.3.1   Selection Of The Preferred Countermeasures

One can perform either an informal or a formal analysis to select the preferred countermeasures. An *informal analysis* can be based on the feelings that the decision-maker has for the various alternatives and then select the preferred one on the basis of which option he or she likes best. This method relies on the expertise and knowledgeable judgment of the analyst.

A formal analysis is explained below. An informal analysis can be performed by the owner or named user simply selecting the desired alternative based on intuitive feelings for the attributes of the countermeasure without actually even defining what the attributes are.

A *formal method* for selecting the preferred countermeasures is to use a multi-attribute utility analysis (MAUA). The MAUA is based on a selected set of attributes that are the same for a given countermeasure, the weights of these attributes, and the values of these attributes for each specific alternative.

Suppose the countermeasure is firewalls. Then there should be a set of attributes (i.e., performance metrics) that can be used to describe and rank order all firewalls. For example, the attributes might be the following: 1) cost, 2) speed, 3) throughput, 4) ease of use, 5) ease of learning, and 6) all other performance parameters that the analyst or owner considers to be important.

The owner or user (hence the preferer) would estimate the weight of each attribute. These weights should sum to one and if they do not they will be normalized so that they do sum to one.

Then, for each firewall, the owner or user would estimate the attribute's value of the firewall in meeting their organization's needs, for each of the attributes. Values should be in the interval [0,1] and can be determined through the use of a strength of preference on the part of the assessor. For example, if a specific attribute for a specific alternative is felt to be of low value, then the assessor may give the attribute a 0.1 value, whereas if the alternative's attribute value is felt to be very high, then the value may be assigned a 0.9 value. Some users may employ a curve like the one shown in Figure 7.6, *Example of a Sigmoid Curve,* for the values of the attributes. This type of curve is often called a "Sigmoid." Using attribute values in the range [0,1] means that the Utility of each alternative is also in the range [0,1], where a larger Utility value means that the alternative is more highly preferred than one with a smaller Utility value.

The various risks for this system can be determined using the probabilities of attack given the threats and assumptions about the potential losses should an attack be successful.

Often, the risk assessment is based on the expectation of some threat occurring during a yearly basis (i.e., annual loss expectancy (ALE)[68]). A utility value for each type of

---

[68] A loss expectancy is actually an annual probabilistic cost, not just the annual probability, however, for this analysis an annual probability is used.

countermeasure can be determined using the following Multi-Attribute Utility Analysis formula (C. Smith 1998):

$$\text{Countermeasure}_i \text{ (utility)} = \Sigma_j \, w_j * v \, (\text{att}_j);$$

where w_j is the weight for the jth attribute, $v \, (\text{att}_j)_i$ is the value of the jth attribute for the ith countermeasure, $w_j \geq 0$, and $\Sigma_j \, w_j = 1$.

It has been shown that the linear form (Dawes 1979) for finding a utility value is of benefit even when the user does not ensure that the attributes are preferentially independent[69], which they should be for a mathematically correct solution. The utility values of the various countermeasures allow the analyst to rank order them so that the one that ranks the highest (i.e., greatest utility value) can be assumed to be the most preferred.

The weights and the attribute values should come from the users so that the final utility values (for all the alternatives) represent the prioritized preferred set of countermeasures when ordered largest to smallest. For the ith countermeasure, if more than one alternative ties for the greatest value, then the user can select one of them on whatever basis is considered, even a whim.

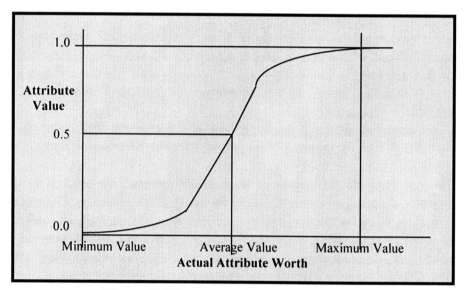

Figure 7.6 Example of a Sigmoid Curve

For each type of countermeasure, rank order the alternatives and select the one with the greatest utility value. Do this for all of the different countermeasures and for each countermeasure select the alternative with the greatest utility value. Then compare the total cost of the countermeasure (acquisition, installation, testing, operation, and maintenance) and if this cost is less than the risk associated with the threat, then the vulnerability can be

---

[69] An attribute Y is *preferentially independent* of attribute X if preferences for different levels of attribute Y do not depend on the level of attribute X. In other words, the value for X does not affect the preference choice between two values of Y.

minimized[70] using the countermeasure. If the lowest total cost countermeasure is not of lower cost than the risk associated with the countermeasure, then none of the countermeasures (for this threat) is cost-effective (assuming their performances are similar and limited to a single threat area) and should not be purchased and installed. That is, if the cost of the countermeasure is more than the potential loss that might occur then it is not cost-effective to purchase, install, and operate this particular countermeasure. This means that the owner will have to accept the risk or risks associated with this particular threat.

If a particular countermeasure is used to counter more than one threat, then the security analyst should sum up the total cost of this countermeasure and compare it with the sum of the potential losses (i.e., risks) to which this countermeasure applies, that is:

Total Cost of Countermeasure$_i$ < $\sum_k$ Relevant Risk$_k$, in order for the ith countermeasure to be cost-effective.

In fact, some organizations may require that the Total Cost of the Countermeasure be much less that the sum of the relevant risks. For example, suppose there are three relevant risks such that Relevant Risk$_1$ is Risk$_3$, Relevant Risk$_2$ is Risk$_5$, and Relevant Risk$_3$ is Risk$_6$, then if the countermeasure total cost is \$100 and the three risks are \$20, \$30, and \$50, respectively, then the owner may not wish to implement the countermeasure (\$100 cost and risk both). However, if the total cost of the countermeasure is \$30, then the owner may very likely wish to purchase and install the countermeasure (less than 1/3 the risk cost).

Owners or users of the system should be able to make this decision best. The priorities and values of the various information system assets can be identified by the appropriate organizational individuals and asking the right questions based on the security analysts' knowledge of the system, the organization, and other relevant issues. The appropriate corporate individuals will likely be the owners and named representatives of the organization.

For example, for some types of organizations (e.g., a research and development organization), the owners might say that the information that their system has produced is of highest value. Whereas another organization (e.g., a consumer-oriented organization) might say that availability of their system to their potential customers is of highest value. As you can plainly see, the value of the information system network assets is highly organizationally dependent and must be performed for each organization.

Although the primary source of this information is the owners, named representatives, and users, should they be unavailable or have no knowledge of this issue, then the analyst can use common sense to ascertain the priorities and values of the information system assets may be based on the organization's mission and what other owners and users have said in similar situations.

7.3.2 Remaining Risks

Once the countermeasures have been selected, and an implementation plan has been developed, there will be some remaining risks that the owners and users believe are

---

[70] Remember that implementing a countermeasure can reduce the relevant vulnerabilities (and effects of associated threats) but it cannot unequivocally cancel them.

Charles L. Smith, Sr.

acceptable. These remaining risks should be clearly listed and the assumptions about the likelihood of occurrence and the potential cost should be explicitly identified.

For the risks that are not countered with some sort of cost-effective security mechanism, then these threats and associated risks will have to be accepted, whatever they are.

7.3.3 Return On Investment

The return on investment (ROI) for the money spent on the countermeasures for your information system is difficult to quantify. Any organization is interested in having some understanding of what their investment in security means in terms of some sort of payoff. It is difficult to quantify the payoff for an investment in security because the return is probabilistic. In case of some catastrophe where an entire system is destroyed, say a tornado that wipes out a facility, then if there is a backup system that has all of the critical information stored online, then the organization can be up and running in a matter of minutes (if it is a hot backup) or hours at most.

This kind of capability (i.e., backup and recovery) can be the difference between staying in business and going out of business. Thus, its worth can be determined in terms of the value of the organization's business, which could be substantial in some cases. Its value after the tornado is the value of the business itself, whereas its value before the tornado could only be probabilistically estimated based on the probability of a tornado (or other catastrophe) affecting the business and totally demolishing the facility. Many organizations will learn the value of having a backup and recovery capability only through actually having endured an actual catastrophe, and then it may be too late if they were not prepared. If they were prepared, then the value of the backup and recovery capability is equal to the value of the business, perhaps many millions of dollars, so the ROI for backup and recovery is considerable.

The ROI for the ith countermeasure (cm$_i$), in general, can be computed using the following equation:

ROI (cm$_i$) = Benefits of the countermeasure—(Total cost of the countermeasure - Operation and maintenance cost of the current, or legacy, capability),

where the *benefits* are in dollars; *total cost* (also in dollars) is the sum of the acquisition cost, installation and unit test cost, test and evaluation cost, modification (if required) cost, and operation and maintenance cost (for some period, usually three years) of the countermeasure; and the *operation and maintenance cost* (in dollars) of the current capability is just that.

The total cost for all (or most all) of the possible countermeasures that one might wish to obtain, can be determined and updated as required so that the data would be available when needed without the analyst having to do research to determine the countermeasures, their attributes and values (on a scale of 1 to 5 with 5 being the best), and their total costs as shown in Table 7.3, *Security Items and Attributes*.

Table 7.3 Security Items and Attributes

Security Item	Performance Attributes	Vendor Attributes	Costs
Firewall, Check Point	Scalability - 4 Interoperability - 5 Portability - 5 Speed - 4 Etc.	Time in business - 4 Dependability - 4 Honesty - 5 Etc.	Acquisition - $10 K Installation - $2 K Test and Evaluation - $2 K Operation and Maintenance - $5 K (3 yrs)
Firewall, Cisco	Scalability - 4 Interoperability - 5 Portability - 5 Speed - 4 Etc.	Time in business - 4 Dependability - 4 Honesty - 5 Etc.	Acquisition - $10 K Installation - $2 K Test and Evaluation - $2 K Operation and Maintenance - $5 K (3 yrs)
Etc.			

Now let us look at an example for an anti-DoS capability. An important measure for anti-DoS software is to know the number of DoS attacks that took place and disabled the system prior to the implementation of the anti-DoS package. Then the analyst can assume that these types of attacks have been blocked completely, or the analyst can collect records over the next few months (say three) to see how many DoS attacks occurred and if any were successful. Of course, if any were successful, then the anti-DoS package would have to be upgraded to ensure that similar attacks are blocked in the future. If they were properly blocked, then a probabilistic computation can be performed to estimate the ROI as follows:

$$ROI = N_a * (R_a - R_p) * A_c - \delta C,$$

where N_a is the number of DoS attacks (over some period, say three months, perhaps even using an extrapolated value based on a trend computation), R is the (number of successful DoS attacks)/(number of all DoS attacks) where R is a ratio and is in the interval [0,1] but could have the maximum value of 1.0 if all DoS attacks were successful, R_p is the ratio prior to the implementation of the anti-DoS package, R_a is the ratio after the implementation of the anti-DoS package, A_c is the average cost of a successful DoS attack, and δC is the total cost of the anti-DoS package less the cost of the current capability.

For this process to work properly, the anti-DoS package must collect data on the number of DoS attempts that were blocked. The security administrator will have to count the ones that it did not block. The ROI can be computed over some period (say three months) after the anti-DoS package is operational. R_a, R_p, and A_c are computed from historical data. If the value for the ROI for the anti-DoS mechanism is positive, then there is a return on the investment else there is no positive return but a loss.

To compute an ROI for any aspect of the countermeasure investment, one needs to identify the correct parameters on which to base the computation. A suggested set of parameters for each of the security areas is shown in Table 7.4, *Security and ROI Parameters*, where C represents the total cost for the countermeasures for each security area.

The computation of ROIs requires the collection of relevant data prior to the installation of the security mechanisms or methods and after their installation. This data must be defined and then accurately collected as the data becomes available. The data can be collected on the basis of the period of a year or shorter term if desired. In many cases, there will be no data on the system prior to the installation of the security mechanism or method, and thus must be estimated.

Each countermeasure should have a positive ROI since that is one of the criteria for purchasing the countermeasure. Thus, the total ROI for all of the countermeasures is:

$$ROI \text{ (Total)} = \Sigma_k \, ROI_k,$$

where k = 1, 2, ..., m where m is the total number of countermeasures. This is a compensatory method for computing ROI, that is, some ROIs may not be positive, but if the total ROI is positive, then all of the countermeasures might be implemented.

The case may be that to get the most "bang for the buck" (i.e., the highest ROI value), the organization may wish to hire a consultant who can provide up-front information on how to get the best products from a variety of vendors, not just one vendor, because the security vendors make heterogeneous products, or should. This is one of the objectives of providing standards for security. It is important that the various security products interoperate.

The decision on which countermeasures and which vendors to contact should come from a formal risk assessment, which the consultant should be able to provide to the organization.

One way to ensure that your organization is getting the most for what it pays for with regard to network security is to get the security at the lowest price, but this assumes that the security quality is the same in all cases, which it may not be.

Another, but more certain approach, is for the organization to hire a reasonably-priced consultant who is very knowledgeable about security and will:

1) Assist in the creation of a security policy and rules base (if required),
2) Assist in identifying the security requirements (if required),
3) Perform a risk assessment,
4) Assist in identifying the countermeasures and vendors,
5) Assist in the acquisition of vendors and countermeasures,
6) Assist in the security test and evaluation process, and
7) Assist in overseeing the operation and maintenance of the mechanisms,

in a timely manner with high-quality results for all of the products listed above.

Table 7.4 Security and ROI Parameters

Security	ROI Parameters	Equation
Anti-DoS Package	Number of DoS attacks (N_a), Average Cost of DoS attacks (A_c), number of successful DoS attacks/number of all DoS attacks (prior to installation (R_p) and after installation (R_a))	$N_a * (R_a - R_p) * A_c - \delta C$ δC = cost of anti-DoS package - current costs
Anti-virus Package	Number of virus attacks (N_v), Average Cost of virus attacks (A_c), number of virus attacks/number of successful virus attacks (prior to installation (R_p) and after installation (R_a))	$N_a * (R_a - R_p) * A_c - \delta C$ δC = cost of anti-Virus package - current costs
Physical Protection	Estimated damage caused by physical intruder per attack (D_e), number of potential intruders denied access (N_b)	$N_b * D_e - \delta C$ δC = cost of physical security - current costs
Training and Awareness	Number of employee errors made prior to training (E_p) and number of employee errors made after training (E_a), average cost per employee error (A_c)	$N_b * (E_a - E_p) * A_c - \delta C$ δC = cost of training and awareness - current costs
Identification and Authentication	Estimated damage caused by electronic intruder per attack (D_e), number of intruders blocked (N_b)	$N_b * D_e - \delta C$ δC = cost of I&A package - current costs
Confidentiality	Estimated damage per message caused by interception of messages (D_m), number of intercepted messages per year (N_m)	$N_m * D_m - \delta C$ δC = encryption cost - currents costs
Data Integrity	Estimated damage caused by modification of messages (D_m), number of modified messages per year (N_m)	$N_m * D_m - \delta C$ δC = data integrity costs - current costs
Non-repudiation	Estimated damage caused by repudiation of valid messages (D_m), number of valid messages repudiated per year (N_m)	$N_m * D_m - \delta C$ δC = non-repudiation costs - current costs
Firewalls	Number of malicious messages allowed through the firewall (N_a), Average Cost of malicious message (A_c), number of successful DoS attacks/number of DoS attacks (prior to installation (R_p) and after installation (R_a))	$N_a * (R_a - R_p) * A_c - \delta C$ δC = cost of firewall - current costs
Availability	$MTTF_{wo}$ and $MTTR_{wo}$ (without clustered servers), $MTTF_w$ and $MTTR_w$ (with clustered servers), number of server failures (N_f), A_c (average cost per failure)	$N_f * A_c - \delta C$ δC = cost of clustered servers - current costs

Security	ROI Parameters	Equation
Backup and Recovery (i.e., contingency)	Probability of occurrence (P_c (j)) of some catastrophe of type j (at the site) and loss (L_c (j)) due to the catastrophe	Σ_j P_c (j) * L_c (j) - δC (j) δC (j) = cost associated with a catastrophe of type j - cost of backup and recovery systems and operations

7.4   Steganography

Steganography is the art of hiding information inside other messages (usually plaintext) so that detection of the message is prevented. Although this craft is quite old (even the ancient Greeks used this method), recently the technology of computers has been used to revive this method. Steganographic techniques based on computer capabilities introduce changes to digital messages so that secret information can be embedded in an innocuous message. This secret information can be communicated in the form of text or binary files to provide information to the receiver who knows how to excise the embedded information (Johnson 2001). The secret information can be plaintext or it can be encrypted. For example, a secret message can be hidden inside an image transmission as the least significant bits.

Steganography used in conjunction with encryption methods can provide a greater amount of security (for the user) than either method alone. However, using encryption with long keys (say 156 bits or more) can ensure that no decryptographic method will be capable of decoding the message in any reasonable time.

A puzzle for cryptanalysts is to discover a method for screening information, especially image information, to discover whether or not the image contains embedded steganographic information.

REFERENCES AND BIBLIOGRAPHY

7.1 NSA (1998). "Network Security Framework," Network Security Group, Technical Directors, Security Solutions Framework, Release 1.1, National Security Agency, 3 December 1998.

7.2 Oppliger, Rolf (1998). "Internet and Intranet Security," Artech House, 1998.

7.3 Dawes, R. (1979). "The Robust Beauty Of Improper Linear Models In Decision Making," American Psychologist, 34, pp. 571-582, 1979.

7.4 Smith, Sr., Charles L. (1998). "Computer-Supported Decision Making: Meeting the Demands of Modern Organizations," Ablex Publishing Corporation, Greenwich, Connecticut and London, England, 1998.

7.5 DITSCAP (1999). "Department Of Defense Information Technology Security Certification And Accreditation Process (DITSCAP)," Application Document, DOD Manual, 5200.40-M, 21 April 1999.

7.6 Bernstein, Terry et al. (1996). "Internet Security for Business," John Wiley & Sons, pp. 150-151, 1996.

7.7 Brenton, Chris (1999). "Mastering Network Security," SYBEX Network Press, 1999.

7.8 Smith, Richard E. (1997). "Internet Cryptography," Addison Wesley, Reading, MA, 1997.

7.9 Krause, Micki and Harold F. Tipton (1999). "Handbook of Information Security Management," Auerbach, Washington, DC, 1999.

7.10 Canavan, John E. (2001). "Fundamentals of Network Security," Artech House, Boston, 2001.

7.11 Russell, Deborah and G. T. Gangemi, Sr. (1992). "Computer Security Basics," O'Reilly, Cambridge, MA, 1992.

7.12 Stillerman, Matthew, et al. (1999). "Intrusion Detection for Distributed Applications," Odyssey Research Associates (ORA), Ithaca, NY, 1999.

7.13 NIST (1995). "An Introduction to Computer Security: The NIST Handbook." NIST Computer Policy, SP 800-12, October 1995.

7.14 Smith, Sr., Charles L. (1999). "A Process for the Development of a Security Architecture for an Enterprise Information Technology System," Command and Control Research and Technology Symposium, U.S. Naval War College, Newport, RI, 29 June - 1 July 1999.

7.15 Swanson, Marianne (1998). "Guide for Developing Security Plans for Information Technology Systems," NIST Special Publication 800-18, Federal Computer Security Program Managers' Forum Working Group, December 1998.

7.16 NIST (1994). "Keeping Your Site Comfortably Secure: An Introduction to Internet Firewalls," NIST, SP 800-10, December 1994.

7.17 Smith, Sr., Charles L. (1997). "A Survey to Determine Federal Agency Needs for a Role-Based Access Control Security Product," *International Symposium on Software Engineering Standards '97*, IEEE Computer Society, Walnut Creek, CA, June 1997.

7.18 CSC (1999). "Evaluation Technical Report for Check Point Software Technologies, LTD, Firewall -1," Version 4.0, CSC, Hanover, MD, October 1999.

7.19 Curtin, Matt (1997). "Introduction to Network Security," Kent Information Services, Inc, March 1997. This document can be found at: (http://www.interhack.net/pubs/network-security.html).

7.20 Stallings, William (2001). "SSL: Foundation for Web Security," Cisco Web Page, http://www.cisco.com/warp/public/759/ipj_1-1/ipj_1-1_SSL2.html, 2001.

7.21 Check Point (2001). "Firewall-1, A Complete Solution for Securing the Internet," Check Point Software Technologies, 2001. This document can be found at: http://www.checkpoint.com/products/downloads/FW1-4.1_Brochure.pdf

7.22 Aladdin (1999). "eSafe Gateway," Aladdin, 1999. This document can be found at: http://www.checkpoint.com/opsec/partners/aladdin.html

7.23 Cisco Cable (2001). "Cisco Cable Web Page," Cisco, 2001. This information can be found at: http://www.cisco.com/cable/products/.

7.24 Zimmerman, Phil (2001). "Why Do You Need PGP?" PGP Web Page, http://www.pgpi.org/doc/whypgp/en/).

7.25 Network Associates (1999), "An Introduction to Cryptography," Network Associates, ftp://ftp.pgpi.org/pub/pgp/6.5/docs/english/IntroToCrypto.pdf, 1999.

7.26 NIST (2001). "NIST Subject List," http://www.nist.go, 2001.

7.27 VeriSign SSL (2001). "Strong Security in Multiple Server Environments," VeriSign, http://www.verisign.com/rsc/wp/onsite/index.html, 2001.

7.28 VeriSign PKI (2001). "Public-Key Infrastructure (PKI) - The VeriSign Difference," VeriSign, http://www.verisign.com/whitepaper/enterprise/difference/difference.html, 2001.

7.29 Carmeli, Buky (2001). "Gap Appliances Enhance Security," Spearhead Security Technologies, Network World, Page 29, August 20, 2001.

7.30 VeriSign Server (2001). "Strong Security in Multiple Server Environments," VeriSign, http://www.versign.com/rsc/wp/onsite/onsite_wp.html, 2001.

7.31 Fisher, Dennis (2001). "Startup Preps Weapon for DDOS Attacks," pp. 1 and 15, eWeek, September 3, 2001.

7.32 Johnson, Neil F. (2001). "Steganography," George Mason University, http://www.jjtc.com/stegdoc/index2.html, 2001.

7.33 Borisov, Nakita, Ian Goldberg, and David Wagner (2001). "Security of the WEP Algorithm," http://www.isaac.cs.berkeley.edu/isaac/wep-faq.html, 2001.

7.34 Proxim (2001). "What is a Wireless LAN?" http://www.proxim.com/learn/library/whitepapers/wp2001-06-what.html 2001.

PART IV. IMPLEMENTING AND TESTING THE PREFERRED SOLUTION

This section contains a process for ensuring that the preferred cost-effective countermeasures are installed in a manner that minimizes the technology, schedule, and performance risks for the countermeasure products. In addition, it also addresses the testing of the installed countermeasures to ensure that they are operating properly as desired by the owner organization.

Charles L. Smith, Sr.

CHAPTER EIGHT

MIGRATION PROCESS

8.1   Introduction

Once an appropriate set of security mechanisms and methods has been identified, in most cases it is necessary to install these mechanisms incrementally so that a migration (or remediation) process needs to be formulated. A *migration process* is a method of upgrading (i.e., correcting) the current system to the target system in a manner that ensures minimal risks (e.g., technical, schedule, and performance) for the complete upgrade as shown in Figure 8.1, *A Migration Process.*

Since the recommended security mechanisms and methods include physical, behavioral, and technical countermeasures, the physical and behavioral methods can be implemented apart from the information system. These countermeasures are not a part of the migration (or upgrade) process intended here; only the technical upgrades are part of the migration process.

In the case shown in the figure, there are anticipated to be **m + 1** upgrades to go from the legacy or current information system to the target system, based on having **m** interim systems, where Interim System #1 is the first upgraded system. If the security analyst so desires, **m** can be 0 so that there is just one upgrade. However, minimizing the various risks will usually require that there be multiple upgrades.

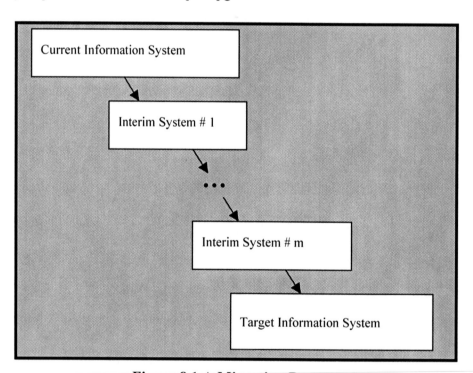

Figure 8.1 A Migration Process

Implementation in the form of multiple upgrades means that the devices and methods do not have to be procured all at once but over some period of time so that only those countermeasures required for moving from the current system to the first interim system need be purchased and installed, initially. This also means that the security test and evaluation (ST&E) process is simpler and probably can be done more quickly, completely, and accurately (but more frequently), than if all of the security measures were undertaken all at once (i.e., within one upgrade cycle).

To greatly diminish the likelihood of a successful attack on their information system, owners can be proactive by identifying and instituting safety measures that provide protection against malicious attacks on their system. Many organizations tend to feel that money spent on computer network security is money diverted from more productive avenues within the organization. However, when a security disaster occurs, such as a denial-of-service (DoS) attack that brings the system to its knees and causes a loss of availability for users, the owners justifiably get upset, perhaps very upset. Some of these disasters can cost the organization a lot of money, such as the DoS disaster in February 2000 that essentially shut down Amazon.com, ebay.com, and yahoo.com and cost them many millions of dollars in lost customer orders. Today, there are software algorithms (i.e., rules) that can be placed in a firewall, which can greatly diminish the possibility of a successful DoS attack. Many vendors sell anti-DoS products.

8.1.1   The Legacy System

The *legacy system* is the existing system. There should be a current architecture for it so that when defining the target system (i.e., the "To Be" system), there is an "As Is" illustration of the current system. If there is no current architecture model available, then the analyst must create one. The architecture of the legacy system should contain a graphical and verbal description of the operational system.

8.1.2   The Target System

The *target system* is the current system with the new countermeasures installed (and perhaps other devices that were installed while the security upgrade was being installed). The security analyst can create the upgraded security architecture (based on the selected countermeasures), modify the overall architecture accordingly, and thereby identify the target architecture.

The security analyst should then decide how the upgrade from the current system to the target system should be accomplished. The manner in which this is done should be such that the various upgrade risks are minimized. The discrete upgrades will define a set of incremental or interim systems that can be defined so that they form a sequential process from the current system to the target system.

Subsequent to each upgrade there should be a security test and evaluation. However, the owner may wish to wait until all of the upgrades are performed, and then do a single ST&E. There are risks associated with this approach (i.e., a single upgrade) if the installed countermeasures are not performing as they should and the owner is not aware of it because the operational personnel are not collecting data that reflects the capabilities of the new devices.

8.2   Migration Plan

The *migration plan* is a roadmap and can be represented by a Gantt chart that shows the various tasks and a schedule for performing the task through its completion and a product due at the end of the task, usually a document but could be a software subsystem. An example of a Gantt or roadmap chart is shown in Figure 8.2, *Simple Example of a Gantt Chart*.

A primary objective of the migration plan is to develop an implementation strategy that minimizes the risks and provides an acceptable schedule for attaining the target system.

The roadmap should do the following:

1) Identify the objective(s) of the program (box),
2) Name the various tasks to be performed,
3) Present a schedule for each task (horizontal line), and
4) Show the product to be generated by each task and the date when it will be delivered (downward arrowhead with product name over it), if there is one.

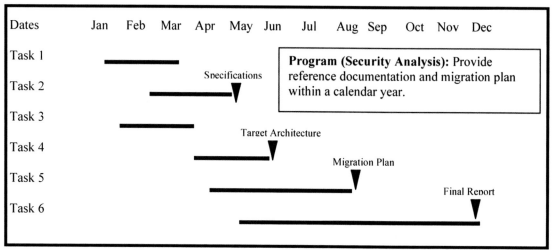

Figure 8.2 Simple Example of a Gantt Chart

Gantt charts are a convenient form for revealing the information needed to view the schedule implications and to keep track of the progress of the migration process.

8.3   Minimizing Non-Security Risks

There are other risks than security risks. Minimizing these other risks (technical, schedule, and performance) can be performed as follows. These other risks are the risks that can be minimized by a migration plan.

The *technical risks* are the risks associated with the technology of the countermeasure. These risks include the following issues:

1) Does the level of technology required by the countermeasure exist in the information system?
2) Can the technical abilities of the operations and maintenance personnel be easily learned?
3) Will the device perform as advertised?

4) Will the device be ready for installation according to the vendor's promised delivery date?

The other two risks, schedule and performance, can occur as part of the technical risk, but it is convenient to keep them separate.

Schedule risks are those risks associated with getting the countermeasure installed, unit tested, security tested, evaluated, and operated and maintained within a planned schedule. If the answer is that a reliable schedule cannot be developed or that the implementation process may lag behind the plan, then the schedule risks may be too severe. If the answers to these concerns are that a reliable schedule can be developed and adhered to by the upgrade process, then the countermeasure has acceptable schedule risks.

Performance risks arise whenever the vendor's reputation is in doubt. If the device cannot be relied on to perform as specified, then there is a vendor issue and the device has a high risk associated with it. The security analyst should consider the credibility of the vendor's specifications for their devices when making any recommendations to purchase and install their products.

Identification and determination of the various risks and the effects of the implementation process is more an art than a science and is quite dependent upon the analyst's knowledge and intuitive capabilities.

8.4   Developing the Migration Plan

The security analyst should ensure that the interim upgrade systems have been determined based on the minimization of the technical, schedule, and performance risks. The schedule should also consider the availability of funds for the upgrades. Subsequent to each upgrade, there should be an ST&E process performed. However, the analyst may decide that a single ST&E following the final upgrade (even if several interim architectures are required) is sufficient.

For each task (which should be a portion of a lifecycle process), a schedule should be developed and a final product (if appropriate) should be indicated. The program name and its objectives should be entered somewhere on the chart.

CHAPTER NINE

SECURITY TEST AND EVALUATION

9.1   Introduction

The objective of a *Security Test and Evaluation (ST&E)* process is to provide *evidence* that can be used to infer that the information system network's countermeasures are (or are not) providing the desired protection. The proof or evidence should be capable of being used to determine that the countermeasures have or have not worked properly. This evidence is arrived at on the basis of a tester performing the following activities:

1) performing some tests that require that all of the countermeasures perform all of the duties required of them,
2) writing down the results of the test, and
3) providing these results to an evaluator.

During these tests, data must be collected and this data should be of the form that a security analyst can decide that each countermeasure has passed or failed the test, and if a test has failed, then the test data should be sufficiently detailed to identify the particular problem so that the appropriate vendor can be notified of the issue that needs to be resolved. In some cases, the test data may not have been defined properly so the test results may be indeterminate, so that there are actually three potential outcomes: PASS, FAIL, or INDETERMINATE.

The tests should be defined so that the desired outcome is a PASS, and any other outcome is considered a FAIL, unless the outcome is indeterminate, in which case the outcome is marked INDETERMINATE. The tests with an INDETERMINATE outcome will usually be tests for which the data collected was either incorrectly defined or the information that was supposed to be collected was not properly collected.

When writing up the test results, it also is a good idea to add comments to the test outcome so that the security analyst is able to recreate the capabilities of the countermeasures, especially in the case of the failed or indeterminate results because the security analyst must define the following:

1) The modification (in general, not detailed since that the actual modification is the responsibility of the vendor) that needs to be made to the software, and
2) An additional test with different test data to be collected in order to properly test the modified countermeasure.

A recommended format for recording the test information is shown in Table 9.1, *Recommended Test Results Format*, below.

The ST&E process can be divided into four separate activities:

1) Test planning,
2) Test operations and data collection,
3) Test analysis and evaluation, and
4) Reporting of Test results.

309

Table 9.1 Recommended Test Results Format

No.	Test Description	Anticipated Outcome	Actual Outcome	PASS/FAIL	Comments
1					
2					
3					
Etc.					

The needed countermeasure performance evidence is determined by executing the following ten actions:

1) Define the evidence required to validate the security operations,
2) Define the system data that should be collected to produce the evidence,
3) Define the system inputs required to produce the data,
4) Define the tests that will generate this data,
5) Define the methods and mechanisms for collecting this data,
6) Set up the system for collecting the data,
7) Run the defined tests and collect the appropriate test data,
8) Analyze and evaluate the test data,
9) Document the evidence, and
10) Publish the ST&E report.

I will refer to the analyst(s) who implements the tests as the *Test Analyst(s)*. There may be one or more Test Analysts. Each Test Analyst must ensure that all of the tests are run and that all of the appropriate test data is collected. The analyst who analyzes and evaluates the recorded information is called the *Security Analyst(s)*. The Test Analyst(s) and Security Analyst(s) may be the same person(s).

The tests should be *repeatable*, that is, if the tester performs the same tests again, then the tester should get exactly the same results. Or, if another tester should perform these same tests, precisely following the plans as described in the Security Test Plan, then the other tester should get the same results as the original tester.

Just like the case of a formal experiment, the security analyst should know what to expect from a properly performing security mechanism, he or she should also have planned the test so that it can be implemented in a precise manner and performed repeatedly with the same results, and the tests should be designed so that the system can fail if the system is not operating properly.

Security tests can be performed in at least three different ways, see Figure 9.1,*The Alternative Architectures for ST&E*:

a) The entire Enterprise system can be tested (Figure 9.1.a),
b) A local area network (i.e., an intranet) can be tested (Figure 9.1.b), and
c) A specific application can be tested (Figure 9.1.c).

If the entire Enterprise system (i.e., all of the intranets (viz., LANs that are part of the Enterprise)) is tested, then any extranets (LANs belonging to other organizations), if trusted, also must be tested to see if they have the same level of security as this organization's level. For this type of testing, both host-based and network-based countermeasures are appropriate. The boundary of this system would be the connections of the LANs to a WAN or the Internet.

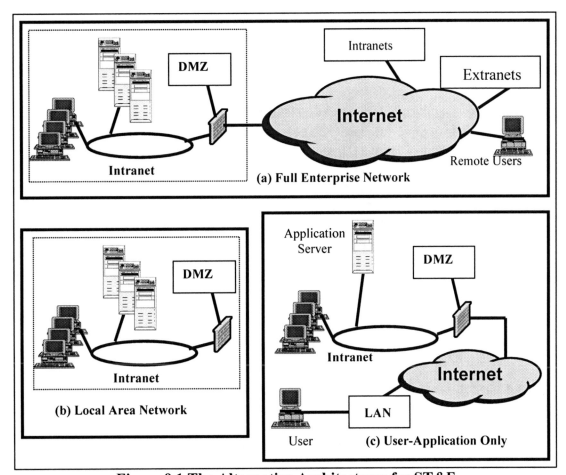

Figure 9.1 The Alternative Architectures for ST&E

For systems with many different LANs (i.e., the intranets), each LAN must be tested to see if it has the proper level of security. For this type of testing, both host-based and network-based countermeasures are appropriate. The boundary of this system would be the gateway firewall or router. The testing may encompass the firewall or it may not. If it does, then the tests should consider that some of the inputs to the intranet must come through the firewall. The various security processes that are performed within the LAN will be subjected to security testing, namely logging on, accessing certain applications, accessing certain databases, the single sign-on process if applicable, and other security capabilities such as confidentiality (i.e., encryption) of transmitted and stored information, data integrity, non-repudiation, access control (e.g., RBAC), anti-virus software, anti-DoS devices and/or software, firewall rules, and so forth.

In the case of testing of a specific application, then the tests must be set up so that only the security capabilities of the specific application are tested. This type of testing would assume host-based security measures where the host in question is the server that contains the specific application being tested. The boundary of this system is anything outside the connection of the application (and any servers it needs to perform its duties) with the PCs used to test the application. If the Enterprise system is Web-enabled, then the user site can be anywhere in the system that is connected to the Internet, otherwise the user must be located on a link that is part of the private communications system that is owned by the organization.

9.2   ST&E Plan

Testing of the specified software system should be done by an independent organization (i.e., one that did not create the specified software). This independent organization will be called the Independent Testing Organization (ITO) comprising the test analysts. The organization that performs the analysis and evaluation is called the Analysis Organization comprising the security analysts. It may be the case that the ITO and Analysis Organization are the same. It is assumed that the ST&E plan is developed by the Analysis Organization.

The purpose of these tests is to test target system software security capabilities to ensure that this software is operating properly with respect to the security requirements. The objective of the ST&E is to assess the extent that each of the Statement of Work (SOW) security requirements is met in the specified software.

An ST&E plan needs to be developed that the ITO can follow in order to perform the security tests.

9.2.1   Testing Approach

Most often the method used for conducting security tests is called "Black Box Testing," see Figure 9.2, *An Illustration of Black Box Testing*. This type of testing is based on considering the target software system as a black box and knowing its security functionality, then identifying the various tests that can be performed to ensure that the security functionality is performing properly.

This is also called "Open Loop" testing because the output of the test is not used as feedback into the system. The system is performing properly if and only if the *actual outcome* (i.e., the results of each test) is the same as the *anticipated outcome*. Of course, the anticipated outcome must be correct and its correctness requires a proper understanding of the system's security functions.

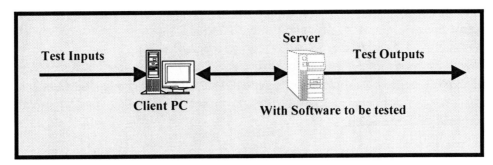

Figure 9.2 An Illustration of Black Box Testing

9.2.2 Test Plan

The tester (i.e., test analyst) needs to be informed of each test that must be performed and the anticipated outcome of the test. Determination of the test outcome will be provided by audit information or by information collected by the tester that clearly illustrates the outcome (i.e., pass or fail) of the test, see Figure 9.3, *The ST&E Plan Development Process.*

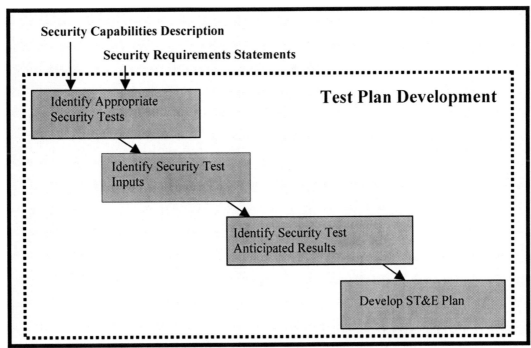

Figure 9.3 The ST&E Plan Development Process

The inputs to the process of test development are the security capabilities description (as deduced from the security description embedded in the specified software) and the SOW concerning the security requirements.

The entire test process is for the tester to:

1) perform the tests following the directions as presented in the ST&E plan (if the plan, at any point, is in error, the tester should modify the plan appropriately and inform the Analysis Organization as to the required modification),
2) during the tests, collect the test performance data for each test, and
3) fill in the blocks (actual test result, pass/fail/indeterminate, and comment), and then
4) turn over the collected data and filled-in tables to the Analysis Organization to be analyzed and evaluated.

The format for a test plan should ensure that the following information (for each test) is presented to the tester:

1) A description of the test that needs to be performed,
2) A description of the steps in the test process that the tester should follow (including the input information required for this test), and
3) The anticipated test results.

The format for the test results that needs to be filled in by the tester and recorded by the auditing mechanism is the following:

1) Any modifications to the test process,
2) A description of the actual results of the test with the tester's selection of pass, fail, or indeterminate for this test, and
3) Any comments that the tester feels are required for this test.

A recommended format is the following:

Recommended Test Plan and Test Execution Format

Testing Process to Be Used:
   A. Step one.
   B. Step two.
   C. Step three.
       …
   N. Step n.
Initials of Tester(s):
Anticipated Test Result:
Actual Test Result: Pass/Fail/Indeterminate (Circle correct one.)
Comments:

9.2.3   When to do ST&E

The time for the ST&E testing is just subsequent to the implementation of security upgrades, or if no upgrades have been installed, then any time that is convenient will suffice. This of course assumes that there is only a single online system that needs to be upgraded and tested. However, if the organization has an offline system that essentially matches the online system, then the countermeasure can be installed and tested using the backup system. Since it is unusual for the organization to have an offline copy of the online system, then the assumption is that there is only the online system that must be upgraded and tested.

9.2.4 Who Should do ST&E

The personnel for the ST&E testing should be the security analyst(s) who prepared the ST&E plan. However, it is a good idea to also have an operations person involved because most of the security mechanisms are performed in response to operations activities. Since the operations person is familiar with the operations processes she or he should have no problems with performing the test activities as they are described in the ST&E plan.

It is a good idea to perform the ST&E activities at the same time that functional testing of the upgraded system is being done. The principal reason for this is that generally the functional test personnel are more knowledgeable about the system than are the ST&E personnel and they are then closely available for answering any questions that the ST&E personnel might have.

9.2.5 How to do ST&E

If the ST&E plan was prepared accurately, then the ST&E testing should be performed by simply following the plan as written. In some cases, the security analyst charged with implementing the security test and evaluation plan will notice that the plan has an error for some test or tests. In this case, the analyst should appropriately modify the plan (in writing on the plan itself) and implement the plan as modified.

9.2.6 Results of the ST&E Effort

The results of the ST&E effort should provide test data that can be collected and evaluated to provide the evidence for whether or not the system's security countermeasures are performing as required.

9.3 Defining the Evidence

The process for *defining the evidence* is required to determine whether or not the countermeasures are operating properly and depends upon an understanding of each countermeasure and how these countermeasures should operate. The method for the operation of each of the countermeasures is explained in an earlier section. Each countermeasure is supposed to provide some measure of security during operations and proof that the countermeasure is operating properly requires that the countermeasure be stimulated with an appropriate input and the output (i.e., from the countermeasure device or software method and all collected test data, e.g., from an auditing mechanism) needs to be examined. To ensure repeatability, often stimulators or simulators are used instead of manual inputs from the keyboard. The test plan should define the mechanisms required to collect the appropriate test data. This means that the tester must ensure that the proper test data collection devices are in place and are operational, including the auditing mechanism.

I will call the vector of evidence **E** (n x 1, where n = number of test) so that **E** = **f** (all relevant test data) where **f** is some vector function that changes the raw test data into a form that is understandable evidence and can be placed in the Test Results table in the "actual outcome" column.

The Test Analyst desires to test the various capabilities of the security countermeasures for at least the following cases:

1) Identification and Authentication,
2) Authorization,
3) Access Control,
4) Confidentiality,
5) Data Integrity,
6) Anti-virus software (i.e., anti-malicious code software),
7) Anti-denial-of-service software,
8) Nonrepudiation,
9) Backup and recovery,
10) Firewalls and intrusion detection systems, and
11) Demilitarized zones.

9.3.1 Identification and Authentication

Identification and Authentication (I&A) is a process of determining that entities on the network are "who they say they are." I&A consists of an entity having a User ID and Password to logon or to access some piece of information or an application. In addition, for strong authentication, some entities may be required to have further evidence such as a token (e.g., a smart card or credit card) or a biometric identifier, such as their fingerprint, retinal pattern, pupil pattern, voice print, facial infrared pattern, handwriting pattern, or other some other specific natural characteristic of a human being (if the entity is a person).

This biometric process requires a specialized hardware device for recording the user's characteristic (such as a fingerprint recorder) and a stored template that is used for comparison by a concomitant software program. If there is a match (or great similarity, e.g., a match based on fuzzy mathematics), then the entity is approved for further activity and if not, then the user's access is blocked. At this point, if the user is blocked, the user should be notified by the system, such as a displayed message like the following:

[Your access is denied because your biometric reading does not compare with the stored template value.]

For certain persons who have unusual access privileges, it is recommended that the system require token or biometric identification. This will enforce a strong authentication for these privileged users, such as system administrators, and make it extremely difficult, if not impossible, for an unauthorized user to enter the system with these unusual privileges (such as GRANT, CREATE, or DELETE).

9.3.2 Authorization

Authorization is a mechanism for controlling access by system entities to certain applications and database information. Access to certain applications and some databases (or database subsets) are available only to some entities. These entities will possess different access privileges than others and this is according to their pedigree (i.e., authorization level) that is determined by the privileges granted to them according to their distinguished name in their certificate (if certificates are used) or in the access control list, and authorization can only occur subsequent to their proper identification and authentication.

For example, some users may only have READ privileges, while others may have READ and WRITE privileges. In addition, some users may have unusual privileges such as GRANT, CREATE, JOIN, REPLACE, APPEND, and DELETE capabilities that enable them to create a database or database line, or to delete a database or database line, or to access an entire database rather than some portion of it.

In many cases, the system users of a database management system (DBMS) do not have direct access to the DBMS, and can only perform database actions by proxy through some software application that actually accesses the database or the DBMS, such as the Cognos Impromptu package. This proxy process is useful in avoiding unpleasant accidents and malicious activities by some users or hackers.

9.3.3 Access Control

Access control is a mechanism that limits access to applications and databases based on an entity's appearance in an Access Control List (ACL) and a permissions or privileges list. It is provided by a software mechanism that limits an entity's access to the set of applications or databases stored in the systems memory. In the case of commercial or civil government organizations, it is most appropriate for the access control to be provided by a role-based access control (RBAC) mechanism. RBAC and the other two primary DOD access control mechanisms were presented and explained in an earlier section.

9.3.4 Confidentiality

Confidentiality is the protection of information from unauthorized entities. It is provided by encrypting the information, either when transmitting it or when storing it. Data or information that is sensitive must be encrypted prior to its transmission over a public communications link (such as the Internet) or prior to its storage, or transmission between servers, such as among servers in an untrusted environment.

Encryption may be provided by either a *symmetric* (a single or private key) or *asymmetric* (two keys, a private and a public key for each entity) encryption capability. Symmetric encryption is much faster than asymmetric encryption but has problems with getting the keys to the users in a confidential manner. However, a hybrid method based on asymmetric encryption to transmit a symmetric session key (which is then used for symmetric encrypted messages) is much faster and is the usual method found in modern systems.

9.3.5   Data Integrity

Data integrity is the process of identifying that transmitted or stored information (e.g., a message or password) has been modified. It can be provided through the use of encrypted messages, however encryption is not required. Data integrity means that the receiver can detect an alteration in a message and can request that the message be retransmitted. The detection of a message change can be through the use of a checksum or a hashed message (i.e., a message digest).

A hashing function is an algorithm that produces a fixed length message digest from a message. The message digest can be used to detect that a message has been altered, but cannot be used to correct the message. So the receiver must request that the original message be retransmitted. The process can be done with an acknowledgement (ACK message) from the receiver to the sender if the message is received in unaltered form and a not-acknowledgement (NACK message) if there is an error (of any sort, intentional or a transmission bit error).

If the sender wishes to ensure both confidentiality and data integrity, then the message must be encrypted and the hash encrypted also.

9.3.6   Anti-Virus Software

Anti-virus software looks for a given set of malicious code characteristics to locate potential viruses. Malicious code characteristics are a unique sequence of bits known to be contained in some particular virus. For every known virus, a unique sequence is known and is embedded in the anti-virus software mechanism for checking for that virus.

Generally, anti-virus software works only if the traffic is not encrypted. It can work for encrypted messages if the anti-virus mechanism is capable of decoding the encrypted messages prior to attempting to detect the virus, or if the anti-virus mechanism can be invoked following message decryption by some other package.

Anti-virus software also protects the system against other attackers (e.g., worms, Trojan horses, etc.) with malicious intent. Some anti-virus software has been created that attempts to check for the characteristics of malicious software based on some way in which malicious software works. This method is not always successful, in that it often calls non-malicious software malicious and malicious software non-malicious.

One of the advantages of current anti-virus packages is that these packages can be updated remotely from the vendors' homesites whenever new viruses have been identified and the malicious characteristics have been identified.

The Symantec AntiVirus Solution 7.5, in both Windows and Netware versions, is available at the following Web address: http://enterprisesecurity.symantec.com/products/.

For organizations that worry about virus threats that may affect costs, customer credibility, and systems uptime, an anti-virus package provides protection for any sized network. A scalable management console can offer real-time communications with clients and servers from a single point, allowing convenient distribution of new virus definition sets. The ability to scan machines, lock down settings, and monitor system activity anywhere on the network can offer protection from various attacks.

Anti-virus software can offer the following advantages:

1) Reduce operating costs by maximizing system uptime.
2) Help system personnel maintain service agreements.
3) Simplify installation and management of servers by centralizing administration across multiple platforms from a single console.
4) Deliver automated, macro-virus protection that can include scheduled updates of programs and automatic delivery of virus definition files.

9.3.7 Anti-Denial-Of-Service Software

In addition to viruses and other malicious code, anti-denial-of-service (Anti-DoS) software packages claim to be capable of detecting and countering denial-of-service (DoS) attacks. Often these attacks are from distributed servers and are called Distributed DoS (DDoS) attacks. However, at present, these packages are fallible. Fortunately, there is a device (called a gap appliance) for defeating the objectives of a DoS or DDoS attack (see section 7.2.16).

One of the problems with creating an anti-DoS package is that intruder tools used by the attackers are constantly changing, especially for distributed attacks, and any anti-DoS existing package will be outmoded in short order.

9.3.8 Non-Repudiation

Non-repudiation requires that the sender cannot deny having sent the message nor can the receiver deny having received the message. To achieve non-repudiation, the message must be accompanied with a *digital signature* or an encrypted message digest with private key.

Digital signatures require a digital certificate and are an encoding of the message from the sender with a unique key encryption in a Public Key Infrastructure (PKI) environment, meaning that the digital signature encryption is asymmetric (sender's private key for encoding and sender's public key for decoding). If both the sender and the receiver keys are used for encryption (dual encryption with the sender's private key and receiver's public key), then the message is also non-repudiated by the receiver, since only the receiver can have decoded the message (using the sender's public key and the receiver's private key).

Non-repudiation is based on the Diffie-Hellman asymmetric keys encryption methodology.

9.3.9 Backup and Recovery

Testing of backup and recovery capabilities requires the following activities:

1) Determining that an acceptable *Contingency Plan* has been developed,
2) Determining that there is a *remote site* that is survivable and that the remote site contains all of the appropriate devices required to record backup data,
3) Determining that a proper *replication period* has been identified,
4) Determining that the software system properly *replicates all critical information* (either in toto or just the information that has changed during the replication period) to the remote site,

5) Determining that the *remote site receives and records all of the critical information* as transmitted, and

6) Determining that the primary site can return to an operational state in case of a catastrophe.

9.3.10 Firewalls And Intrusion Detection Systems

Firewalls are computers with a minimal operating system that have a high-speed throughput, similar to routers. They operate on the basis of one of two philosophies: 1) allow everything to pass except what is specifically to be blocked, or 2) block everything except what is specifically allowed to pass. Both are based on a set of rules for passing or blocking all traffic through the firewall. In Case I, the final rule is "Pass all remaining messages." All preceding rules are "Blocking Rules." In Case II, the final rule is "Block all remaining messages." All preceding rules are "Passing Rules."

Intrusion detection systems (IDSs) are also computers with a minimal operating system that can evaluate the audit log trails either online (i.e., in real-time) or offline to identify any event that appears to be an intrusion. Each intrusion event that the IDS detects is recorded and reported to some administration officer (such as an information system security officer).

There are three types of attacks that avoid the inspection usually provided by IDSs, insertion attacks, evasion attacks, and denial-of-service attacks. DoS attacks have been discussed elsewhere (Ptacek and Newsham 1998). *Insertion attacks* cause the IDS to improperly accept and process packets that should be rejected (false positives) and thereby cause the insertion of data into the IDS that should be rejected. Insertion attacks are attacks in which an IDS accepts a packet that an end-system[71] rejects. The end-system rejects the packet because of an incorrect checksum (lack of data integrity), improper sequence number, invalid flag, or some other inconsistency that causes the implementation of the TCP/IP stack on the end-system to drop the packet.

Evasion attacks cause the IDS to improperly reject packets that should be accepted (false negatives). Evasion attacks are attacks in which an IDS rejects a packet that an end-system accepts. Evasion attacks thwart malicious packet pattern matching in a manner quite similar to insertion attacks (Ptacek and Newsham 1998). The information that the IDS misses is critical to the detection of an attack. *Denial-of-service* attacks are discussed elsewhere.

Modern IDSs should protect against both insertion and evasion attacks. Two of the best IDS products are the Cisco Secure IDS and ISS's RealSecure (Yocom et al. 2001).

9.3.11 Demilitarized Zones

The *demilitarized zone (DMZ)* is a special area where server information is stored that can be accessed by the public (or untrusted users). The DMZ can house Web-servers, mail servers, caching servers, or FTP[72] servers. If some untrusted user should destroy any

---

[71] An *end-system* is either the source of a TCP/IP packet or the destination of the packet.

[72] FTP is a protocol for transferring files to (from) a local hard drive from (to) an FTP server located on another TCP/IP-based network. It should not be used for transferring sensitive information, because user names and passwords are transferred in plaintext (so malicious sniffers can easily obtain them), unless FTP access is constrained to circuits created over VPN connections that are end-to-end transmissions.

information in the DMZ, then the system can detect the damaged area and replace it with information that is stored in an internal server that is not accessible by the public.

A DMZ, see Figure 9.4, *An Illustration of a DMZ,* is an area between a router and a local area network where a server can be placed, such as an e-mail server, that is separate from the LAN but can respond to the public e-mail traffic. Any database info that this e-mail server, should it be destroyed or modified, can be replicated from LAN internal info either offline or in an out-of-band mode. This offers protection to internal LAN systems while offering proper e-mail responses (Wack 1994).

A screened subnet can provide a DMZ and can be used to locate each component of the firewall (dual-homed gateway and screened host) according to the following rules:

a) Application traffic from the application gateway to Internet sites gets routed,
b) E-mail traffic from the e-mail server to Internet sites gets routed,
c) Application traffic from Internet sites to the application gateway gets routed,
d) E-mail traffic from Internet sites to the e-mail server gets routed,
e) FTP, gopher, rlogin, Telnet[73], etc traffic from Internet sites to the information server gets routed, and
f) All other traffic gets rejected.

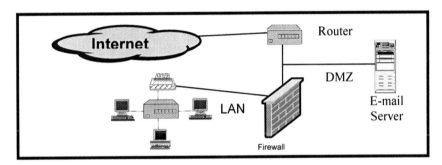

Figure 9.4 An Illustration of a DMZ

9.4   Internet Router Security

Companies using a firewall to isolate corporate IP networks from the Internet may have a false sense of security. Although using a firewall to isolate corporate IP networks from the Internet is a good security practice, many organizations unwittingly believe a firewall is their first and only line of defense. A firewall system is definitely recommended for any company connected to the Internet, but some organizations fail to include the router used to connect to their ISP in their arsenal of countermeasures.

A router not used in conjunction with a firewall can result in the DMZ being penetrated, possibly leading to a compromise in front of the firewall that may go undetected. Depending on the configuration of the DMZ, a remote attacker can cause a number of disruptions, including monitoring Internet traffic by using a packet sniffer program or gaining access past the firewall.

---

[73] "Telnet" is a remote terminal emulation application that has its own protocol for transport.

Even if the DMZ isn't vulnerable, the Internet boundary router itself can be a target for an attacker. Published flaws in outdated or router firmware that has bugs may allow specifically tailored attacks to reboot the router or cause problems with its operation. The result could be a Denial of Service or intermittent and confusing behavior of the router.

Routers are used to connect logical networks by stripping off the header that contains the destination address and providing a "nearest router address" for relaying the message. Routers use static, distance vector, or link state entries to create a blueprint of the network known as a routing table.

Routers are multiport devices that decide how to handle the contents of a frame, based on protocol and network information. IP routing can be done by a specified route (viz., Source Routing) or by an active method where the next router is determined dynamically. Usually, routers assume that the *identified source address* is in fact the IP address where the message began and hence are subject to spoofing (Wack 1994). However, modern routers have probably overcome this difficulty.

A router should reject any application traffic originating from inside the intranet (i.e., the local LAN) unless it came from the application gateway.

Some Internet routers have no administrative access controls or traffic filters. The lack of administrative access controls enables hackers to use "telnet" to gain access to the command prompt or use the Simple Network Management Protocol (SNMP) to monitor and control a router without authentication. Without filters, an attacker could possibly spoof the router into passing traffic that looks like it originated from the router or flood the router with garbage IP traffic (Brenton 1999).

Clearly it is critical for organizations to consider their boundary Internet router when implementing Internet security. Ensure that the firmware of the router is not outdated. The router operating system should be updated at least once each year. The Cisco Internet operating system (IOS) is improved much more often than that.

An Internet boundary router should always be configured to process IP traffic for IP networks assigned to your connection. This keeps garbage traffic from entering the DMZ and can greatly reduce the work of the firewall. A few older routers do not directly support traffic filtering. If any of your routers do not support filtering, it would be wise to invest in one that does. Cisco routers have excellent access control list features with excellent documentation.

A router without administrative access controls, simple administrative usernames and passwords, or default administrative logon information is an invitation for abuse. For example, do not use the vendor's name as a password. This occurs because many vendors pre-configure their routers that way, and then their customers never change the username and password. Unsecured SNMP access can be used to monitor and even change settings on the router remotely.

Administrative access control should always be used to secure an Internet router. When a router is using the latest firmware and traffic filters and administrative access controls are in place, the router, DMZ, and firewall are better protected. If your router supports it, you can filter IP traffic even further by protocol (such as dropping or limiting Internet Control Message Protocol (ICMP) packets from the Internet), disabling protocols that aren't needed (the Internet only uses IP), and disabling other routing protocols (single-homed Internet connections can usually only use static routing). This further screens your router and firewall from unwanted traffic and improves the performance of both.

In some situations, routers have been the object of DoS attacks where the attacker uses vendor-supplied default passwords that enable the intruder to modify a router's configuration and protocol information to misdirect traffic over the Internet (Vijayan 2001). This vulnerability of routers is well known among vendors. Compromised routers can be used by attackers to scan networks for vulnerable systems and as launch points for more traditional DoS attacks such as flooding a network with useless data. Breaking into routers usually requires some insider information and fairly sophisticated investigative work, but once accomplished, attacking the routers in a network can be devastating (Vijayan 2001). Router vendors may be aware of methods to counter this type of DoS attack.

9.5   Defining the System Data

Defining the system data that should be collected to produce the set {E} of **n** items of evidence {E_j: j = 1,n}, can be represented by a set {P} of **m** system parameters, {P_i: i = 1,m}, where $E_j = f_i (Q_k)$, where f_i is some function that transforms some subset Q_k of the system parameters set (i.e., Q_k is contained in {P}) into some piece of evidence E_j for some particular j.

For example, suppose we wish to determine whether or not the Identification and Authentication (I&A) process is operating properly. The I&A is a software process (unless a token card or biometric process is used in which case there must be a hardware mechanism that "reads" the token or identifier but a software subsystem is used to determine if there is a "match" where match means that the read token compares exactly (or very closely in a fuzzy system) with the stored template of the identifier) identifies and authenticates a particular entity.

Assume that the I&A process operates on the following four premises:

1) Each authorized test entity (user, application, or server) has a user ID,
2) Each authorized test entity has a password,
3) User IDs are permanent (not necessarily always the case in actual systems),
4) Passwords are at least eight characters in length, have at least three non-alphabetical characters, have at least one special character, and have a lifetime of 100 days.

Thus, the I&A tests should be capable of assessing each of the above capabilities:

1) Test the correctness of an authorized entity ID,
2) Test the correctness of an entity ID that is not authorized,
3) Test the correctness of the entity's authorized password,
4) Test the correctness of the entity's password that is not authorized,
5) Test the correctness of the number of password characters,
6) Test the correctness of the number of numerical characters in the password,
7) Test the correctness of the number of special characters (e.g., %, ^, &, *, or !) in the password, and

8) Test the lifetime of the password (e.g., set the life of the password at 99 days) and implement a password update reminder when the password has less than two days of remaining lifetime.

If a biometric is being used, then there will be more tests. There are at least eight tests (as shown above), but may be more to ensure that the I&A countermeasure is behaving properly. In the I&A case, for the first test the set {P} consists of the system output to the user, e.g.,

User ID is Acceptable

The function that transforms this test information to a PASS or FAIL is a comparison with the anticipated output which should be that for an authorized user, the system will accept the correct User's ID, so $f_1(P_1)$ = PASS.

By filling in the entries in Table 9.1, the analyst can identify the various tests to be performed and define the anticipated results (for a successful test), then when the tests are run and the appropriate data is collected that infers the actual test results, then the analyst can conclude that the system passed or failed the test, or that the test was indeterminate.

To decide whether the list in the test table (i.e., Table 9.1) is complete requires that the analyst be knowledgeable enough to determine that a tabular list that has omitted some tests forms an incomplete testing process. The omitted tests must be identified and placed in the table before the actual test process begins. Although it will mean an additional testing effort, it is better to have too many tests (some tests may be duplicative) than to have too few tests.

9.5.1 Defining the System Inputs

In order to run the tests, a set of inputs is required. These inputs are required to produce the data that will be collected. For each test, the data needed to determine that the system passed or failed the test must be defined so that prior arrangements can be made to ensure that the relevant data is collected.

For example, for the first test just described above, the input will be the correct User ID for an authorized user. Of course, if the tester makes a mistake during the keyboard input of the User ID, then the system will deny access (unless the tester accidentally types in an authorized user's identification). However, the test requires that the tester type in the correct identification.

When the system inputs for all the tests have been identified, a table should be created that lists the test number, the test descriptions, and the anticipated outcomes. These inputs become part of the table described in Table 9.1 above. When the tests are being run, the tester can fill in the rest of the table entries.

9.5.2 Defining the Tests for Data Generation

The tests should be defined so that they are as simple as they can be. The more complex the analyst makes the test, the more data is required to ascertain the test outcome, and the more encompassing will be the test because a PASS means that all elements of the test were positive and a FAIL means that at least one of the elements of the test did not happen properly. Thus the best tests are ones that have a single event outcome, such as the number of character test for the I&A "user ID." Hence a PASS or FAIL is a determination of the capability of the countermeasure to perform or not perform a single action.

Defining the tests that will generate this data is the next requirement.

9.5.3 Defining the Methods for Collecting the Data

The next step is to define the methods and mechanisms for collecting this test data. This is the data that will be used to infer the capabilities of the various countermeasures.

Most of the ST&E data can be determined from audit trails. This means that the existence of audit data is always a test of at least two countermeasure capabilities, 1) the auditing capability of the system and 2) the capability of some countermeasure. Other test data includes the following:

1) Sniffer outputs,
2) Special software coding, and
3) The system's reaction (e.g., block messages).

For example, suppose the security analyst provides an unauthorized User ID as an input for a logon process. Assume that the system responds to the input with the displayed block message shown in Figure 9.5, *User ID Error Message*. The parameters for this test can be the recording by the test analyst writing down the system response as the block message is displayed.

For testing the User ID capability, provide five randomly chosen entity IDs as input. During the test process, record the audit trails to ensure that these entities are allowed to enter the system. The relevant audit trails would be the auditing data that corresponds to the same time period during which the security tests were run.

**This is an inappropriate User Identification.
Please try again.**

Figure 9.5 User ID Error Message

Much of the actions by these mechanisms are transparent to the user and it is a good choice that the testing reflects this transparency.

Tests of the firewall and IDS activities can be determined by sniffing the input links to the firewall or IDS to include actions as:

1) remote logon,
2) encrypted messaging,
3) attempts by malicious hackers to logon with multiple attempts at User IDs or Passwords, and
4) firewall responses to internal access (or Web access) by authorized users (or public users).

The test should be defined in complete sentences and they should be uniquely numbered so that the tests can be numbered with a corresponding number and correlated so that the test analyst can ascertain that each of the ST&E tests were run with a corresponding test-data recording with results as shown in Table 9.2, *Test Plan Information.*

Table 9.2 Test Plan Information

No.	Test Description	Data to be Collected	Anticipated Outcome	Pass/Fail Criteria
1				
2				
3				
Etc.				

9.6 Setting Up The Test System

Prior to the ST&E tests, the test analyst should ensure that the preparation information is provided to the system. Examples of this preparation information are as follows:

1) Logon User IDs for the tests,
2) Logon Passwords for the tests, and
3) A listing of each of the tests with anticipated outcomes.

9.6.1 Performing the Tests

The next step is the running of the defined test and collecting the appropriate test data. This means that the test analyst must ensure that the auditing data is in operation and that the auditing data will be accessible at the end of the test process.

9.6.2 Analyzing And Evaluating The Collected Data

The analysis and evaluation of the test data are to determine the countermeasure operations performance outcome (pass, fail, or indeterminate - for those tests for which the test data was not defined properly or for some reason the test process was not completed).

9.7 Documenting the Evidence

The test results should be a principal part of the final ST&E report. A format for the test results is shown in Table 9.3, *Test Results Format*.

Table 9.3 Test Results Format

No.	Test Description	Pass/Fail	Comments	Action Required
1				
2				
3				
Etc.				

9.8 Publishing the ST&E Document

The final report should have the following contents:

1) Introduction
2) ST&E Plan
3) Tests to be run
4) Test Run
5) Data Collected
6) Analysis and Evaluation
7) Results
8) Remaining Actions

9.9 Performing Security Tests

To ensure that your information system network is really secure requires regular assessment to expose any security problems, misconfigurations, and architecture flaws (Raikow 2001). This is one reason why many organizations outsource their security activities. Unfortunately, most organizations tend to ignore the reports they get concerning the flaws in their networks.

When a problem occurs that infers that the organization's policy is inadequate, then the policy should be changed immediately. More often, the security problem is ignored, but if the policy is appropriately changed then the organization should make some modification to its systems to ensure that the security flaw (i.e., system vulnerability) is not simply ignored or forgotten. Raikow believes that most network security vulnerabilities are the result of "nonexistent, poorly developed, or unenforced security policies." If this is true, then an axiom of security, namely that policy should change slowly is not true. However, this axiom was based on the organization writing a good security policy, not one that is severely flawed. Thus, if the organization is careful and thorough in creating its security policy, then many of the system vulnerabilities will be alleviated before they occur.

The NIST has a program to improve the state-of-the-art for automatic test generation by developing a computer process that will automatically generate software security tests given

327

a set of formal specifications for the software (Black 2001). An objective of this program is to improve the quality of software by greatly reducing the cost of software testing. However, the automatic test process greatly depends upon having a set of specifications that are complete and follow a standard process for their storage. The development of the software must follow a standard set of procedures and a particular life cycle process. Hopefully, the software products produced in this manner will experience fewer failures and be less prone to failures once they are placed on the market or implemented in a system.

If the project is successful, then time-to-market for new software products will be reduced and the benefits of automatic test production will include reducing costs of producing the software vis-à-vis current methods.

REFERENCES AND BIBLIOGRAPHY

9.1 Raikow, David (2001). "Expert Advice Only Helps If You Listen," Sm@rtPartner, May 28, 2001, p. 40.

9.2 DITSCAP (1999). "Department Of Defense Information Technology Security Certification And Accreditation Process (DITSCAP)," Application Document, DOD Manual, 5200.40-M, 21 April 1999.

9.3 Wack, John (1994). "Keeping Your Site Comfortably Secure, An Introduction to Internet Firewalls," National Institute of Standards and Technology, SP 800-10, December 1994.

9.4 Brenton, Chris (1999). "Mastering Network Security," SYBEX Network Press, 1999.

9.5 Black, Paul E. (2001). "Automatic Generation of Tests From Formal Specifications," NIST, http://hissa.nist.gov/~black/FTG/autotest.html.

9.6 Ptacek, Thomas H. and Timothy N. Newsham (1998). "Insertion, Evasion, and Denial of Service: Eluding Network Intrusion Detection," Secure Networks, Inc., http://secinf.net/info/ids/idspaper/idspaper.html, January 1998.

9.7 Yocom, Betsy, et al. (2001). "IDS Products Grow Up," Network World, pp. 54-62, October 8, 2001.

9.8 Vijayan, Jaikumar (2001). "Network Routers Vulnerable to Denial-of-Service Attacks," Computer World, p. 8, November 5, 2001.

Charles L. Smith, Sr.

PART V. CONCLUSIONS AND RECOMMENDATIONS

This section contains the conclusions from the content of the various previous chapters. It also contains some recommendations for implementing a security capability in any information system.

Charles L. Smith, Sr.

CHAPTER TEN

CONCLUSIONS AND RECOMMENDATIONS

10.1 Conclusions
10.2 Recommendations
10.3 Future Activities
10.4 Storage Evolution

10.1   Conclusions

The contents of this book were designed to present a prescriptive model that describes the steps that one should follow for performing a comprehensive security analysis, from the development or identification of a security policy to the determination of the performance of the countermeasures based on a security test and evaluation of the implemented security mechanisms that were recommended by a risk assessment, to making the indicated modifications to ensure proper operation of the security mechanisms, as required, and operating and maintaining the upgraded (with security mechanisms) system.

10.2   Recommendations

For those who are in charge of an information system, especially systems that encompass several distributed LANs connected by at least one public WAN, it is a good idea to do at least one of the following:

1) Acquire enough security knowledge to ensure that the proper security vendors can be selected and that they are doing the proper job, or
2) Hire a security consultant who can provide sufficient comprehensive security awareness and offer accurate advice to the information system owner so that he or she can acquire the appropriate countermeasures and vendors to ensure adequate security for his or her system.

Some recommended countermeasure actions for safeguarding an information system network are to:

1) Ensure that all *critical information is replicated* at a remote location that is not accessible other than for read-only privileges. Ensure that this information is protected against modification (data integrity checksums) or reading by unauthorized individuals (perhaps through encryption).
2) Ensure that the *overall architecture* (that includes the security architecture) is kept current.
3) Ensure that there is *no single point-of-failure* in the network.
4) Ensure that the *contingency plan* for the system is easily accessible, is practiced by managerial and technical personnel, and is kept current.
5) Ensure that the *backup and recovery mechanisms* are installed and operating properly.
6) Ensure that *catastrophic recovery operations are periodically tested* and that relevant personnel are knowledgeable of the procedures.
7) *Remove all unneeded services.*
8) Ensure that all networks that are connected to your network provide the *same or better security* or that you have a firewall between your system and theirs.
9) Ensure that *security administrators are knowledgeable* of current security capabilities.
10) Ensure that managerial, user, administrative, and technical personnel are trained and aware of security policies.

11) Ensure that the organization has a current *security policy and security rules base*. Ensure that the *firewall rules accurately* reflect these.

12) Ensure that the organization's *living document of security requirements* is kept current.

13) Ensure that all *vendor patches* for relevant COTS software are implemented immediately upon arrival.

14) Ensure that a *formal risk assessment* is performed to identify the preferred countermeasures for this organization's information system.

15) Ensure that a *proper security test and evaluation* is periodically performed to assure that the installed countermeasures are operating properly, and if not, are appropriately modified or upgraded.

16) Ensure that a *DMZ has been installed* to protect public-accessed servers. Ensure that all organizational information that is accessible by the public is placed on DMZ servers.

17) Ensure that *perimeter routers and firewalls* have the proper set of rules and cannot be accessed to provide rules changes by external users.

18) Ensure that a list of *phone numbers of critical personnel* is kept current.

19) Ensure that *anti-virus software* is kept current and is operating properly.

20) Ensure that *anti-DoS software and devices* are kept current and are operating properly.

21) Ensure that there is a *plan for contacting customers if* the organization's Web site is inactive.

22) Ensure that there is a *plan for identifying customer versus organizational systems* that are affected by any malicious code (e.g., a worm) in the system.

23) Ensure that *hardware availability is maintained* through the use of redundant or clustered servers.

24) Ensure that *applications and information storage capabilities* are properly maintained.

25) Only allow *read-only privileges* when possible.

A list of vendors that can offer security services is contained in Table 10.1, *Security Service Vendors*. There are many more since this list is certainly not exhaustive. Also, over time, some may disappear and some new ones will appear.

Just go to www.yahoo.com or www.google.com and then do a search on "Network Security" or "Network Information." The names of nearly 600 vendors and security information sources will pop up. Some of the names appear in the tables below. In addition, some sources of security information (590 vendors) on the Web are listed in Table 10.2, *Web Sources of Security Information*.

Table 10.1 Security Service Vendors

No.	Vendor	Services	URL Address
1	Symantec	Security software	www.symantec.com
2	VeriSign	Certificates and Security services	www.verisign.com
3	SANS	Security services	www.sans.com
4	SEI	Security services	www.cert.org/octave/omig.html
5	Arbor Networks	Security services	www.arbornetworks.com
6	Oracle	Security services	www.oracle.com
7	Novell	Security services	www.novell.com
8	Sun	Security services	www.sun.com
9	Microsoft	Security services	www.microsoft.com
10	Netscape	Security services	www.netscape.com
11	MITRE Corp.	Certificates and Security services	www.mitre.org
12	IBM	Security services	www.ibm.com
13	Axent	Security services	www.axent.com
14	Yahoo	Security services	www.yahoo.com
15	Cabal Network Security	Security services	www.cabal.net
16	Network Security Int'l Association	Security services	www.netsec-intl.com
17	Check Point	Security services	www.checkpoint.com
18	Network-1 Security	Security services	www.network-1.com
19	Network Associates	Security services	http://www.nai.com/
20	Apple	Security services	www.apple.com
21	Network Security Assurance Group	Security services	www.nsag.net
22	Firewall Network Security	Security services	www.firewall-security.com
23	Innovative Security Network	Security services	www.innovativesecuritynet.com
24	Firedoor Network Security	Security services	www.firedoor.net
25	Architecture Technology Corp.	Security services	www.arcorp.com

Table 10.2 Web Sources of Security Information

No.	Vendor	URL Address
1	International Relations and Security Network	www.isn.ethz.ch
2	TimeStep Network Security	www.timestep.com
3	Network Security Buyer's Guide	www.netsecurityguide.com
4	Network and IT Security Policies	www.network-and-it-security-policies.com
5	ClearStar Security Network	www.clearstar.com
6	Mac Network Security	www.macintouch.com/macattack.html

No.	Vendor	URL Address
7	Computer Security Information	http://www.alw.nih.gov/Security/security.html
8	British American Security Information	http://www.basicint.org/
9	eMailman	http://www.emailman.com/encryption/index.html
10	Group for Research and Information on Peace and Security	http://www.ib.be/grip/
11	AT&T Information Security Center	http://www.att.com/isc/
12	Information Security and Privacy	http://www.stanford.edu/group/tdr-security/
13	Laboratory for Information Security Technology	http://www.list.gmu.edu/
14	Information Systems Security Associate	http://www.uhsa.uh.edu/issa/
15	Canadian Centre for Information Technology Security	http://www.ccits.org/
16	American Society for Industrial Security	http://www.asisonline.org/
17	Windows IT Security	http://www.windowsitsecurity.com/
18	Microsoft Excel Virus Information	http://www.microsoft.com/excel/productinfo/vbavirus/emvolc.htm
19	Access Control & Security Systems Integration	http://www.securitysolutions.com/
20	Information Assurance and Technology Center	http://iac.dtic.mil/iatac/
21	Security Informer	http://www.security-informer.com/
22	Kerberos General Information	http://web.mit.edu/kerberos/www/
23	The Security Net	http://www.the-security-net.co.uk/
24	NIST Computer Security Resource Clearinghouse	http://csrc.nist.gov/
25	The Clipper Chip	http://epic.org/crypto/clipper/

The decision process for performing a complete risk analysis is to formulate a hypothesis, find a solution to this issue, and then implement, test, and operate the identified solution. The process is divided into five parts and is explained below.

I. The Hypothesis Process:

Identify all of the potential threats.
Identify those that are relevant to this organization's system.
Identify the vulnerabilities using a scanner and/or architecture model.
Identify the assets that could be attacked and estimate their values.

Identify the threat scenarios (an exhaustive set of size n) that could exploit the identified vulnerabilities.

Formulate the Hypothesis.

Explanation: For each threat, identify the vulnerabilities that could be exploited by the threat. Determine the scenarios that could be used to perform an attack for this threat. For each scenario, estimate the potential losses that might occur. Of these, select the largest loss (i.e., perform a worst case analysis). Then, perform an exhaustive analysis by creating scenarios for all of the relevant threats, identified vulnerabilities, and identified assets (and their values) that are affected by the threats, so that the complete hypothesis will consist of the following statements (n of them):

Describe threat scenario 1 (TS_1) that exploits an identified vulnerability and could result in a loss to the organization of X_1 dollars.

Describe threat scenario 2 (TS_2) that exploits an identified vulnerability and could result in a loss to the organization of X_2 dollars.

...

Describe threat scenario n (TS_n) that exploits an identified vulnerability and could result in a loss to the organization of X_n dollars.

II. The Resolution Process:

Identify a comprehensive list of the alternative countermeasures (required to counter each scenario) and identify the attributes of these countermeasures (including performance attributes, vendor attributes, and total cost).

Compute the risk (for each of the n scenarios and associated maximum loss).

For each computed risk (R_i), identify the relevant countermeasures that could reduce the success probability of the ith threat scenario (thereby lessening the ith risk).

Rank order the identified countermeasures (for the ith risk) if there are more than one countermeasure (and repeat for all of the n risks).

Formulate a list of the cost-effective countermeasures.

Explanation: For each threat scenario (TS_n), compute the risk (R_i). For each risk (R_i), determine the countermeasures that could be used to reduce the ith threat scenario's success probability. If there is more than one countermeasure for this threat scenario, rank order them using a multi-attribute utility analysis. Select the one with the highest utility value. Do for all of the n threat scenarios. Delete any replications. Place all of the identified unique cost-effective countermeasures in a table. Identify the locations in the organization's network where each type of countermeasure must be located (e.g., firewalls may be required at each entry point in the organization's network, and there may be two of them at each point to ensure no single point-of-failure).

III. Implement the Solution:

Create a remediation or migration plan that minimizes the upgrade risks (technical, schedule, and performance).

Follow the plan and upgrade the information system, accordingly.

Explanation: For each of the upgrade risks, determine an upgrade process (remediation) that minimizes these risks. For example, if a particular countermeasure has been difficult to implement in other environments (which means that it probably should not have been selected in the first place), then this countermeasure should be implemented in an offline location and tested prior to installation in the online environment. Using this process, develop a plan for installing all of the countermeasures. After the plan has been approved, follow the plan during the implementation process.

IV. Perform Security Test and Evaluation of the Upgrade:

Create a Security Test and Evaluation (ST&E) Plan.

Implement the ST&E Plan.

Modify or replace, if required, the security mechanisms that are not operating properly.

Explanation: Create an ST&E plan. Get the ST&E plan approved. After all of the countermeasures have been installed, implement the ST&E plan and document the results. If any countermeasure did not work properly, either modify the device or replace it. If it is replaced, then the replaced mechanism must be tested and evaluated.

V. Operate the System:

Operate the system in its environment.

Maintain and upgrade the system, as required.

Respond properly to all security incidents.

Explanation: If the tested system is an offline system that is a replica of the actual online system, then subsequent to the test and evaluation process, the offline system should be installed as the online system with, perhaps, the original online system remaining as a hot backup in case any aspect of the new system is deemed non-operable.

To guard against malicious internal users, some recommended actions are presented in Table 10.3, *Recommended Actions to Minimize Internal Threats* (Verton 2001).

Table 10.3 Recommended Actions to Minimize Internal Threats

No.	Action
1	Ensure that no one has root access to the information system except those who are highly trusted and have a need for such access.
2	Ensure that security policies are provided to all relevant personnel.
3	Ensure that all relevant personnel are trained and educated relative to security issues.
4	Ensure that proper physical security is provided for the facility and the areas containing critical equipment.
5	Ensure that all personnel with sensitive positions, such as security administrators, are submitted to strict background investigations.
6	Ensure that all critical organizational information is declared sensitive and provided with the appropriate protection for confidentiality and data integrity.
7	Ensure that the organizational information system network is segmented to facilitate security.
8	Ensure that all auditing and intrusion detection systems are operating properly.
9	Request all organizations, with extranets connected to the organization's network, have sufficient security.
10	Ensure that security policies for internal users are strictly enforced.
11	Ensure that the information system provides sufficient identification and authentication capability for all internal personnel.
12	Ensure that the information system provides an appropriate level of authorization for employee access to applications and database information.

10.3   Future Activities

A logical follow-on to this effort is a book on enterprise information system architecture design and security control from a well-designed security command and control center as shown in Figure 10.1, *Architecture Design with Security Control*. In the figure, the enterprise information system is generic and consists of a single WAN with several LANs, with each LAN having a local storage area network (SAN). A set of sensor systems is attached to the WAN but these could also have been separated into several groups with a subset attached to each LAN.

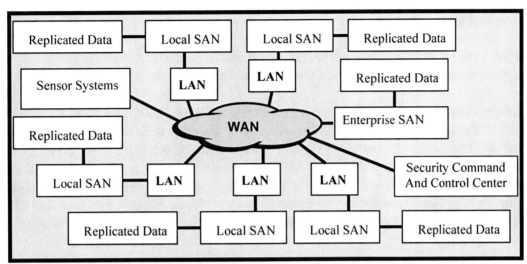

Figure 10.1 Architecture Design with Security Control

Each local SAN is replicated at a remote site as is the enterprise SAN. These replications contain the latest data on the information, applications, and middleware located at the local SAN or the Enterprise SAN.

The security command and control center has inputs required to support configuration management and control, auditing, intrusion detection, and health status of the sensors, devices, and software systems located throughout the enterprise network.

This information system's security capability is similar to a command, control, and communications system and is treated similarly. It attempts to:

1) Scan the relevant environment collecting all data on potential issues,
2) Identifies issues that need to be corrected,
3) Identifies a solution for correcting the situation,
4) Sends out a command to correct the situation, and
5) Then monitors the situation to determine if the command was properly performed.

For example, a server may be down and although it has been automatically replaced by another server in its cluster (using automatic failover software), the security administrator must be aware of the failure and provide a command to a technician to replace the failed server component.

Much of the configuration management may be automatically performed by smart software programs, however a security administrator or system administrator must be aware of the configuration status in order to understand the potential situations that may occur that will require the administrator to make a decision regarding the enterprise system. The relevant information on system status will be displayed in a manner that is readily understandable by the administrator.

Scanning of the relevant environment will require auditing information, intrusion detection information, status information, sensor information, and any other information relevant to understanding the potential security situations for an enterprise network system. Sensor information might be:

1) Aircraft traffic information (for an air traffic control environment), or
2) Enemy locations and activities (for a military environment).

The primary objective of the security command and control system is to provide commands that are accurate and timely resolutions to the correct issue.

The effort to present the material for designing an enterprise architecture and performing the activities required for providing security for the system is not dissimilar to the design of a large complex modern skyscraper. This is the objective of the follow-on book. Although the efforts to provide security for the enterprise network are similar, in many ways, to the efforts to provide overall control for the network devices and software, the focus of the book will be on security.

For those owners and administrators who wish to contract the responsibility for designing a security network to a vendor, there is a Cisco white paper (titled, "SAFE: A Security Blueprint for Enterprise Networks") that provides best practices for designing and implementing a secure network (Convery and Trudel 2001). Unfortunately, this paper does not provide any analyses for identifying and using the organization's security policy, security rules base, or security requirements, so there is no way of factually determining if the resulting architecture (from the paper) consists of a set of cost-effective security mechanisms (for this organization), which should result from a formal risk assessment that requires a full understanding of the threats, vulnerabilities, threat scenarios, and alternative security mechanisms (plus the cost and performance of each one) that are available from the various vendors at the time of the analysis.

10.4   Storage Evolution

The five basic phases of LAN storage evolution are shown in Figure 10.2, *LAN Storage Evolution.* Figure 10.2.a illustrates the initial phase of client/server storage where the data is stored in the server. This figure presents the basic elements of a client/server LAN system, namely the clients (or PCs), the servers, and an Ethernet network. Applications and databases may be stored in the PC or on the server. Most often databases are stored in the server since the database management system is located on the server. For this form, the LAN is adequate for a small office or home environment (Lais 2001a).

For a more capable LAN, one might install direct-attached devices, such as a redundant array of independent discs (RAID), as shown in Figure 10.2.b, by adding severs with tape backups as an interface between the Ethernet and the RAID devices. Also, there may be an additional server with tape backup. Accessing tape backups for information can take a long time and this form of backup is not recommended for situations where the stored information may be required for online usage. A third architecture for storage is the implementation of network-attached storage (NAS) as shown in Figure 10.2.c. In this case, an NAS server is the interface between the Ethernet and the RAID devices, with mobile users, other servers with tape backups, and clients also connected. The mobile users would be connected via modems through an Internet Service Provider.

The fourth step (Figure 10.2.d) in the evolution is to install a storage-area network (SAN) subsystem that interfaces with the Ethernet via an NAS server, as in the previous architecture, where the SAN subsystem (RAID and tape library connected to a fiber channel switch) replaces the RAID system (Lais 2001a). The most advanced storage capability is the fiber channel SAN shown in Figure 10.2.e (Lais 2001). This version is based on a SAN fabric[74] that is connected to the LAN (with communications provided by fiber distributed data interface, Ethernet, or token network) by servers. Storage is provided by: 1) small computer system interface (SCSI) discs with a fiber SCSI bridge, 2) a RAID device, and 3) a SCSI tape library via another fiber SCSI bridge. The fiber channel is a high-speed data transmission technology used to connect multiple hosts to dedicated storage systems over copper or fiber optic links (e.g., gigabit networks).

Steps to ensure that the needed storage capabilities for your system are met by the potential vendor are as follows (Lais 2001b):

1) Write a high quality request for proposal (RFP) for your system, even it you have to hire a consultant to do this,

2) Request that the selected vendor provide an explanation of the storage architecture that they are providing and how the vendor's storage devices will meet the organization's requirements, and

3) Establish a partnership with the selected vendor to ensure that the vendor is committed to your organization's information technology goals.

---

[74] A *fabric* is a combination of interconnected switches that perform as a unified routing infrastructure by allowing multiple connectors among devices on a SAN and allows new devices to enter inconspicuously by logon.

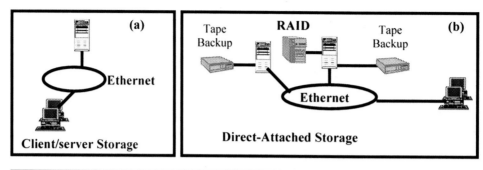

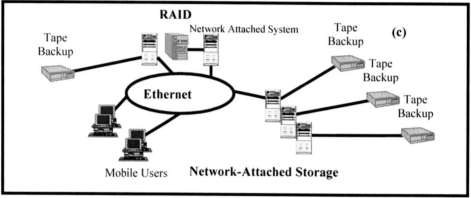

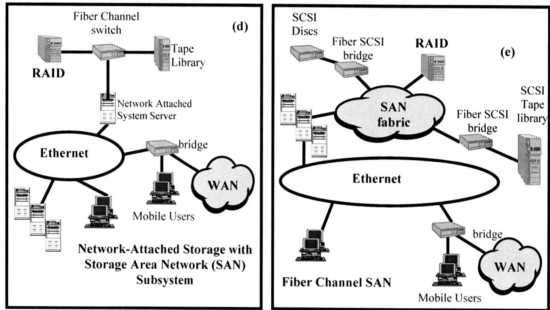

Figure 10.2 LAN Storage Evolution

REFERENCES AND BIBLIOGRAPHY

10.1   Verton, Dan (2001). "Analysts: Insiders May Pose Security Threat," Computerworld, p. 6, October 15, 2001.

10.2   Lais, Sami (2001a). "A Storage Sketchbook," Computer World, pp. 42-43, October 15, 2001.

10.3   Lais, Sami (2001b). "The 12 Most Costly Storage Mistakes, and How to Avoid Them," Computer World, pp. 43-46, October 15, 2001.

10.4   Convery, Sean and Bernie Trudel (2001). "SAFE: A Security Blueprint for Enterprise Networks," Cisco, http://www.cisco.com/warp/public/cc/so/cuso/epso/sqfr/safe_wp.htm, 2001.

Charles L. Smith, Sr.

APPENDIX A

GLOSSARY

A good glossary for information technology terms is the *Novell Glossary*, which can be found at: http://www.novell.com/documentation/lg/glossary/index.html. Also, another fine glossary is the *Free Online Dictionary of Computing* that can be accessed at http://foldoc.doc.ic.ac.uk/. The SANS Institute has the *NSA Glossary of Terms Used in Security and Intrusion Detection* at http://www.sans.org/newlook/resources/glossary.htm.

Another source is the Storage Networking Industry Association (SNIA) at http://www.snia.org/, then click on the Resource Center (Dictionary). Another source of definitions for Web and Internet terms is: http://www.webopedia.com/.

Abend - This is used with an agent, to bring a server to an abrupt halt or to crash a server (transitive). If used without an agent, it means to crash (intransitive).

Acceptable level of risk - This is a judicious and carefully considered assessment by the appropriate Designated Approving Authority that an information system meets the minimum requirements of applicable security directives. The assessment should take into account the sensitivity and criticality of information; threats and vulnerabilities; safeguards and their effectiveness in compensating for vulnerabilities; and operation requirements.

Access - This means to view, retrieve, or otherwise invoke objects, which include files, properties, aliases, user lists, file structures, and so on.

Access control - This is the process of limiting entities to their appropriate permissions. Administrating access control is based on organizational policy, employee job descriptions and their tasks to be performed, information sensitivity, user need-to-know, and other factors. Administration can be *centralized* (e.g., a single individual or office) or *distributed* (e.g., owners or file or functional manager) or *hybrid* (i.e., both central and distributed). Access controls can be provided in three categories: 1) physical (e.g., fences and gates), 2) operating system (e.g., by the Windows NT OS), or 3) application (e.g., by an Oracle DBMS).

Access control list (ACL) - This is a list of users or applications together with the specific information that they have access to and what they may do with the accessed information (read-only, read and modify, create, or delete).

Access point - This is the point in a wireless LAN where a mobile user can connect to the wired network.

Accident - This is any situation that occurs due to an inadvertent action.

Accountability - This is a security policy that states that every entity is identified and authenticated, and any activity by the user or application is audited and uniquely traced back to the entity that committed the action. This supports non-repudiation, deterrence, fault isolation, intrusion detection and prevention, and after-action recovery and illegal action.

Accreditation - This is the official authorization and approval, granted to a computer system or network, to process sensitive information in a particular environment. Accreditation is performed by specific technical personnel after a security evaluation of the system's hardware, software, configuration, and security controls. It is a formal declaration by the DAA concerning who has fiscal and operational responsibility that an information system is approved to operate in a particular security mode using a prescribed set of safeguards (i.e., countermeasures). This is the official management declaration for system operation and is based on information provided in the System Security Authorization Agreement (SSAA) and Security Certification and Authorization Package (SCAP) documents as well as other management considerations. The accreditation statement affixes security responsibility with the DAA and shows that due care has been taken for security.

Address - This means to specify the location of something. For example, in electronic mail, it means to specify the e-mail location of a user, resource, or group. In computer usage, it means to uniquely identify and specify the location of an element in a computer or computer network, for example: a location in memory or disk storage, a network or portion of a network, a station or other device on a network, and so forth.

Address resolution protocol (ARP) - This is a method for finding a host's Ethernet address from its Internet address. The sender broadcasts an ARP packet containing the Internet address of another host and waits for it (or some other host) to send back its Ethernet address. Each host maintains a cache of address translations to reduce delay and loading. ARP allows the Internet address to be independent of the Ethernet address but it only works if all hosts support it.

Adequate protection (OMB Circular A-130) - This is protection that is commensurate with the risk and magnitude of the potential harm resulting from the loss, misuse, or unauthorized access to or modification of information resources. This protection includes ensuring that systems and applications operate effectively and provide appropriate confidentiality, data integrity, availability, and accountability by using cost-effective management, personnel, operational, physical, and technical controls commensurate with an information system's sensitivity level.

Agent - This is software that processes queries and returns replies on behalf of an application. Typically, an agent is a small and well-defined task. In Network Management Systems, an agent is a process that resides in all managed devices and reports the values of specified variables to management stations. Although the theory behind agents has been around for some time, agents have become more prominent with the recent growth of the Internet.

Aggregation - This is the result of assembling or combining distinct units of data when handling sensitive information. Aggregation of data at one sensitivity level may result in the total data being designated at a higher sensitivity level.

Air gap - This is any point in the transfer of information which requires personal handling or administrative delays, e.g., physical transport of magnetic tape or floppy disk, intermediate paper document, or message traveling by courier.

Alarm - This is an audible signal from the computer to notify the user of the condition it was configured to specify. For example, an alarm can warn a user of an error condition, or it can notify the user that a certain program is being started or shut down.

American national standards institute (ANSI) - This is the organization that sets the standards for many technical fields and provides the most common standard for computer terminals.

Anti-virus programs - These programs are designed to search for viruses, notify users when they are found, and remove the virus from the infected disks or files.

Appendix III to OMB A-130 - This Appendix establishes a minimum set of controls to be included in Federal automated information security programs; assigns Federal agency responsibilities for the security of automated information; and links agency automated information security programs and agency management control systems established in accordance with OMB Circular No. A-123. It incorporates requirements of the Computer Security Act of 1987 (Public Law 100-235) and responsibilities assigned in applicable national security directives.

Applet - This is a small application that performs a specific task, such as the Calculator in Microsoft Windows.

Application - This is a computer program designed to perform a task for the user, such as payroll processing or general ledger entry.

Application entity - This is the part of an application process that interacts with another application process.

Application-level gateway - This is a firewall technology that proxies traffic. By doing so, the firewall become the client, typically opening two real sessions for every session—one to the LAN client and the other to the Internet host. This technology actively inspects all of the data in a packet before approving or denying the packet.

Application-level virus scanners - These scanners are responsible for securing a specific service throughout an organization. For example, there are products for scanning e-mail to identify viruses in the e-mail messages.

Application program interface (API) - This is a message and language format that allows programmers to use functions within another program.

Architecture - This is the configuration of computers, network devices, network linkages, and software that provides the entire support needed for an information technology system within an organization. There are at least two types of architectures:

1) **Enterprise architecture** - This is the overall architecture for an entire organization and may be spread around the country or even across the world.
2) **Security architecture** - This is the subset of the enterprise architecture that provides the security mechanisms that protect the system and its information from unauthorized viewing, loss, or damage.

An architecture may be the actual information technology system or it may be a model of that system. In most cases, when one speaks of an "architecture," they may mean the "actual system" or they may mean the "model." The speaker should say "architecture model" when speaking of the model.

An architecture model may be divided into three categories:

1) **Operational architecture** - This provides a description of the tasks and activities, operational elements, and information flows to accomplish or support an operation. It is the total aggregation of missions, functions, tasks, information requirements, and business rules for an organization.
2) **Technical architecture** - This provides a minimal set of rules governing the arrangement, interaction, and interdependence of system parts or elements, whose purpose is to ensure that the system as defined satisfies a specified set of requirements stipulated by the users. It consists of the "building codes" upon which systems are based.
3 **Systems architecture** - This is a description, including graphics, of physical systems and interconnections (and their attributes) providing for, or supporting, physical systems and functions. It is the physical implementation of the operational architecture. It describes the layout and relationship of systems and communications.

Archiving - This is the copying of files (applications or data) to a storage device (such as a diskette, magnetic tape, or optical disc) for long-term storage or backup purposes.

Assurance - This is the set of policies and procedures that protect and defend information and information systems by ensuring their availability, data integrity, confidentiality, identification and authentication, authorization, accountability, access control, no object reuse, and non-repudiation. It refers to a basis for believing that the defined functionality will be achieved; it includes tamper resistance, verifiability, and resistance against circumvention or bypass.

Asynchronous - This term refers to processes in which data streams can be broken by random intervals.

Asynchronous transfer mode (ATM) - This is a communications system with high performance capable of transporting data, digital voice, and video economically. It is capable of using available network bandwidth by clumping information in small blocks that can be transmitted separately at high data rates but not contiguously. ATM is based on transferring data in cells or packets of a fixed size and is sometimes referred to as "Cell Relay." ATM uses a fixed route between source and destination. Risks to ATM information include physical security, internal attacks, and international connections.

Attack sensing, warning, and response (ASW&R) - This provides protection of the centralized portion of the PKI. ASW&R is sometimes called intrusion detection.

Attribute - This is a property of an entity that reveals the value (quantitively and qualitatively) of some characteristic of the entity. For example, the attributes of a computer system may include the values of its reliability, availability, and mean-time-to-failure and its color.

Audit - This is the recording of network activity and later examination of the records to identify user or application network access and activities. An audit means collecting information in an *audit trail* that is a chronological set of records that provide evidence of systems activity and can be used to reconstruct, review, and examine any transaction from inception to output of final results. **Auditing** is the activity of conducting an audit.

Audit record or trail or log - This is a record of computer events, about an operating system, an application, or entity activities. A computer system may have *multiple audit records*, one for each type of activity. Audit records can be used to *provide user accountability*. They can be analyzed to *discern flaws* in the system. They can be used in a *system-specific sense* such as determining access to specific files or applications. *Application audit records* can be used to determine the specifics of user activities for a particular application. An entity *audit record* can be used to track the activities of a specific user. An audit record should include the following:

a) Date and time of the event,
b) The identity of the user or process,
c) The type of event(s) that occurred,
d) The origin of the request or the name of the object's security level,
e) The file or application that was accessed, and
f) The success or failure of the event.

Audit record information should be protected from viewing, modifying, or deleting by any unauthorized user. The integrity and confidentiality of the audit record should be ensured.

Audit record tools - These tools can perform the following: 1) audit reduction (i.e., cull out clearly legitimate events); 2) trends/variance-detection (detect anomalies in user or system behavior); and 3) attack signature-detection (e.g., failed logon attempts).

Authentication - This is the process of an entity providing proof that the entity attempting to logon to the system is "who they say they are." Strong authentication is the process of providing almost "foolproof" evidence, such as a one-time password, smarter card, or a biometric identifier. It is the process of proving that an entity is who or what they claim to be. It is a measure used to verify the eligibility of a subject and the ability of that subject to access certain information. It protects against the fraudulent user of a system to the fraudulent transmission of information. There are three classic ways to authenticate oneself: something that you know, something you have, or something you are.

Authentication header (AH) - This provides integrity-checking information (cryptographic checksum) so that one can detect if the packet's contents were forged or modified while in route over the Internet.

Authenticator - This is a record sent using a ticket to a server to certify the entity's knowledge of the encryption key in the ticket, to help the server detect replays, and to help choose a "true session key" to use with the particular session.

Authenticity - This is a measure designed to provide protection against fraudulent transmission by establishing the validity of a transmission, message, station, or originator.

Authorization - This is the granting of rights to a user or application and includes the granting of access based on specific access rights. The provider of authorization answers the question "Are you allowed to do that?" This is often referred to as the entity's "pedigree."

Availability - This is the security goal that generates the requirement for protection against intentional or accidental attempts to: a) perform unauthorized deletion of data or b) otherwise cause a denial of either service or data to authorized users. It is the property that ensures that a resource is accessible and usable upon demand by an authorized principal. Availability is dependent upon the following four capabilities:

1) *Protection from attack* - This is partly achieved by closing holes in operating systems, firmware, and network configurations. It is also achieved by protecting the system against Denial-Of-Service (DoS) attacks. Anti-virus software also protects the system against attackers (viruses, worms, etc.) with malicious intent.

2) *Protection from unauthorized use* - This is achieved by ensuring that no unauthorized entity (person or process) gets access to an application or data such as by improperly using legitimate User IDs and Passwords.

3) *Resistance to failures* - There are two types of failures, routine and catastrophic. Routine failures, such as power outages, can be minimized through "loss of support" protection such as uninterruptible power supplies (UPSs) and redundant subsystems such as clustered servers with load balancing and automatic fail-over software. Protection against catastrophic failures, such as earthquakes and floods, can be provided through backup and recovery capabilities where critical data is stored in off-site backup servers and can be used to replace lost data in a timely manner.

4) *Resistance to undue latent response times* - Examples of these excessive times are: a) logging onto a system that takes an excessive amount of time to conclude that the user is authenticated; b) excessive time to connect following a single sign-on request for another application or database; and c) excessive time to respond to any action by the user. These types of availability problems can be addressed through the use of top quality programmers and programming techniques for operating system, middleware, and application software.

Thus, improved availability can be achieved through system adjustments that include encryption and authentication improvements, software patches, DoS detection and anti-DoS software, as well as the use of backup servers and associated software.

Backbone - This is the central part of a network, which carries the heaviest traffic, transmits at the fastest rate, and connects smaller networks that have lower data-transfer rates.

Backend systems - These are systems that comprise servers, which provide database management and network operating system capabilities.

Backplane - This is the reverse side of a panel or board that contains interconnecting wires. Or, it is a printed circuit board or device containing slots or sockets for plugging in boards or cables.

Bandwidth - This is the carrying capacity of a circuit, usually measured in bits per second (bps) for digital circuits, or hertz (Hz) or cycles per second (cps) for analog circuits. It is used to mean either how fast data flows on a given transmission path or somewhat more technically, the width of the range of frequencies that an electronic signal occupies on a given transmission medium. Any digital or analog signal has a bandwidth. Generally speaking, bandwidth is directly proportional to the amount of data transmitted or received per unit time.

Baseband - This is a communication technique in which digital signals are placed onto the transmission line without change in modulation. It is usually limited to a few miles and does not require the complex modems used in broadband transmission. Command baseband LAN techniques are token ring and Ethernet. In baseband, the full bandwidth of the channel is accomplished by interleaving pulses using TDM. Contrast with broadband transmission, which transmits data, voice, and video simultaneously by modulating each signal onto a different frequency using FDM.

Baseband transmission - This is a technique for transmitting digital data up to approximately 10Mbps using low frequency transmission over twisted pair or coaxial cable. Baseband transmission distance is usually limited to a couple of miles.

Basic input/output system (BIOS) - This is a set of low-level routines in a computer's ROM that enables application programs and operating systems to read characters from the keyboard, output characters to printers, and interact with the hardware in other ways. Many plug-in adapters include their own BIOS modules that work in conjunction with the BIOS on the system board.

Bastion host - This is a computer system that is part of a firewall configuration and hosts one or several application gateways. A bastion host must be highly secure in order to resist direct attacks from the Internet.

Biometric identification - This is the use of a characteristic of a person (e.g., fingerprint, voiceprint, facial infrared pattern, handwriting pattern, pupil pattern, retinal pattern, or other) that involves a reading device, a comparison program, and a template of the user's characteristic.

Biometric identifier - The particular personal characteristic used in a biometric identification process.

Birthday attack - This attack on hash functions is defined as follows. Let the hash function be a 64-bit hash. The birthday function states that it takes only 2^{32} hashes (about 4 billion) to find a collision, not 2^{64} (about [(4 billion) squared]). Thus, the attacker formulates two messages, one that he thinks is the proper message and one that is the bogus message. He then computes a hash for the proper message and computes 2^{32} hashes of the bogus message. He then compares them until he finds two that match (a collision). He can then cut out the original message and replace it with the bogus message (leaving the hash value alone) and send the modified message to the victim. It will be considered legal since the hash functions will match.

Blades - These are servers that are thinner so that they can be stacked together. They take up less space and use less power. They do not have a standard backplane design so they are not compatible in a heterogeneous environment.

Block encryption - This is when a message is split up into a number of blocks of equal size and then each block is encrypted.

Bootp - This package supports automatic address assignment. *Dynamic Host Configuration Protocol (DHCP)* supports both automatic and dynamic IP address assignments. Both **bootp** and DHCP use UDP as their communications transport. The amount of time a host retains a specific IP address is referred to as the *lease period*. A short lease period ensures that only systrems requiring an IP address have one assigned.

Border gateway protocol (BGP) - This is a routing method protocol that involves two basic activities: 1) determination of optimal routing paths and the 2) transport of information groups (typically called packets) through an internetwork. The transport of packets through an internetwork is relatively straightforward. Path determination, on the other hand, can be very complex. BGP performs inter-domain routing in Transmission-Control Protocol/Internet Protocol (TCP/IP) networks. BGP is an exterior gateway protocol (EGP), which means that it performs routing between multiple autonomous systems or domains and exchanges routing and reachability information with other BGP systems. BGP was developed to replace its predecessor, the now obsolete *Exterior Gateway Protocol* (EGP), as the standard exterior gateway-routing protocol used in the global Internet. BGP solves serious problems with EGP and scales to Internet growth more efficiently. EGP is a particular instance of an exterior gateway protocol (also EGP), and the two should not be confused.

Bot - Short for robot, a computer program that runs automatically

Bounded media - Twisted-pair and fiber optic cabling are *bounded media*, whereas the atmosphere is an *unbounded medium*. Unbounded media have a host of security problems. *Light transmissions* through the atmosphere use lasers to transmit and receive network signals but bedause precisse alignment and a focused signal, these transmissions are relatively secure. However, *radio waves* broadcast using microwave signals can be more easily intercepted. *Fixed frequency signals*, such as radio station signals, use a single frequency as a carrier wave and can easily be monitored once an attacker knows the carrier frequency. *Spread spectrum signals* use multiple frequencies for transmission and are more difficult to monitor. Most transmissions rely on encryption to scramble the signal so that it cannot be monitored by outside parties. *Terrestrial transmissions* are completely land-based radio signals and are usually line-of-sight systems and can only be received locally. *Space-based transmissions* are signals that originate from a land-based system but are then bounced off one or more satellites that orbit the earth above the atmosphere. The greatest benefit of these systems is the range. But the larger the range, the more likely the signal will be monitored.

Bridge - This is a small box with two network connectors that attach to two separate portions of the network. It is nearly identical to a repeater but it looks at frames of data. It is a device that connects two LAN segments togrether, which may be of similar or dissimilar types, such as Ethernet and Token Ring. Bridges are inserted into a network to improve performance by keeping traffic contained within smaller segments. Bridges maintain address tables of the nodes on the network through experience. They learn which addresses were successfully received through which output ports by monitoring which station acknowledged receipt of the address. Bridges work at the data link layer (OSI layer 2), whereas routers work at the network layer (OSI layer 3). Bridges are faster than routers, because they do not have to read the protocol to glean routing information.

Broadband - This is used to reference a type of Internet connection. A broadband connection allows for extremely high speeds. It supports data, voice, and sometimes even video information.

Broadband-integrated services digital network (B-ISDN) - This extends the ISDN cocept further and is envisioned to meet the total communication needs of customers, ranging form a few bits per second to 150 Mbps. This bandwidth spectrum covers voice, data, image, video, and High Definition TV services. In addition, the broadband network will provide interactive and multimedia conmmunications.

Broadband transmission - This is a technique for transmitting large amounts of data, voice, and video over long distances. Using high frequency transmission over coaxial cable or optical fibers, broadband transmission requires modems for connecting terminals and computers to the network.

Buffer overflow - This is when more data is received by a process than the programmer ever expected the process to see, and no contingency exists for dealing with an excessive amount of data except a modification the software package.

Bus - A bus is a path through which a device sends its data so that it can communicate with the CPU and/or other devices. Typically, each bus has a uniquely shaped interface to prevent you from damaging devices by plugging them into the wrong ports. PCs usually have three or more buses.

C2 requirements (DOD controlled access protection) - C2 requirements include the following (from the DOD Orange Book):

1) A Discretionary Access Control policy enforced in the network to prevent unauthorized users from reading the sensitive information entrusted to the network
2) No Object reuse
3) Accountability (identification, authentication, and audit)
4) Assurance
5) System integrity
6) Confidentiality and data integrity
7) Life-cycle assurance (i.e., security testing)
8) Documentation (Security User's Guide, Trusted Facility Manual, Test Documentation, and Design Documentation)

Cable modem - At the cable company's head end controller, IP traffic is modulated, or converted to an analog signal, for delivery over a spare downstream television channel. When the signal reaches the user site, it reaches a splitter. The splitter sends normal cable TV traffic to the user's set-top box. The Internet traffic is shunted off to the user's "cable modem" where it is demodulated, or converted back into Ethernet packets, and fed into the user's computer using a 10Base-T Ethernet card. The process is reversed for upstream traffic, but on a channel with less bandwidth, resulting in slower upload speeds. Users also have to share bandwidth with other cable modems in their neighborhood. Cable modems are a good choice in a residential area that already has a cable TV infrastructure. It then becomes a popular choice for telecommuters or people who have home offices. Cable modem services also are inexpensive, and are frequently bundled with cable TV packages. But there are disadvantages, too. The Ethernet card limits download speeds to just 4Mbps.

Cache - This is a place where temporary copies of objects are kept. If a server is used to store temporary copies of objects, for example Web pages, then the server is called a *Caching Server*.

Carnivore - This is the e-mail sniffing software, similar to a wiretap, which the FBI uses to monitor alleged criminals. Carnivore is a program that monitors packets of data passing through an Internet service provider's network, so this software can track users surfing on the Web. Critics have pointed out that Carnivore can be configured to capture all traffic on the network, not only the e-mail to and from the person on the warrant (as issued by a judge to the FBI), and that there is effectively no audit trail.

Certificate - This is a computer-based record (usually a template) that binds a subscriber's identity with the subscriber's public key in a trustable association.

Certificate authority (CA) - This is an organization that issues and manages security credentials and public keys for message encryption and decryption. Often it is a part of the Public Key Infrastructure because it manages the process of issuing and verifying the certificates used to grant entities access to other systems.

Certificate management - This involves the generation, production, distribution, control, tracking, and destruction of public/private keys and associated public key certificates. Certificate management is composed of Certificate Authority and Directory Services. Certification Authorities serve as trusted thrid parties and provide digitally signed-certificates for users and components. The certificate management process is responsible for:

1) Digitally signing each certificate,
2) Managing the revocation of certificates by publishing a certificate revocation list or by providing a mechanisms for a real-time check fo the revocation status,
3) Archiving required certificate managmeent information to support non-repudiation of digital signatures, and
4) Providing tools and procedures for personnel resopnosible for user registration.

Certificate obligations - There are responsibilities when certificates are used.
Certificate Authority (CA) obligations:

1) Accuracy of representations - The CA is obligated to all who reasonably rely on the information contained in the certificate that it has issued the certificate to the named subscriber and that subscriber has accepted the certificate.

2) Notification of certificate issuance - The CA is obligated to ensure that the subscriber who is the subject of the certificate and others who reasonably rely on that certificate are notified of the certificate issuance in accordance with this policy.

3) Notification of revocation or suspension of a certificate—The CA is obligated to ensure that the subscriber who is the subject of the certificate and others who reasonably rely on that certificate are notified of the certificate revocation or suspension in accordance with this policy.

4) Maintain certificate information - The CA is obligated to maintain records necessary to support requests concerning its operation, including audit files and archives. CAs may be required to account for certificate related user equipment.

Registration Authority (RA) obligations:

1) Accuracy of representations - The RA is obligated to accurately represent the information it prepares for a CA, to process requests and responses timely and securely.

2) Maintain certificate application information - The RA is obligated to keep supporting evidence for any certificate request made to a CA (e.g., certificate request forms).

3) An RA who is authorized to assume other CA functions may have obligations commensurate with a CA role; this situation will be described on a case-by-case basis.

Subscriber obligations:

1) Accuracy of representations in certificate applications - Subscribers are obligated to accurately represent the information required of them in a certificate request.

2) Protection of subscriber private key - Subscribers are obligated to protect their private keys at all times, in accordance with this policy and local procedures.

3) Notification of CA upon private key compromise - Subscribers are obligated to notify the CA that issued their certificates upon suspicion that their private keys are compromised.

4) Proper use of certificate - Subscribers are obligated to abide by all restrictions levied upon the use of their private keys and certificates.

A subscriber who is found to have acted in a manner counter to these obligations will have its certificate revoked, and will have no claim against the CMI in the event of a dispute.

Relying party obligations:

1) Proper use of certificates - Relying parties are obligated to use the certificate for the purpose for which it was issued.

2) Revocation or suspension checking responsibilities - Relying parties are obligated to check each certificate for validity, as described in the X.509 standard, before use.

3) Digital signature verification responsibilities - Relying parties are obligated to verify the digital signature of the CA who issued the certificate they are about to use.

4) Establishing trust in CA - Relying parties are obligated to establish trust in the CA who issued the certificates they are about to use by verifying the chain of certificates at root of which a trusted CA exists. The path processing should be based on the guidelines set by the X.509 v3 Amendment.

A relying party who is found to have acted in a manner inconsistent with these obligations will have no claim against the CMI in the event of a dispute.

Certificate registration - This process to register certificates requires Registration Authorities (RAs) with software tools recognized by the infrastructure. An RA is an agent of a Certificate Authority (CA). An RA is usually established to collect the information that is to be entered into a certificate. The CA may also assign other duties or responsibilities to an RA; in the extreme case, the CA signs any request made by an authorized RA. Security requirements imposed on the CA are likewise imposed on any RAs to the extent that the RAs are responsible for the information collected.

Certificate revocation list (CRL) - This is a list of user certificates for which the begin-and-end dates that appear on the certificates are no longer valid, or which for any other reason, the certificates have been revoked.

Certification - This is a technical evaluation performed as part of, and in support of, the accreditation process that establishes the extent to which a particular computer system or network design and implementation meet a specified set of requirements.

Change management - This is a formal process for reviewing, modifying, and documenting changes to system software. It can also be defined for changes to the system hardware.

Checksum or cyclical redundancy check (CRC) - This is a mathematical method that permits errors in long runs of data to be detected with a very high degree of accuracy. It is a mathematical verification of the data within a file. If files have a checksum attached (for ensuring data integrity), then it is far more difficult to replace the file with a Trojan horse or modify it without the receiver being able to detect that a change was made.

Cipher Suite - This is a set of authentication, encryption, and data integrity algorithms used for exchanging messages between network nodes. During an SSL handshake, for example, the two nodes negotiate to see which cipher suite they will use when transmitting messages back and forth.

Classified information system - This is an information system designed in accordance with Department of Defense standards to handle Confidential, Secret, or Top Secret information.

CLIPPER and CAPSTONE chips - These chips enable a computer to comply with the escrowed encryption standard using the law enforcement access feld mechanism.

Closed router - This is a router that peforms for internal clients only.

Clustering - This is where two or more servers are connected into a single package (*clustered servers*) and is similar to redundant servers except that all systems take part in processing service requests. The reason that clustering is more attractive that server redundancy is that the secondary systems are actually providing processing time ensuring the highest level of system utilization.

Cold spare - This is a secondary device that must be manually installed to take over for a failed primary device.

Common criteria - This is a multi-part standard (ISO/IEC 15408) that defines criteria that are to be used as the basis for evaluating security properties of information technology products and systems. By establishing such a common criteria base, the results of an information technology security evaluation are meaningful to a wider audience.

Competent - This is a person who has some experience with a particular situation and environment, but is not an expert, and will likely use a rule-based process for resolving a situation.

Computer Oracle and password system (COPS) - This package was written by Dan Farmer at Purdue University. It is a collection of tools that can be used to check for common configuration problem, on UNIX systems. It checks for items such as weak passwords, anonymous FTP, or just FTP, and inappropriate permissions.

Computer Security Act - This law requires federal agencies to provide for the mandatory periodic training in computer security awareness and accepted computer security practices for all employees who are involved with the management, use, or operation of a federal computer system within or under the supervision of the federal agency. This includes contractors as well as employees of the agency. Hospitals, health services, banks, and companies that do the majority of their business over the Internet are especially fearful of private information being divulged.

Common criteria - These are the multi-part standards (ISO/IEC 15408) that define criteria that are to be used as the basis for evaluating security properties of information technology products and systems. By establishing such a common criteria base, the results of an information technology security evaluation are meaningful to a wider audience.

Common vulnerabilities and exposures (CVE) - This is a list of standardized names and descriptions for vulnerabilities and other information concerning security exposures.

Compression - This is the downsizing of information to a smaller set using some sort of algorithm where the information can be returned to its original set by a decompression algorithm. Compression can be exact or approximate, so that the decompressed information is either exactly reproduced or approximately reproduced. Approximate reproductions are fine in many picture contexts. Data must be returned to its exact original state.

Confidentiality - This is the security process of ensuring that information (either stored or transmitted) is protected from being revealed to any unauthorized entities.

Configuration - This is the combination of elements, either software or hardware or both, that form a computer system, and the way they are set up for working together based on a number of possible choices.

Configuration control - This is the use of a tool for controlling the configuration of a system. It can be automatic or manual or both.

Configuration management - This is a collection of processes and tools that promote network consistency, track network change, and provide up-to-date network documentation and visibility. By building and maintaining configuration management best-practices, you can expect several benefits such as improved network availability and lower costs. These include:

1) Lower support costs due to a decrease in reactive support issues.
2) Lower network costs due to device, circuit, and user tracking tools and processes that identify unused network components.
3) Improved network availability due to a decrease in reactive support costs and improved time to resolve problems.

Conformance - This is the capability of a system element to meet necessary standards or requirements. By running standard test scripts, conformance testing can ensure that a product meets the relevant standards.

Connection - In data communications terminology, it is a logical link established between application processes that enables these processes to exchange information. In the OSI Reference Model, it is an association established by one layer with two or more entities of the next higher layer for the transfer of data. In TCP/IP, it is a logical TCP communications path identified by a pair of sockets, one for each side of the path.

Connection-oriented communication - These links exchange control information referred to as a *handshake* prior to transmitting data. It also requires acknowledgements after each packet transmission.

Connectionless communication - These links require neither a handshake nor an acknowledgement. They rely on the stablity of the underlying layers. The User

Datagram Protocol and the NetWare Core Protocol are examples of connectionless transports.

Consistent - This is a specification that infers that no requirements subset is in conflict with any other subset.

Constraint - This is a limitation on some entity, such as an architecture or some element of an architecture.

Constructable - This is a situation for a system such that there exists a systematic approach to formulating requirements for the system.

Content delivery network (CDN) - A *content delivery network* is a network of servers that delivers a Web page to a user based on the geographic locations of the user, the origin of the Web page and a content delivery server. A CDN copies the pages of a Web site to a network of servers that are dispersed at geographically different locations, caching the contents of the page. When a user requests a Web page that is part of a CDN, the CDN will redirect the request from the originating site's server to a server in the CDN that is closest to the user and deliver the cached content. The CDN will also communicate with the originating server to deliver any content that has not been previously cached.

This service is effective in speeding the delivery of content of Web sites with high traffic and Web sites that have global reach. The closer the CDN server is to the user geographically, the faster the content will be delivered to the user. CDNs also provide protection from large surges in traffic

The process of bouncing through a CDN is nearly transparent to the user. The only way a user would know if a CDN has been accessed is if the delivered URL (i.e., IP address) is different from the URL that has been requested.

Contingency measures - These are measures for emergency response, backup operations, and post-disaster recovery for an information system, to ensure the availability of critical resources and to facilitate the continuity of operations in an emergency situation. *Disaster prevention* is the set of precautionary steps one takes to ensure that any disruption of an organization's information system does not affect the day-to-day operations. *Disaster recovery* is the set of precautionary steps one takes to ensure that any disruption of an organization's information system can be overcome in a reasonable amount of time. This requires *contingency planning*, that is, defining:

1) The required actions needed for how an organization can recover from a disaster,
2) Who is responsible for managing and performing these required actions,
3) What backup systems are needed for the organization to promptly recover,

Disaster solutions fall into two categories: 1) maintaining or restoring a service, and 2) protecting or restoring lost, corrupted, or deleted information. *Mirroring* requires that a system have a backup system that is performing the exact same actions at the exact same time as the primary system and can immediately replace the primary system should it fail for any reason. *Good disaster recovery procedures* start with appropriate network cabling. One of the biggest improvements one can make to improve network availability is to replace Thinnet and Thicknet cabling with a more modern solution. Category 5 (CAT5) cabling is the current cabling standard for most network installations. Even though gigabit Ethernet is based on fiber, concessions have been made to include CAT5 cabling for short cable runs (50 to 75 meters). CAT5 cabling can support up to 100Mbps bandwidths.

Cookies - These are packets of information that Web sties can put on a user's computer and are often used to personalize a Web site by identifying the actions that a particular user performs on the site. Many persons are fearful of the possible invasion of personal privacy from these cookies.

Core - This is the set of minimal essential elements of a system without which the system could not operate. A core architecture is the simplest descriptive model, or set of models, representing the principal elements of a system architecture.

Countermeasure - This is the specific safeguard required to minimize the potential negative effects from attacks based on a group of threats (one or more threat scenarios) and can be either a method (e.g., training) or a product (e.g., a firewall or anti-virus software). It is any proactive measure taken to protect an information system against physical, behavioral, or technical threats.

Credentials - This is a record that is needed to establish the claimed identity of a principal.

Critical infrastructure protection remediation plan - This is the annually updated plan that details who the CIPP will be implemented.

Critical system - This is a system that is crucial to the functioning of an organization and the accomplishment of its mission.

Criticality - This is a measure of how important the correct and uninterrupted functioning of the system is to national security, human life, safety, or the mission of the using organization.

Cryptanalysis - This is the methodology used by experts to break cryptographic encoding algorithms.

Cryptography - This is the transformation of plaintext information into information that can only be read by authorized users. Cryptography can provide for *confidentiality, data integrity, non-repudiation* (i.e., digital signatures), and *advanced user authentication*. It can be provided by secret key systems (symmetric key cryptography), public key systems (asymmetric key cryptography), or hybrid systems (symmetric session key). It depends on an *algorithm* (i.e., a transformation function) and *one or more keys*. Mathematically, a plaintext message, P, can be transformed to an encrypted message, C, using an encryption algorithm, E, with a cryptographic key, K, as follows: $C = E_K (P)$.

Cyberattack - This is an attack on the information technology capabilities of an organization, usually through its Internet connection.

Cyberspace - This is the Internet or public wide area network domain.

Cybersecurity - This is a term to denote the security aspects of public wide area networks.

Cybersystems - This is a term for the information technology systems used by commercial and government organizations.

Cyber-terrorism - This is the use of information technology systems as the object of terrorist activities.

Daemon - This is a program that is constantly running as a background task looking for a request for a specific service.

Dark fiber - This is optical fiber infrastructure (cabling and repeaters) that is currently in place but is not being used.

Database - This is a storage array of information within computer memory.

Database management system (DBMS) - This is the middleware that accepts inputs and controls the information provided to, and stored in, the organizational databases.

Database server - In a client/server system, this is the computer that houses the DBMS and the database information.

Data dictionary - This is a database that contains information about the stored data in the database and is often referred to as "metadata.".

Data element - This is a basic unit of information having a meaning with a distinct unit and value. It is uniquely named and defined with a "cell" into which it is placed.

Data encryption standard (DES) - This is the encryption standard used by the U.S. government for protecting sensitive, but not classified data. For export, a 40-bit key is used and a 56-bit key is used for U.S. organizations. Triple DES can also be used for U.S. organizations. DES is a symmetric key (i.e., single key) algorithm.

Data entity - This is the representation of a set of people, objects, places, events, or ideas that share the same characteristic relationships.

Data integrity - This is a requirement that information (either stored or transmitted) can only be changed by authorized entities.

Data link - This is the physical interconnection between two points in a network. It may also refer to the modems, protocols, and all required hardware and software to perform a transmission.

Data link layer - This is layer 2 of the OSI model.

Data management - This provides management of information within an organization. It is software that allows for the creation, storage, retrieval, and manipulation of files interactively from a terminal or personal computer in a network.

Data-over-voice technology - This is the network technology to provide simultaneous transmission of voice and data or slow-scan video images from one computer to other computers.

Data separation - This requires prevention of disclosure to either authorized or unauthorized entities by preventing an adversary from getting to data or even knowing that it exists or through prevention of access to conflicting data for authorized entities. This can be achieved using Access Control Lists (ACLs) through the role-based access control (RBAC) mechanism provided by the Oracle DBMS.

Decision rules - These are rules that a person (i.e., decision-maker) uses to select an option from among several alternatives. There are three types of decision rules:

1) *Comparative elimination* - Comparing two alternatives and selecting the preferred one. Then continuing until there is a single option remaining.
2) *Systematic evaluation* - Rank ordering the alternatives and selecting the one with the highest ranking, and
3) *Personal judgment* - Selecting the preferred alternative, even if there is only one alternative, based on an intuitive feel for the preferred option.

Decryption - Whenever a message sent over a public network is encrypted, the receiver must be able to decrypt the message in order to read it. Decryption is the process of decrypting a message. Decryption is also called *Decipherment*.

De facto standard - This is a method or product that enjoys widespread acceptance in the industry but is not generally recognized by one of the official standards-setting bodies.

Defense data network (DDN) - This system provides a common-user packet switched network for all DoD long haul data communications.

Defense information infrastructure (DII) - This is a seamless network of communications, computers, software, databases, applications, and other capabilities that meet the information processing and transport needs of DoD users during peacetime and during all crises, conflicts, humanitarian support situations, and wartime.

Defense information system network (DISN) - This a subset of the DII and is DoD's consolidated world wide enterprise-level telecommunications infrastructure that provides the end-to-end information transfer network for supporting military operations.

Defense-in-depth - This is the security approach whereby layers of information assurance are used to establish an adequate security posture. Implementation of this strategy also recognizes that due to the highly interactive nature of the various systems and networks, information assurance solutions must be considered within the context of the shared risk environment and that any single system cannot be adequately secured unless all interconnected systems are adequately secured.

Defense messaging service (DMS) - This is all the hardware, software, procedures, standards, facilities, and personnel used to exchange messages electronically between organizations and individuals within the DoD.

Defense switched network (DSN) - This is a telecommunications system, which provides end-to-end common user and dedicated telephone service for the DoD.

De jure standard - This is a method or product that was developed and is recognized by one of the official standards-setting bodies.

Demilitarized zone (DMZ) - This is a separate segment (interface) on a firewall that typically has less stringent security policies than the secure internal LAN segment. It is an area between a router and a local area network where a server can be placed, such as an email server, that is separate from the LAN but can respond to the public email traffic. Any database info that this email server, should it be destoryed or modified, can be replicated from LAN internal info either offline or in an out-of-band mode. This offers protection to internal LAN systems while offering proper email responses.

Denial of service (DoS) - This is a situation where a malicious attacker floods a server with commands that the server cannot handle thus preventing the server from providing services that it should render and denying its services to authentic users. When DoS is deployed using a distributed set of hosts, then this is called distributed DoS (DDoS).

Descriptive process - This is a description of how a real person in a real situation actually performs some task or set of tasks.

Designated approving authority (DAA) - This is a senior management official, appointed in writing by a member of the Management Board, who determines whether or not to authorize a system for operation or to remove that authorization.

Developer or developing organization - This is the organization that has primary responsibility for developing or acquiring an information system. If a contractor develops a system, the organization responsible for that contract is the developing organization.

Diffie-Hellman encryption - This is the first encryption that used two separate keys for encrypting information based on exponentials.

Digital certificate - Using the X.509 format, it can provide a very strong method for authenticating a user's identity. It can also be used for storing the user's public key and for performing single sign-on and advanced access control such as fine-grained access. It can also contain the user's *distinguished name (DN); a* string of characters identifying the user in significant detail. Certificates are usually stored on a server, often designated as a *digital certificate server.*

Digital communication - This is analgous to the telegraph in that certain patterns of pulses are used to represent different characters during transmission. Digital communication is usually much more efficient than anallog circuits, which require a larger amount of overhead in order to detect and correct errors in noisy transmissions.

Digital encryption standard (DES) - This is a private key algorithm adopted as the federal standard for the protection of sensitive unclassified information and used for the protection of commercial data.

Digital encryption standard encryption algorithm - This is the only publicly available cryptographic algorithm that has been endorsed by the U. S. Government. Plaintext is encrypted in blocks of 64 bits, yielding 64 bits of cipher text. The algorithm, which is parameterized by a 56-bit key, has 19 distinct stages. The algorithm was designed to allow encryption and decryption to be done with the same key.

Digital signature - This is a group of bits that uniquely identify the source. It is based on the user entity's distinguished name or some random set of bits.

Digital subscriber line (DSL) - The foundation of this technology is the four-wire telephone cable that is standard throughout North America. Until recently, only two wires were needed for voice calls. The other two were largely unused until the mid-90s, when explosive demand for Internet bandwidth led several firms to look beyond the 56K bit/sec limit on analog dial-up connections. Early DSL standards required installing a signal splitter at the customer premises to separate the voice and data as well as a "DSL modem" to connect the user's PC to the data-stream. The two extra wires in standard telephone cable were employed to carry digital data to and from the local telephone office, which had to be located no more than 2.5 miles away. At the Telco office, a digital subscriber line access multiplexer (DSLAM) serves as the connection between the individual DSL streams and the backbone network. Top speeds are 1.5Mbps but at a much lower cost than a T-1 (1.544Mbps) or ISDN (128Kbps) connection.

DSL is often known as "xDSL" to account for the many versions of DSL that have arrived on the scene. A common variant is asymmetric DSL (ADSL). Asymmetric refers to the unequal upstream and downstream speeds—typically, there is much higher bandwidth for downloading than uploading. ADSL technology is aimed at home users and small offices that are mostly interested in high-speed access to Web content, and even streaming media. Download speeds can top 8Mbps, depending on the distance to the local Telco office. Upload speeds are much slower - typically a maximum of 640Kbps is possible. The faster the downstream DSL connection then the higher the monthly price.

Symmetric DSL (SDSL) has equal upstream and downstream speeds of up to 1.5Mbps in both directions. Very high bit rate DSL (VDSL) is a developing asymmetric alternative that can reach downstream speeds of 55Mbps, but only at a distance of up to 1000 feet from the local Telco office. Beyond that, VDSL transmission rates drop dramatically, which severely limits its customer base.

Also gaining in popularity is DSL Lite, or G.Lite. This is yet another asymmetric variant, but what makes it different is that it does not require a splitter at the customer premises. The line is instead split at the Telco office, which cuts down installation costs. Speeds range from 1.5Mbps to 6Mbps downstream, and about one-tenth those rates for upstream.

Digitized signature - This is a computerized image (viz., a scanner image or bitmap) of a person's actual signature.

Directory - This is a collection of files and other subdirectories. Directories are information containers, like files. However, instead of text or other data, directories contain files and other directories. In addition, directories are hierarchically organized; that is, a directory has a parent directory "above" and may also have subdirectories "below." Similarly, each subdirectory can contain other files and also can have more subdirectories. Because they are hierarchically organized, directories provide a logical way to organize files.

Directory services - There are flexible, special-purpose distributed databases designed to enable the rapid storage and retrieval of entry-oriented information for a wide range of applications. Directory services have the following attributes: 1) flexible, 2) special-purpose, 3) distributed, 4) manage entry-oriented information, and 5) support a wide range of applications. A *directory service* is a database that can inventory and control every directory-enabled resource on the nework. *True directory services* provide secure, location-independent access to the widely dispersed Internet/intranet/extranet resources today's increasingly mobile end users need. *Directory services* are good for providing access control lists and for managing security credentials, such as that required for single sign-on. Relational databases and directory services have the differences noted in the table below.

Directory Services	Relational Databases
Entry-oriented	Table-oriented
Global data	Location-specific data (e.g., a table)
Usually requires distributed access	May require distributed access
Globally-consistent naming policies	Application-consistent naming policies
High access-to-update ratio	Low access-to-update ratio
Performance emphasis on retrieval	Performance emphasis on transactions

Discretionary access control (DAC) - This is the situation where a user or application has access to particular information that is granted based on their appearance in an access control list.

Dispatcher - The Dispatcher in the SNMP engine sends and receives SNMP messages. It also dispatches SNMP protocol description units (PDUs) to SNMP applications. When an SNMP message needs to be prepared or when data needs to be extracted from an SNMP message, the Dispatcher delegates these tasks to a message version-specific Message Processing Model within the Message Processing Subsystem.

Distinguished name (DN) - This is the name in the digital certificate that includes the complete set of attributes of an entity's identity.

Distributed processing - This is a system where information resources and responsibilities are allocated over the entire network to levels consistent with mission needs, required interoperability, and applicable architectures.

Distribution list - This is a list containing the names of mail users and/or other distribution lists. It is used to send the same message to multiple mail users and can be private or public.

Domain - This is a group of workstations and servers associated by a single security policy. Domain information is stored on *domain controllers*. Domains can be configured to be trusting. When a domain trusts another domain, it allows users of that *trusted domain* to retain the same level of access they have in the trusted domain. Domains can be trusted in a uni-directional or bi-directional manner. You cannot check your trust relationships from a central location; you must to each primary server in each domain to see what trust relationships have been set up. *Trusts can be used to enhance security*, but the number one rule is to keep it simple, such as by trying to keep the number of trust relationships to only one or two.

Domain name system or service (DNS) - This is a distributed database system that provides name-to-IP address mapping for hosts on a network such as the Internet. It enables local users to give an easy-to-remember name to a local host that is then mapped to an IP address by the DNS server. This means that the user does not have to remember numbers of the form xxx.xxx.xxx.xxx for each host on the Internet.

Domain name system poisoning (or cache poisoning) - This is the process of handing out incorrect IP address information for a specific host with the intent to divert traffic from its true destination.

Dual active - This is a solution that enables two units that can perform the same duties (they may be unequal in the quantity of duties, but are both as busy through load balancing).

Due diligence - This is the meeting of standard controls of acting responsibly in acquiring and implementing security controls to protect the information system. After identifying and adhering to the standards of due diligence, one should:

1) Identify areas of particularly elevated risk
2) Address security risks not covered in the particular set of baseline controls that you use,
3) Obtain consensus when different sets of baseline controls are contradictory, and
4) Provide business justification for employing one or more security controls when the justification is less than compelling.

Dumpster diving - This is when an attacker rummages through an organization's trash (e.g., in a dumpster) in an effort to find company private information.

Dying gasp - This is a message from a router to its nearest routers that it is about to go offline.

Dynamic data exchange (DDE) - This is a mechanism used in the MS Windows OS to transfer data between tow applications or two separate instances of the same application.

Dynamic entity -This is an entity that is subject to change with time.

Dynamic host configuration protocol (DHCP) - Whenever the client and server exist on two separate network segments, the DHCP relay agent acts as a proxy between the two systems.

Dynamic link library (DLL) - This is a special type of Windows program containing functions that other programs can call, has resources (such as icons) that other programs can use, or both. DLL links are defined at runtime, hence the adjective "dynamic."

Dynamic random access memory (DRAM) - This is a read-write memory used to store data in PCs and must be refreshed each time the computer is turned off and then back on.

Electromagnetic interference (EMI) - This is produced by circuits that use an alternating signal, like analog or digital communications (referred to as an alternating current or AC circuit). EMI is not produced in DC circuits. While twisted-wire cabling has become very popular due to its low cost, *twisted-pair cabling is also extremely insecure* due to its EMI radiation. *Fiber Optic Cabling* consists of a cylindrical glass thread center core 62.5 microns in diameter wrapped in cladding that protects the central core and reflects the light back into the glass conductor and is encapsulated in a jacket of tough KEVLAR fiber and sheathed in PVC or Plenum with a total diameter of 125 microns and is often referred to as 62.5/125 cabling. Fiber optic cabling does not produce EMI and hence is much more secure.

Electronic data interchange (EDI) - This is the computer-to-computer exchange of routine business information in a standard format. The *Electronic Commerce/EDI (EC/EDI) system* is a store-and-forward system that handles business transaction information much like an electronic mail (e-mail) system. It looks at the header information and routes the data packet, ignoring the body of the message. If an intruder were to compromise the switching feature of the system, based on information in the header, he could probably obtain a copy of the whole message. Technical equipment such as network analyzer devices (i.e., sniffers) could be used to capture individual message packets based on header information. Sniffers can also be used for record field analysis, data manipulation and insertion, and messages replayed based on information in the packet header.

Electronic mail (e-mail) - This is a computer network application that allows messages to be passed between computers on the network.

Emacs - This is a popular screen editor for UNIX and most other operating systems.
Emacs has an entire LISP system inside it. Emacs contains modes for assisting in editing most well-known programming languages. Most of these extra functions are configured to load automatically on first use, reducing start-up time and memory consumption.

Emerging standard - This is a standard that is still in formal review and has not yet gone through the balloting process.

Encapsulation - This is the process of ensuring that data inside an object is protected from the outside world and can only be seen provided a message is sent to the object requesting specific information.

Encapsulating security payload (ESP) - This encrypts the data contents of the remainder of the packet so that the contents cannot be extracted while in route over the Internet. The encryption information is selected using the security parameter index.

Encryption - This is the process of changing a message in plaintext into unreadable text in a manner that the receiver can change the message back to plaintext. *End-to-end encryption* is the case where a message is encrypted at the source and not decrypted until it reaches its destination. Routing data is not encrypted. *Link-to-link encryption* is the case where a message (including routing data) is encrypted at the first link then decrypted, routed, and then encrypted at each link node thereafter, and then decrypted at the last link node prior to its reception at the destination node. The *RSA encryption algorithm* is considered the *de facto* standard in public/private key encryption. Reversing the encryption of a message is called *decryption*. Encryption is also called *Encipherment*.

End node - This is the machine or unit (node) that serves as an originator and as the final destination of network traffic. The end node does not relay traffic originated by other nodes.

End system - In the OSI model, it is the computer containing application processes that can communicate through all seven layers. It is equivalent to an end node. An *end-system* is either the *source* of a TCP/IP packet or the *destination* of the packet.

End user - This is an end-entity that represents the source of inputs to, or destination of outputs from, the network PC.

Enterprise - This is the entire organization, which may consist of a single site or of distributed sites, either all over the country or all over the world.

Enterprise JavaBeans (EJB) - This is a server-side component architecture for writing reusable business logic and portable enterprise applications. EJB is the basis of J2EE.

Enterprise JavaBean components are written entirely in Java and run on any EJB compliant server. They are operating system, platform and middleware independent, preventing vendor lock-in. EJB servers provide system-level services (the "plumbing") such as transactions, security, threading, and persistence. The EJB architecture is inherently transactional, distributed, multi-tier, scalable, secure, and wire protocol neutral, so that any protocol can be used, namely, Internet Inter-ORB (Object Request Broker) Protocol (IIOP), Java Remote Method Protocol (JRMP), Hypertext Transfer Protocol (HTTP), Distributed Component Object Model (DCOM), and others. EJB 1.1 requires the Remote Method Innovation (RMI) for communication with components.

Enterprise network - This is the aggregation of a system consisting of many intranets and the links connecting them.

Entity - An entity is a person (e.g., user), application (e.g., some computer program that performs some organizational function), or service (e.g., middleware subsystem or domain name service) or server.

Escrowed encryption - This is the system by which secret keys are stored for the purpose of key recovery. The secret keys are held in escrow until an authorized entity requests access to one. The entity then uses the escrowed key to recover the actual key used to encrypt a particular message.

Ethernet - By far, the most popular local area network (LAN) topology is the Ethernet. Ethernet's communication rules are called "carrier sense multiple access and collision detection (CSMA/CD)":

1) *Carrier sense* means that all Ethernet stations are required to listen to the wire at all times (even when transmitting).Multiple access means that more than two stations can be connected to the same network and all stations are allowed to transmit whenever the network is free.

2) *Multiple access* means that more than two stations can be connected to the same network and that all stations are allowed to transmit whenever the network is free

3) *Collision detection* resolves the issue of a simultaneous transmission by two or more stations.

It is possible to configure a system to read all information that it receives (*promiscous mode*). This capability is the biggest flaw with Ethernet. A network analyzer is effectively a computer operating in a promiscous mode.

Expert - This is a person who has much experience with a particular situation and environment and can use an intuitive process for resolving situations.

Extensible markup language (XML) - This is a grammatical system for constructing custom markup languages. XML, like HTML, can be written with any text editor or word processor, including the very basic TeachText or SimpleText on the Macintosh and Note pad or Wordpad for Windows. Some text editors will provide a test of the document as it is being typed. XML uses the same building blocks as HTML does, namely elements, attributes, and values.

Extension point - This is an augmentation to access points in a wireless network to extend the area of user roaming.

Fabric - This is the backplane (a hard-wired device) with the capacity of a hardware device (e.g., 5Gbps) for switching electronic information traffic.

Fault management - This allows the network manager to detect, isolate, and find the cause of network faults.

Fault tolerance - This is the capability of a system component to provide a capability to deal with network failures and to maintain continuity of operations of a network. It includes the following features: a) error/fault detection, b) fault treament, c) damage assessment, d) error/failure recovery, e) component/segment crash recovery, f) archive creation and access, and g) whole network crash recovery.

Federal acquisition reform act (FARA) - This is part of the Defense Authorization Act of Fiscal Year 1996. The FARA and ITMRA form the Clinger-Cohen Act of 1996.

Fiber channel - This is a networking technology that send data over copper or fiber-optic cable at speeds up to 2Gbps. It supports distances of up to 1 km over fiber with a design that is primarily for block mode high-performance data transfers.

Fiber distributed data interface (FDDI) - This is a ring topology but a second ring has been added to rectify many of the problems found in Token Rings. FDDI is considered to be a dying technology since no effort has been made to increase speeds beyond 100Mbps. FDDI can be run in a full duplex mode that allows both rings to be active at all times. However, the redundancy of the second ring is not in force in this case. A FDDI network can have either a ring or star topology.

File server - This is a computer in a network, with a hard drive, used to process network commands and to store network files.

File transfer - This is the capability, within a network, to move a file from point A to point B via any transmission system or network.

File transfer and management (FTAM) - This is a standard communications protocol for the transfer of files between systems of different vendors.

File transport protocol (FTP) - This is a protocol that is used to transfer information from one system to another using the TCP as its transport and ports 20 and 21 for communications. *Passive FTP (PASV FTP)* is the mode supported by most Web browsers.

Finger - This is a service that allows the source to identify an account by user name in order to collect information about the identified user. This is one of those services that are typically overlooked but can provide a major security hole.

Fingerprint - This is a unique sequence of bytes known to be contained in some particular virus. *Anti-virus software* looks for a given set of fingerprints to locate potential viruses. Anti-virus software works only if the traffic is not encrypted.

Firewall - This device represents an approach to security; it helps implement a larger security policy that defines the services and access to be permitted, and it is an implementation of that policy in terms of a network configuration, one or more host systems, and routers, and other security measures such as advanced authentication in place of static passwords. The main purpose of a firewall is to control access to/from a protected network. A firewall can be a router, a personal computer, a host, or a collection of hosts that are set up specifically to shield a site or subnet from protocols and services that can be abused from hosts outside the subnet. The primary components of a firewall are: 1) network policy, 2) advanced authentication, 3) packet filtering, and 4) application gateways. Firewalls operate in one of two basic modes, 1) allow all to pass except what is specifically to be blocked, and 2) block all except what is specifically allowed to pass. The newer firewalls include the capabiilty to perform as a VPN encryption using an SSL capabiilty. Firewalls can be divided into three categories:

1) *Packet filters,*
2) *Stateful inspection* or dynamic filters, and
3) *Application-level gateway* or *proxies* (requires the opening of two sessions, one to the client and the other to the Internet host).

Firewalls should provide the following capabilities:

a) Automatic failover,
b) Scalability,
c) Be compatible with current security policies,
d) Resource optimization, that is, provide load balancing and load sharing for clusterd servers, if required,
e) The option to provide virtual private network encryption using the secure sockets layer protocol,
f) No single point of failure, and
g) Automatic health checking capabilities.

Firewalls provide little protection against insider attacks uness the firewall sysetm is appropriately arranged. Other problems not addressed by firewalls are:

a) *Multicast IP Transmission (MBONE)* for video and voice are encapsulated in other packets which are passed through the firewall.
b) Firewalls offer little or no protection against *Viruses.*
c) *Throughput* may go down if the firewall is overloaded.
d) Firewalls offer a *single point of failure* unless properly configured.

Firewall administration - A guide for administrating firewall operations should include the following:

1) The organization's Internet poliocy
2) How the grant/revoke user access to the Internet
3) How to modify existing rules as necessary
4) What to do in the event of an intrusion
5) How to secure new machines as they are installed
6) Procedures for reading and reviewing logs
7) Procedures for distributing token cards to employees.

Firewall verification - Verifying the implementation of a firewall can be accomplished with a checklist of questions:

1) *Policy* - Does the firewall's implementation adequately reflect the organization's Internet policy?
2) *Packet filter configuration* - Are the routers or other packet filters configured to use the simplest set of rules necessary?
3) *Acess control to systems* - Is access to every bastion host system controlled through something more than a simple user ID/Password?
4) *Access control to routers* - Is access control to routers limited to terminal access? Alternatively, is an authentication protocol (such as TACACS or RADIUS) used to control router access? Does the router place limitations on what can send it routing updates? SNMP? ICMP?
5) *Internet services configuration* - Are all externally available services configured securely?

Firmware - This is software (programs or data) that has been written onto read-only memory (ROM). Firmware is a combination of software and hardware. ROMs, Programmable Read Only Memory, and Erasable Programmable Read Only Memory that have data or programs recorded on them are firmware.

Flaming - This is the process of sending e-mail messages that are designed to cause the receiver to get very angry.

Footprint analysis - This is an analysis to identify the various devices that are attached to some network. A diagram of the devices and how they are connected is called a "footprint."

Formal risk assessment - This means that the assessment must be procedurally, mathematically, and logically rigorous.

Frame relay - This is a widely implemented packet-switching protocol that offers an alternative to virutal private network lines or leased lines. It is primarily used for data communications and is not recommommended for voice links. Frame relay is an inexpensive alternative to leased line networks. X.25 and frame relay are packet-switched technologies. Because data on a packet-switched network is capable of following any available circuit path, such networks are represented by clouds in graphical presentations. Both X.25 and frame relay must be configured as permanent virtual circuits, meaning that all data entering the cloud at some point A is automatically forwarded to its designated destination point B. For large WAN environments, frame relay can be far more cost effective than dedicated circuits. Frame Relay uses the same basic data link layer framing and Frame Check Sequence so current X.25 hardware works. It adds addressing (viz., a 10-bit data link connection identifier) and a few control bits but does not include retransmissions, link establishment, or error recovery.

Free software foundation (FSF) - This is an organization devoted to the creation and dissemination of free software, i.e., software that is free from licensing fees or restrictions on use. The Foundation's main work is supporting the GNU project, partly to support the position that information is community property and that all software sources should be shared.

Front-end systems - These are the clients that provide the direct computer system capability to the user, such as Web access via the browser.

Gateway - This is a device for converting one network's message protocol to the format used by another network's protocol. It can be implemented in hardware or software.

General support system - This is an interconnected set of information resources that share a common functionality under the same direct management control. These systems, which include software, host computers (mainframes, minis, and workstations), and networks (LANs and WANs), provide support for a variety of users and applications.

GNU - This is a recursive acronym, "GNU's Not Unix!" It is the Free Software Foundation's project to provide a freely distributable replacement for UNIX. The GNU C compiler and Emacs (gcc) are two tools designed for this project and have become very popular.

Gopher - This is a powerful yet simple file retrieval tool. It uses TCP as a transport and rests on port 70.

Government information - This is information that is created, collected, processed, disseminated, or disposed of by or for the federal government.

Granularity - This is the degree to which access to objects can be restricted. Granularity can be applied to both the actions allowable on objects, as well as to the users allowed to perform those actions on the object.

Hacker - This is anyone who uses the Internet or public wide area network to increase his or her knowledge or enter a local area network or computer in an unauthorized manner. *Malicious hackers* are those hackers who perform some sort of damage to the victim's system. A *cracker* is a special sort of hacker who enters a local area network and may or may not do some damage to the system.

Handshake - This is the exchange of signals between two data communications systems prior to, and during, data transmission to coordinate and control each phase of transmission over a serial connection.

Hardware - This represents the physical aspects of a network, such as computers, firewalls, and routers.

Hash function - This is any mathematical function that can be used to encode a message in such a way as to be computationally infeasible to decode unless one knows the coding method used.

Hashing - This is the process of using the hash algorithm in order to access a file on a large volume. The algorithm calculates a file address both in cache memory and on the hard disk and predicts the address on a hash table, which is much more efficient than searching for the file sequentially.

Header - This is the information at the beginning of a packet that defines control parameters such as size, memory requirements, and entry point of a program, as well as the locations in the program of absolute segment address references.

Heterogeneous system - This is a distributed system that consists of more than one vendor's products.

Heuristic scanners - These scanners perform a statistical analysis to determine the likelihood that a file contains program code that may indicate a virus. One of the biggest benefits of heuristic scanners is that they do not require updating so that they can detect both known and unknown viruses. Unfortunately they often report false positives.

High capacity storage system (HCSS) - This is a data storage system that extends the storage capacity of a NetWare server by integrating an optical disk library, or jukebox, into the NetWare file system. HCSS moves files between faster low-capacity storage devices (the server's hard disk) and slower high-capacity storage devices (such as optical discs in a jukebox).

High-level data link control (HDLC) - This is a bit-oriented synchronous data link layer protocol developed by ISO. Derived from Synchronous Data Link Control, HDLC specifies a data encapsulation method on synchronous serial links using frame characters and checksums.

Hole - A hole in a program, operating system, firmware, or application, is a location in the code where an intruder is able to enter the software system and perform malicious activities. Holes can be created accidentally during the development of the software or it can be intentionally created for later exploitation. Holes can be located by software mechanisms called "sniffers" or "scanners." In most cases, vendors will publish patches to fix holes in their software.

Home directory - This is a directory where you keep personal files and additional directories. By default, the File Manager and Terminal Emulator windows are set to the home directory when you first open them.

Honey pot - This is a fake computer system on the Internet designed to lure malicious hackers who attempt to penetrate other people's computer systems. To set up a honey pot, it is recommended that you: 1) install the operating system without patches and use typical defaults and options, 2) make sure that there is no data on the system that cannot safely be destroyed, and 3) add an application designed to record the activities of the intruder. Maintaining a honey pot may require a considerable amount of attention and its value may be nothing more than a learning experience because you may not catch any intruders.

Host - A host is any mechanism on your network that has an IP address. This includes client computers, servers, printers, image scanners, routers, and firewalls. However, in many cases when "host" is used, it is intended to mean only the client computers or even servers. The reader must somehow determine the author's intention.

Host name - This is the logical name given to a host device.

Hot spare - This is a secondary device that is on the network and waiting to enter service should the primary device fail. Excellent for "no single point of failure prevention."

Hubs - These are boxes of varying sizes that have multiple female connectors designed to accept one twisted-pair cable outfitted with a male connector for connecting a workstation or other device to the hub.

Hurd - This is the foundation of the whole GNU system. The GNU C Library will provide the UNIX system call interface, and will call the Hurd for needed services it can't provide itself. One goal of the Hurd is to establish a framework for shared development and maintenance. The Hurd is like GNU Emacs in that it will allow a broad range of users to create and share useful projects without knowing much about the internal workings of the system, namely projects that might never have been attempted without freely available source, a well-designed interface, and a multi-server-based design.

Hybrid cryptography - This offers the advantages of symmetric cryptography (such as speed, 1,000 to 10,000 times faster) while offering the advantages of asymmetric cryptography (such as ensured protection).

Hypertext transfer protocol (HTTP) - This protocol is used in communciations between Web browsers and Web servers. It includes the Multimedia Internet Mail Extensions (MIMEs) to support the negotiation of data types.

Hypertext transfer protocol secure (HTTPS) - This is the Hypertext Transfer Protocol using the Secure Sockets Layer as a sub-layer under the HTTP application layer.

In-band - A device is in-band if it is accessible from anywhere in the network on which it resides. Contrast with out-of-band.

In-band communication - This is any path within a network for normal information transmission.

Incident - This is any action that happens on a computer system that is considered to be negative or malicious.

Incident handling - This is a capability to detect and react quickly and efficiently to disruptions in normal processing caused by malicious threats.

Industrial espionage - This is the act of gathering proprietary data from Government or commercial organizations for the purpose of aiding some commercial organization.

Identification - This is the process of providing some indication of who you are such as your name and/or your password.

Inference - This is the derivation of new information from known information. The inference problem refers to the fact that the derived information may be classified at a level for which the user is not cleared. The inference problem is that of users deducing unauthorized information from the legitimate information they acquire.

Information - This is any communication or representation of knowledge such as facts, data, or opinions in any medium or form, including textual, numerical, graphic, cartographic, narrative, or audiovisual forms. Information is categorized as follows:
Classified information - This is information that, in the interest of national security, requires protection against unauthorized disclosure as determined under Executive Order 12356. Such information is classified as Top Secret, Secret, or Confidential.
Other controlled or sensitive information - This is information whose unauthorized disclosure, modification, or unavailability would harm the agency. Sensitive information includes:

a) Essential or critical air traffic control information and any other information that must be protected in performance of an organizational mission to ensure confidentiality, data integrity, or availability of information.
b) Information identified in Executive Order 12958, requiring protection under the provision of the Privacy Act of 1974.
c) Information designated "For Official Use Only"
d) Information whose disclosure or modification might affect the contractual or resource management function.
e) Proprietary information.
f) Information identified in the Information Technology Management Reform Act of 1996.
g) Information falling under the auspices of the Computer Security Act of 1987.
h) Information protected under the Sensitive Security Information rule, 14 CFR Part 191.
i) Financial management information.
j) Information exempt from disclosure under the Freedom of Information Act.

Information assurance - This establishes grounds for confidence that the five security goals (viz., data integrity, availability, confidentiality, accountability, and non-repudiation) have been adequately met by a specific implementation. "Adequately met" includes: 1) system security functionality that performs correctly, 2) sufficient protection against unintentional errors (by users or software), and 3) sufficient resistance to intentional penetration or bypass.

Information owner - For information originating within the organization, the information owner is the manager responsible for establishing the rules for the use and protection of the subject information. For information originating elsewhere, the information owner is the originating entity, represented by a designated individual. The information owner retains responsibility for its security even when the information is shared with other organizations.

Information system - This is a finite and well-defined set of information resources either in stand-alone or networked configurations, that is organized for the collection, processing, maintenance, transmission, and dissemination of information in accordance with defined procedures, whether automated or manual. Information systems are of two types:

1) *General support systems* - These are interconnected information resources that are under the same direct management control and share common functionality, e.g., telecommunications and networks.
2) *Major application systems* - These are systems that require special management attention because of their importance to the agency's mission; their high-maintenance, development, or operating costs; or their significant role in dealing with the agency's programs, finances, property, or other resources.

Information system network - This is a set of information subsystems (e.g., routers, firewalls, cables, computers, hubs, printers, operating systems, applications, middleware, database management systems, databases, etc.) that are linked (i.e., connected) together to form a single system to achieving some organizational purpose.

Information system owner - See **System owner**.

Information systems security (ISS) - This is the set of collected attributes that describe the security measures taken to protect information systems and the organizational information resources, either individually or collectively.

Information system security certification (ISSC) - This is a comprehensive evaluation of the technical and non-technical security features of an information system and other safeguards, made in support of the authorization process to establish the extent to which a particular design and implementation meets a set of specified security requirements and that risk ahs been mitigated commensurate with magnitude of potential harm.

Information system security certification agent (ISSCA) - This is a senior level manger responsible for ensuring an impartial, quality control review of the SCAP, who makes recommendations to the ISS certification official and the DAA as appropriate. The certification agent signs the final ISS certification document before forwarding to the ISSM for review prior to the DAA deciding whether or not to authorize the system.

Information system security certification official - This is an employee in the organization that owns the information system who is responsible for certifying that system security technical controls are present and functional, management and physical controls are described and in place, and risk has been mitigated commensurate with magnitude of potential harm.

Information system security certifier (ISSC) - This is a senior manager in the development or operational organization that owns the information system and is responsible for certifying that the system security technical controls are present and functional, management and physical controls are described and in place, and risk has been mitigated commensurate with magnitude of potential harm.

Information system security manager (ISSM) - This is a Federal employee who is responsible for implementing the Agency's ISS program within a single line of business or staff office.

Information system security plan (ISSP) - This is a document that identifies the information system components; operational environment; sensitivity and risks; and detailed, cost-effective measures to protect a system or group of systems. The ISS plan must be maintained throughout the system lifecycle and is complete when selected controls are tested and the responsible organizational official signs the SCAP.

Information system security policy - This is the set of objectives, rules, and practices that are used to provide control over the manner in which sensitive or critical assets (including information) are managed, protected, and distributed within a network system and at interfaces with other systems.

Information sharing and analysis centers (ISACs) were formed by nineteen information technology companies, e.g., IBM, CSC, Cisco, Intel, Microsoft, Entrust, VeriSign, HP, Oracle, AT&T, CAI, EDS, KPMG, RSA, and Titan. The ISACs were called for under Presidential Decision Directive 63. Issued by Clinton in May 1998, PDD 63 set the requirements for critical infrastructure protection. The centers are intended to provide a mechanism for companies within the eight infrastructure sectors to share information about cyberthreats, vulnerabilities and solutions.

Information technology (IT) - This is the hardware, software, and networks that process information, regardless of the technology involved, whether computers, telecommunications, or other.

Information technology management reform act (ITMRA) - This is part of the Defense Authorization Act of Fiscal Year 1996. The FARA and ITMRA form the Clinger-Cohen Act of 1996. It repealed the provisions of the Brooks Act and eliminated the centralized authority of the General Service Administration. It provided the head of each executive agency with full authority to acquire and manage IT under existing laws. It required each executive agency to appoint a Chief Information Officer (CIO) and establish a process to acquire and manage IT investments in accordance with the ITMRA.

Infrastructure - This is the composite of computer and network devices, software, people, procedures, materials, tools, equipment, and facilities that provides the means to operate in a specific environment in order to perform some specific tasks, or achieve specific purposes, or support a specific mission.

Infrastructure subarchitecture - This model defines the communication network and the interaction and interdependence of the nodes in the network. It is used to standardize communication protocols between the various nodes so that they can share communications.

Integrated services digital network (ISDN) - This line can be used to obtain bandwidth up to 128Kbps. Often, dial-up connections make use of protocols, such as the Serial Line Internet Protocol (SLIP) and the Point-to-Point Protocol (PPP), that provide IP connectivity over standard telephone lines. SLIP and PPP offer almost full connectivity to the Internet. The only real difference between a PPP/SLIP connection to the Internet and a dedicated connection to the Internet, other than the difference in bandwidth, is the somewhat lower availability of PPP/SLIP links.

Integration - This is the effort that joins two or more similar products such as individual system elements, components, modules, processes, databases, or other entities (products) into a unified framework or architecture in a seamless manner. IEEE STD 610.12 defines an *integration architecture* as a framework for combining software components, hardware components, or both into an overall system.

Integrity - This is the quality of being uncorrupted. *Message integrity* refers to the state of a message not being modified while in transit. *File integrity* refers to the state of the files not being modified while in storage. *System integrity* is the state when a system is capable of operating in its proper modes. *Data integrity* (whether in storage or in transit) refers to the state of data not being modified by an unauthorized entity.

Integrity checker - This is a security tool that tries to determine whether data on a particular host has been improperly modified.

Interconnection - This is the manual, electronic, or optical communications path/link between systems and subsystems. It includes the circuits, networks, routers, bridges, gateways, relay platforms, switches, and other necessary equipment for effective communications.

Interface - This is a connecting link between two systems. In the OSI Reference Model, it is the boundary between adjacent layers. An interface not only refers to the physical port that devices plug into but also to the electrical operating parameters and the communications format as well.

International data encryption algorithm (IDEA) - This is an encryption method that uses a 128-bit key, is resitant to many forms of cryptanalysis, is very strong, and can be efficiently computed.

International standards organization (ISO) (or International organization for standards) - This is the organization that establishes international standards for computer network architectures. Its OSI Reference Model divides network functions into seven layers. Membership is by country with over 90 countries currently participating.

Internet (or World Wide Web) - This is a web of interconnected networks that use the TCP/IP protocols for communications. It is a collection of regional and backbone networks with associated switching devices and interconnecting links. It allows any user with the appropriate hardware and software to connect to, and communicate with, any other hardware/software system that is also connected to the Internet.

Internet control message protocol (ICMP) - This protocol is used for flow control, detecting unreachable destinations and re-direction routes, and checking remote hosts. A *ping* command is used to send these messages.

Internet daemon (inetd) - This program monitors each of the listed ports (sometimes called *sockets*) on a UNIX system and is responsible for *waking up* the application that provides services to that port. Well-known ports are de facto standards used to ensure that everyone can access services on other machines without needing to guess which port number is used by the service. Port numbers below 1024 make up the bulk of Internet communications. When a system requests information, it not only specifies the port it wishes to access but also which port should be used when returning the requested information. Port numbers from 1024 to 65535 (upper port numbers) are used for this task.

Internet message access protocol (IMAP) - IMAP v4 is designed to be the next evolutionary step from the Post Office Protocol. IMAP suport three different modes, online (online messages), offline (offline messages), and disconnected (only receives a copy of messages).

Internet protocol (IP) - This is a *packet-based protocol* used to exchange data over computer networks. IP handles addressing, fragmentation, re-assembly, and protocol demultiplexing. It is the foundation on which all other Internet protocols, collectively referred to as the Internet Protocol suite, are built. There are two newer versions of the IP, namely IPv6 (version 6) and IPng (next generation). IPv6 provides a new format that increases the number of IP addresses.

Internet protocol address assignment - There are three methods for assigning IP addresses to host systems:

1) *Manual:* The user manually configures an IP host to use a specific address.
2) *Automatic:* A server automatically assigns a specific address to a host during startup.
3) *Dynamic:* A server dynamically assigns free addresses from a pool to hosts during startup.

Internet protocol (IP) multicast - This is a bandwidth-efficient method for broadcasting data or streaming media across a network. Fundamentally, there are two methods of transmitting streaming data across the Internet: *Unicast* and *Multicast*. Unicast is the most common but uses the most bandwidth. In the unicast model, a server must deliver a stream of data for each client that wants to participate in the broadcast. For instance, a video that requires 100Kbps of bandwidth for a single user would require 100 times that amount for 100 users, or a total of 10Mbps of bandwidth. This can lead to network congestion and increased server load that will decrease the overall quality of the broadcast. IP Multicast sends a single stream from the server across the network that serves all users who want to receive the broadcast. In the prior example, this would mean the server could send out a single 100Kbps stream that can be picked up by each of the 100 clients, reducing the load on the network and servers significantly. The work is done by IP Multicast enabled routers, which send one stream to each router connected to it. This is done all the way through the network until the signal reaches the clients that want to receive the broadcast.

Internet protocol security (IPSEC) - This is a public/private key encryption algorithm that offers a set of open standards. It is implemented at the Transport or IP layer or layer 4. It is convenient to use and has been implemented by Cisco on its routers. IPSEC is an obvious VPN solution. At present, IPSEC is limited to a 40-bit key encryption.

Internet protocol version 6 (IPv6) - This protocol, under development by the IETF, is designed to expand the number of IP addresses and create an easier system for managing network-connected devices.

Internet small computer system interface (iSCSI) - This is a proposed standard for mapping SCSI commands over TCP/IP to extend storage connectivity to any distance over standard IP networks.

Internetwork - This is a network that connects other networks.

Interoperability - This is the capability of two or more products from different vendors to operate properly within a single network.

Intranet - This is a LAN, MAN, or WAN that belongs to a single organization (i.e., enterprise).

Intrusion detection - This is the process of monitoring computer and telecommunications traffic and reporting whenever the security of a network is jeopardized. Intrusion detection may be done in real-time or offline. Intrusion detection systems may collect traffic information for auditing and in some cases may respond to an intruder automatically.

Intrusion detection system (IDS) - This device can review the log entries in real time or offline to identify potential attackers. The complete log should contain audit trails, firewall logs, and any other pertinent log entries. All IDSs are capable of logging suspicious events. An IDS comprises two parts:

1) The *engine*, which is responsible for capturing and analyzing the traffic, and
2) The *console*, from which the engine can be managed and all reports are run.

Intrusion detection system countermeasures - These countermeasurees include 1) *session disruption* and 2) *filter rule manipulation*. *Session disruption* is the easiest to implement. It cannot prevent the attacker from launching an attack but it can prevent further damage. For *filter rule manipulation*, the IDS modifies the rules base to include a rule that prevents the attacker from transmitting additional traffic the the traget host. Unfortunately, a firewall that uses filter rule manipulation can be duped into a denial-of-service attack when it becomes so busy changing its rules base that is stops passing traffic.

Intrusion detection system authentication - There are two ways to authenticate an IDS:

a) *Weak authentication* requires the right public key during connection setup, and
b) *Strong authentication* requires a challenge based on the console's pubic key, private key, and a shared secret.

IP over ATM - This is a multiprotocol encapsulation over ATM Adaptation Layer 5.

IP packet - This is a 53-bytes portion of a complete message that may concompass many packets. Each packet comprises the following: 5-bytes header and a 48-bytes body.

ISO X.25 - This is the International Standard Organization's standard protocol for packet-switched communications.

ISO X.400 - This is the International Standard Organization's standard for electronic mail.

ISO X.500 - This is the International Standard Organization's standard for directory services.

ISO X.509 - This is the proposed International Standard Organization's standard for user certificates.

ISO X.800 - This is the International Standard Organization's standard for the Open Systems Interconnect Security Architecture, which outlines measures that can be used to secure data in a communicating open system by providing appropriate security service in each layer. X.800 defines a digital signature as a relatively low-level cryptographic mechanism. However, the term "digital signature" usually means an extension of that mechanism to enable use and/or reliance upon certificate validation during signature verification.

Isochronous - This term refers to processes where data must be delivered within certain time constraints. For example, multimedia streams require an isochronous transport mechanism to ensure that data is delivered as fast as it is displayed and to ensure that the audio is synchronized with the video. Isochronous can be contrasted with asynchronous or synchronous. Certain types of networks, such as ATM, are said to be isochronous because they can guarantee a specified throughput. Similarly, new bus architectures, such as IEEE 1394, support isochronous delivery. *Isochronous* service is not as rigid as *synchronous* service, but not as lenient as *asynchronous* service.

Java 2 platform, enterprise edition (J2EE) - This is a platform-independent, Java-centric environment from Sun for developing, building, and deploying Web-based enterprise applications online. The J2EE platform consists of a set of services, namely, Application Programming Interfaces (APIs) and protocols that provide the functionality for developing multi-tiered, Web-based applications. Some of the key features and services of J2EE are the following:

1) At the client tier, J2EE supports pure HTML, as well as Java applets or applications. It relies on Java Server Pages (JSPs) and servlet code to create HTML or other formatted data for the client.
2) Enterprise JavaBeans (EJBs) provide another layer where the platform's logic is stored. An EJB server provides functions such as threading, concurrency, security, and memory management. These services are transparent to the author.
3) Java Database Connectivity (JDBC), which is the Java equivalent to ODBC, is the standard interface for Java databases.
4) The Java servlet API enhances consistency for developers without requiring a graphical user interface (GUI).

Java - This is a portable object-oriented language. *Portable* means that the language can be run on any computer and its operating system. Java programs for clients are called *applets*. Java programs that provide services and run on servers are called *servlets*.

Java naming and directory interface (JNDI) - This system provides a standard way for Java applications to use directory and naming services. The key to writing LDAP directory-enabled applications is by using the LDAP *software development kit (SDK)*. There are SDKs for C, C++, and Java. The Secure Sockets Layer, version 3 (SSLv3, i.e., transport level security (TLS)) provides LDAP security over non-secure links. The most common operations that one can perform with a client SDK are connecting, authenticating (or binding), searching, and updating. Authentication can be performed using access control lists (ACLs) or access control instructions (ACIs). ACLs are simply lists of entity names with their identified privileges. ACIs are more complex in that they reside in the database as instructions that define an entity's privilege, or fine-grained access control.

Java sandbox - This is a special area in a Java Virtual Machine (JVM) where Java code can be executed but where the code can do no harm should it be malicious. The code cannot disrupt or affect any other sandboxes, even though the code may wreak havoc in its own sandbox. The JVM can allow untrusted application to execute in a trusted environment without fear of corruption or subterfuge. Additionally, if the *code is signed* and the signature is by a trusted source, then the JVM can choose to execute the code in a normal environment where it can interface with whatever objects it wishes.

Java server pages (JSPs) - These are a text-based, presentation-centric way to develop servlets. JSPs offer all of the benefits of servlets and when combined with a JavaBeans class, give you an easy way to keep content and display logic separate.

Kerberos - This is an authentication and key distribution system developed at the MIT. It is a centralized login authorization mechanism, which authenticates requests, such as login, file access, or peripheral usage, from Kerberos-equipped agents. A Kerberos server issues a ticket to a client or server who has successfully logged on. The ticket is used to grant certain privileges to the user or server and is good for a limited period.

Kernel - This a fundamental part of an operating system that resides in memory at all times and provides the basic services required by a user or application. It is the part of the OS that is closest to the machine and may activate the hardware directly or interface to another software layer that drives the hardware.

Key - This is a random number used by an encryption/decryption algorithm for encoding/decoding a message. There may be just one key (as in the case of a private key symmetric system) or two keys (as in the case of a public and private key asymmetric system).

Key distribution - This is the secure controlled movement of keys from the point of generation to the point of use.

Key management - This is the set of procedures and protocols, both manual and automated, that are used throughout the entire lifecycle of the keys. This includes the generation, distribution, storage, entry, use, destruction, and archiving of the cryptographic keys.

Key pair - In an asymmetric cryptosystem, it is the set of keys (public and private) for an entity.

Key recovery - This is a mechamism by which the keys that encrypt the data traveling between two users can be recovered by someone else, probably without the other's awareness. This mechanism is generally intended to allow eavesdroppimg by a third party.

Knowledge base - This is a set of information stored in a computer that contains a set of rules relative to some area of expertise such as repairing computers.

Knowledge management - This is software that makes it easy for customers and agents to look up information about your company.

Label switched path (LSP) - This is the path that a message (packet) takes based on the switch label provided by a Multi-Protocol Label Switching (MPLS) system.

Laptop - This is a portable computer that has a flat screen, is battery operated, and usually weighs less than ten pounds.

Latency - This is the delay inherent between the time of starting to send data and the time that data begins to arrive at its destination.

Laws - These are statutes that define the limits of legality. There are three levels of laws for information systems: 1) *Criminal Laws* require the perpetrator to go to jail or be fined if they are violated. Laws may be federal or state or local. 2) *Civil Laws* require that the perpetrator pay a plaintiff some amount to offset losses incurred due to actions by the plaintiff (i.e., the perpetrator). Laws may be federal or state. Civil law includes tort, contract, or intellectual property. An example of a tort law break is to commit denial of service on another person's network. An example of a contract law break is the breaking of a promise to do something. Intellectual law covers patent and trademark rights. 3) *Federal laws* define the limits of legality for the federal government such as:

- Defamation - Libel and Slander
- Sexual Harassment
- Privacy
- U.S. Export laws on export of cryptography methods.

Layer - This is a level of the OSI Reference Model.

LEAP - Numerous papers have been written regarding vulnerabilities with the implementation of RC4 in the wired-equivalent-privacy (WEP) encryption framework. Cisco Systems has been shipping a security scheme known as LEAP since November 2000. Based on the 802.1x authentication framework, LEAP mitigates several of the weaknesses by utilizing dynamic WEP and sophisticated key management. Also, Cisco has incorporated Media Access Control (MAC) address authentication for third-party clients to provide simple and easy management of device authentication.

Least privilege - This is the condition that a user has access to only the data the user needs to perform his or her job functions, and no more. Application of this principal limits the damage that can result from accident, error, or unauthorized use.

Legacy system - This is the current system.

Level of concern - This is a rating assigned to an information system that indicates the extent to which protective measures, techniques, and procedures must be applied. Usually there are three levels:

1) *Basic* is for an information system that requires implementation of the minimum standard.
2) *Medium* is for an IS that requires layering of additional safeguards above the minimum standards.
3) *High* is for an IS that requires the most stringent protection measures and rigorous countermeasures.

Lifecycle - The lifecycle, a description of all of the phases of a system, is divided into two categories: 1) *Information* - These are the phases through which information passes typically characterized as creation or collection, processing, dissemination, use, storage, and disposition. 2) *System* - These are the phases through which a system (e.g., an information system) passes and is sometimes characterized as initiation, development, operation, termination, and decommissioning.

Lifecycle assurance - This is the basis for believing that the various phases of a system's life cycle are being properly planned, managed, implemented, and performed. Intrusion detection systems collect traffic information for auditing.

Light emitting diode (LED) - The light in a fiber optic cable is produced by a light emitting diode and another diode also receives the signal. Light transmission can take two forms:

1) *Single-mode transmissions* consist of an LED that produces a single frequency of light that is faster and can travel long distances but is expensive and installation can be difficult.

2) *Multimode transmissions* consist of multiple light frequencies and are less expensive than single mode but are subject to light dispersion (the tendency of light rays to spread out as they travel).

Lightweight Directory Access Protocol (LDAP) - This is an enterprisewide protocol service for managing information about users, groups, and roles and is used to get the user's credentials.

LDAP data interchange format (LDIF) - This is a widely used file format that describes directory information or modification operations that can be performed on a directory. LDIF is completely independent of the storage format used within any specific directory implementation and is therefore implementation-neutral. Typically, it is used to export directory information from, and import data to, LDAP servers.

Link - This is the physical realization of connectivity between two system nodes.

Link access procedure, balanced (LAPB) - This is a data link layer protocol in the X.25 protocol stack. LAPB is a bit-oriented protocol derived from High-Level Data Link Control.

Link encryption - This is the individual data encryption on one or all individual links of a data communications system.

Link state protocol (LSP) - This is a message from a router to its nearest routers that it is up and running.

Linux - This is an open systems operating system based on UNIX. It is an implementation of the UNIX kernel originally written from scratch with no proprietary code. The Sun SPARC, IBM PowerPC, Apple PowerMAC, and 68k ports all support shells, X, and networking. The Intel and SPARC versions have reliable symmetric multiprocessing.

Work on the kernel is coordinated by Linus Torvalds, who holds the copyright on a large part of it. The rest of the copyright is held by a large number of other contributors and employers. Regardless of the copyright ownerships, the Linux kernel is available under the GNU General Public Release. The GNU project will support Linux as its kernel until the research Hurd kernel is completed. This kernel would be no use without application programs. The GNU project has provided large numbers of quality tools, and together with other public domain software is a UNIX environment. A compilation of the Linux kernel and these tools is known as a Linux distribution. Compatibility modules and/or emulators exist for dozens of other computing environments.

LISP (List processing language) - This is artificial intelligence's mother tongue, a symbolic, functional, and recursive language based on the ideas of lambda-calculus, variable-length lists, and trees as fundamental data types and the interpretation of code as data and vice-versa. Data objects in LISP are lists and atoms. Lists may contain files and atoms. Atoms are either numbers or symbols. Programs in LISP are themselves lists of symbols, which can be treated as data. All LISP functions and programs are expressions that return values.

List - In computer jargon, a list usually means a table such as a directory of user names, their identifications, and access rights to appilcations and databases (e.g., an access control list).

Listener - A listener is a process that resides on the server whose responsibility is to listen for incoming client connection requests and manage the traffic to the server.

Load balancing - This is the distribution of traffic to devices for the purposes of high availability and resource optimization. Distributed decisions are dynamically made based on a load-balancing algorithm.

Load balancing algorithm - This algorithm uses the parameters upon which traffic is distributed to resources to identify the appropriate traffic loads for all relevant devices.

Load sharing - This is when two or more devices share responsibility for a single task, such as an organization's Internet security. Load sharing does not respond to a single firewall in the farm becoming overloaded. Typically, load sharing is achieved by network segmentation.

Local area network (LAN) - This is a network located within a small area or common environment, such as in a building or a building complex.

Local registration authorities (LRAs) - These are an emerging class of certificate-based security management solutions. They assist corporations secure their intranets, extranets, and electronic commerce solutions by tapping into the potential of digital certificates

Logical access controls - These are controls that are computer-based, that is, those that are provided by computer software.

Logical network - This is a group of systems assigned a common network address by the network administrator.

Logon - This is the process whereby a user identifies himself or herself to the computer.

Long haul system - This is the communications system that transfers information over long distances.

Loop - This is either a closed circuit or a single connection from a switching center to an individual communications device

Loop back - This is a method of performing transmission tests on a circuit that does not require the assistance of personnel at the far end.

Loop start - This is a method of calling a central office by applying a closed direct current loop across the line.

Loop-start trunk - This is a two-wire central-office trunk or dial-tone link that recognizes an off-hook situation by putting a 1000-ohm short across the tip and ring leads when the handset is lifted. This is the most common type of line, also called a POTS line. See also off-hook.

Loop test - This is a method of testing a link from a single point.

Macro - This is a set of code that provides special programs for a user. It is like having a personal version of some software that provides specific capabilities for this particular user. In some cases, viruses are developed to infect certain macros and these are called *macro-viruses*. If a virus infects a macro that enables a certain document, say an attachment to an e-mail message, then this is the case of a *macro-enabled document virus*. Should the client open a virus infected macro-enabled document (that is attached to an e-mail message) then the virus will be activated and can perform malicious actions, such as erasing the hard disk.

Major application - This is an application that requires special attention to security because of the risk and magnitude of the potential harm that could result from the loss, misuse, or unauthorized access to, or modification of, the information in the application. Such a system might actually comprise many individual application programs and hardware, software, and telecommunications components.

Malicious code - There are several kinds of malicious code. A *worm* is an application that can replicate itself via a permanent or dial-up network connection. A *virus* is a set of computer code that can reproduce itself. It is usually attached to some piece of software and may damage or erase databases and other software. A *bomb* is a virus that waits for some specific event to begin running and do its malicious handiwork. A *Trojan horse* is an application which hides a nasty surprise and is a stand-alone application which had its bomb included from its original source code.

Malicious host - This is a network server that attempts to do damage of some sort to one or more devices or packages on the network system.

Management information base (MIB) - This is a database of network management information about the configuration and status of nodes on a TCP/IP-based internetwork. MIB is used by the Common Management Information Protocol (CMIP) and the Simple Network Management Protocol (SNMP).

Mandatory access control - This access control method (developed by DOD for clissifed information access) is used when the objects (e.g., files and databases) as sell as the subjects (e.g., users and applications) have classified labels so that only objects with appropriate labels can be accessed by subjects with their labels. The use of the mandatory access control requires that all objects (e.g., information) and subjects (e.g., people) be appropriately labeled.

Man-in-the-middle attack (or session hijacking) - This attack begins with the attacker forcing the client to crash by sending the client a Ping of Death or through a utility such as WinNuke. Now the attacker can communicate with the server as if it were the client. This can be done by capturing the server's replies in order to formulate an appropriate response. These kinds of attaacks can most easily be countered by using encrypted transmissions.

Marshalling - This is the process of gathering data from one or more applications or non-contiguous sources in computer storage, putting the data pieces into a message buffer, and organizing or converting the data into a format that is prescribed for a particular receiver or programming interface. Marshalling is usually required when passing the output parameters of a program written in one language as input to a program written in another language. *Demarshalling* is the reverse of the marshalling process, that is, it undoes what a marshalling process does.

Media access control (MAC) - In the OSI reference model, it is the data-link layer protocol that governs communication between the data-link and physical layers for controlling the use of the network hardware. The MAC includes numbers used by all the systems attached to a network (PCs and Macs included) to uniquely identify themselves.

Media control interface (MCI) - This is the Windows OS component that allows multimedia devices such as a CD-ROM drive or videodisk player to be programmmed using high-level function calls that insulate the software from the nuances of the hardware.

Message - This is a group of words transmitted over a network from a source to a destination that conveys some information.

Message digest 5 (MD5) - This is an algorithm that assures data integrity by generating a unique, 128-bit cryptographic message digest value from the contents of a file. If a little as a signal bit value in the file is modified, the MD5 checksum for the file will change. Forgery of a file in a way that will cause MD5 to generate the same result as that for the original file is considered extremely difficult.

Messaging application programming interface (MAPI) - This is the API developed by Microsoft and other computer vendors that provides Windows applications with an implementation-independent interface to various messaging systems such as Microsoft's, Novel's, or IBM's e-mail system.

Meta-data - This is data about the data. That is, data that describes data. In general, any term (e.g., meta-physics) that begins with the prefix "meta" means the "use of the <u>second term</u> to describe or study the <u>second term</u>." Thus, meta-physics is use of physics to describe or study physics. Similarly, a meta-language is a language used to describe languages.

Microcomputer - This is a personal computer that usually adheres to the standards of IBM's or Apple's personal computers or both.

Microsoft Internet information server (IIS) - This system adds Web, FTP, and Gopher functionality to the NT Server. Both the NT Server and the IIS have vulnerabilities. The Network Monitor Tool installs a network analyzer similar to Novell's Analyzer or Network General's Sniffer, except that it can only capture broadcast frames or traffic traveling to and from the NT server.

Microsoft Windows NT - This is an operating system for PCs that allows two types of permission, 1) share permissions and 2) file permissions. *Share permissions* are enforced when users remotely attach to a shared file system. The share permissions are checked to see if the user is allowed access. *File permissions* are access rights that are assigned directly to the files and directories. File permissions are enforced regardless of method used to access the file system. This means that while a user would not be subjected to share permissions if he accessed the file system locally, he would still be challenged by the file-level permissions. Access permissions for a Web server are only regulated by file-level permissions; share permissions have no effect. NT accesses include: no access, list, read, add, add and read, change, full control, and special access. The auditing button allows you to monitor who is accessing each of the files on hour sever. All NT events are reported through Event Viewer. You should plan on reviewing your logs on a regular basis. NT events include: logon and logoff, file and object access, use of user rights, user and group management, security policy changes, restart and shutdown and system, and process tracking. Windows NT supports static packet filtering of IP traffic. The *Distributed Component Object Model (DCOM)* is an object-oriented approach to making Remote Procedure Calls. *Supporting the DCOM application across a firewall is a severe security threat.* One of the caveats about DCOM is that raw IP address information is passed between them. This means that network address translation cannot be used. NAT is typically used to translate private IP address space into legal IP address space for the purpose of communicating on the Internet. If the DCOM server is sitting behind a device performing NAT, then DCOM will not work. You can use DCOM applications across the Internet using private address space if the data stream will be traveling along a VPN tunnel. This is because a tunnel supports the use of private address space without performing NAT.

Middleware - This is the software that resides between the user (client) and the applications or other software such as database management systems that reside on the servers or mainframes.

Migration - This is the process of upgrading the legacy system until it becomes some defined target system.

Minimal specification - This is a specification that does not overly constrain system design.

Mirror - This means to duplicate a disk, partition, server, or other device.

Mission - This is a succinct statement of an organization's objectives together with the purpose of the implied actions.

Mission area - This is the general class to which an operational mission belongs.

Mission support systems - These are systems that are not used for operational air traffic control services, but are unique to the performance of the organization's mission.

Mistake - This is an error that happens due to a misunderstanding of the proper principles involved (e.g., a programming error either due to an erroneous requirement or poor programming).

Modem - This stands for modulator/demodulator and is a device that can convert electronic traffic to digital bits and vice versa. It is a device that receives and transmits digital computer traffic.

Modular - This describes a program that is written in component pieces that can be run independently. Object-oriented programming is used to develop self-contained modules that run independently, but that work together when plugged in.

Multihoming - This is the case where a client has two or more upstream Internet providers. These service providers could be other ISPs that the client exchanges backup service with (e.g., large backbones such as MCI, Sprint, UUNet, or Agis/Net99; regional backbone providers; or other local ISPs). Multihoming is a good idea since the provider that is best today may be quite poor tomorrow and vice versa. The best way to achieve redundancy is to obtain multihoming.

Multilevel security - This is security that is provided within a single network where data (objects) and users (subjects) can be assigned any security level (Super Classified, Top Secret, Secret, Confidential, and Unclassified). This type of security is not considered in this book.

Multilevel security system - This is a system that has multilevel security capabilities.

Multiplexer - This is a device that divides a channel's bandwidth by frequency or time to allow multiple devices to share a channel.

Multiprotocol label switching (MPLS) - This is a standards-based approach to applying label switching to large-scale networks being defined by the IETF since early 1997. MPLS defines protocols and procedures that enable the fast switching capabilities of ATM and Frame Relay to be used in IP networks. It represents the next level of standards-based evolution in combining layer 2 (data link layer) switching with layer 3 (network layer) routing in order to create flexible, faster, and more scalable networking. This includes traffic-engineering capabilities, which provide, for example, aspects of Quality of Service and facilitate the use of VPNs.

Multithreading - This is the simultaneous execution of more than one thread within a process or application. Multithreading and task are similar and are often confused. Today's computers can only execute one program instruction at a time, but because they operate so fast, they appear to run many programs and serve many users simultaneously. The computer operating system (for example, Windows 95) gives each program a "turn" at running, and then requires it to wait while another program gets a turn. Each of these programs is viewed by the operating system as a "task" for which certain resources are identified and kept track of. The operating system manages each application program in your PC system (spreadsheet, word processor, Web browser) as a separate task and lets you look at and control items on a "task list." If the program initiates an I/O request, such as reading a file or writing to a printer, it creates a thread so that the program will be reentered at the right place when the I/O operation completes. Meanwhile, other concurrent uses of the program are maintained on other threads. Most of today's operating systems provide support for both multitasking and multithreading. They also allow multithreading within program processes so that the system is saved the overhead of creating a new process for each thread.

Mutual recognition arrangement (MRA) - This is an agreement between two or more entities to accept Common Criteria certifications and validations completed by the other members of the arrangement.

Named representative - Often, the owner of an organization will name a person to represent him/her in certain situations, such as discussing technical issues. This person is called a "named representative."

National institute of standards and technology (NIST) - This is the division of the U. S. Department of Commerce that ensures standardization within Government agencies and for all manufactured goods and services in the U. S.

Net8 - This is Oracle's remote data access software that enables both client-to-server and server-to-sever communications across any network. Net8 supports distributed processing and distributed database capability. Net8 runs over, and interconnects, many communications protocols. Net8 is backward compatible with SQL*Net (version 2). A feature of Oracle Connection Manager, *Net8 access control*, sets rules for denying or allowing certain clients to access designated servers. Net8 access control is also known as firewall support. The *Net8 assistant* is a graphical user interface tool that combines configuration abilities with component control to provide an integrated environment for configuring and managing Net8. It can be used on either the client or server. Net8 assistant can be used to configure the following network three components: naming, naming methods, and listeners. *Net8 configuration assistant* is a post-installation tool that configures basic network components after installation, including: 1) listener names and protocol addresses, 2) naming methods, 3) net service names, and 4) directory server access. *Net8 open* is an application program interface to Net8 that enables programmers to develop both database and non-database applications that make use of the Net8 network already deployed in their environment. Net8 open provides, for any application, a single common interface to all industry standard network protocols. Net8 consists of three layers:

1) The Network Interface (NI) layer that hides the underlying network protocol and media from the client application.
2) The Network Routing (NR)/Network Naming (NN)/Network Authentication (NA) layer that takes care of routing the data to its final destination.
3) The Transparent Network Substrate (TNS) layer that takes care of generic communications such as sending and receiving data.

Network - This is the communications hardware and software that allow one user or system to connect to another user or system and can be part of a system or a separate system. Examples of networks include local area networks (LANs) or wide area networks (WANs) and public networks such as the Internet.

Network administrator - This is the person responsible for ensuring that a network is operating and maintained properly.

Network address translation (NAT) - This is the translation of an Internet Protocol (IP) address used within one network to a different IP address known within another network. *Stateful inspection* firewalls can use NAT to protect an internal LAN segment by hiding client's IP addresses from the Internet.

Network-attached storage (NAS) - This is a file server that attaches directly to the network, allowing file-level access by heterogeneous clients and servers without a separate storage network. These products can implement one or more distributed file system protocols to allow both clients and servers to access files that are stored (usually in RAID products) in a common shared storage pool. NAS is a thing whereas a Storage-Area Network (SAN) is a process. NAS servers are turnkey devices a) that accommodate a cluster of hard drives, b) the hardware needed to connect to a network, and c) the hardware and software to make it work.

1) NAS is a term used to refer to storage elements that connect to a network and provide file access services to computer systems. An NAS Storage Element consists of an engine, which implements the file services, and one or more devices, on which data is stored. NAS elements may be attached to any type of network. When attached to SANs, NAS elements may be considered to be members of the SAN Attached Storage (SAS) class of storage elements.

2) NAS can also mean a class of systems that provide file services to host computers. A host system that uses network-attached storage uses a *file system device driver* to access data using file access protocols such as Network File System (NFS) or Common Internet File System (CIFS). NAS systems interpret these commands and perform the internal file and device I/O operations necessary to execute them.

Network authentication service - This is a means for authenticating clients to servers, servers to servers, and user to both clients and servers in distributed environments. It is a repository for storing information about users and the services on different servers to which they have access, as well as information about clients and servers on the network, and can be physically separate machine, or a facility co-located on another server within the system.

Network backbone - This is the communications network that provides the transmission infrastructure upon which other capabilities are built.

Network basic input/output system (NetBIOS) - This is a protocol that provides the underlying communication mechanism for some basic NT functions, such as browsing and communication between network servers.

Network file system (NFS) - This is a distributed file system that allows data to be shared across a network regardless of machine vendor.

Network interface card (NIC) - This is a device (usually a card in a PC or other device) that allows communications over a local area network.

NetWare directory service (NDS) - This is a 32-bit, multitasking, multithreaded kernel. This kernel also provides support for symmetrical multiprocessors. The kernel is designed to be modular so that applications and support drivers can be loaded or unloaded on the fly. Most of the changes can be made without rebooting the system. The *Java Virtual Machine (JVM)* allows the server to support the execution of Java scripts. Users are arranged in a tree format.

Network file system (NFS) - This provides access to remote file systems. It is a distributed file system developed by Sun Microsystems that allows a set of computers to access each other's files cooperatively in a transparent manner.

Network layer - This is the third layer of the OSI Reference Model.

Network management system (NMS) - This system is responsible for managing at least part of a network. An NMS is generally a reasonably powerful and well-equipped computer such as an engineering workstation. NMSs communicate with agents to help keep track of network statistics and resources.

Network operating system (NOS) - This is an operating system for providing operational support for controlling network devices.

Network security - This is the protection of networks and their services from unauthorized modification, destruction, or disclosure.

Network security policy - This is the set of statements that an organization makes regarding its information system.

Network topology - This is the generic physical or electronic configuration describing a local communications network, such as a bus, ring, or star configuration.

Neural network - A neural network is a program (or hardware device) that learns from various examples of some specific phenomena. These examples express a variety of relationships so that the neural net discovers the relationships among the inputs. Because of its capability to learn, a neural network is insulated from the shortcomings that plague expert systems in terms of adapting to changing conditions.

Nmap - This is a scanner that is designed to allow system administrators and security analysts to scan an information system network to determine which hosts are up and what services they are offering. **Nmap** supports many different scanning techniques such as: UDP, TCP connect(), TCP SYN (half open), FTP proxy (bounce attack), Reverse-ident, ICMP (ping sweep), FIN, ACK sweep, Xmas Tree, SYN sweep, IP Protocol, and Null scan. It also offers a number of advanced features such as remote OS detection via TCP/IP fingerprinting, stealth scanning, dynamic delay and retransmission calculations, parallel scanning, detection of down hosts via parallel pings, decoy scanning, port filtering detection, direct (non-portmapper) Remote Procedure Call (RPC) scanning, fragmentation scanning, and flexible target and port specification.

Node - This is an element of an architecture where information is produced or consumed.

Nonce - This is a random number generated and used once.

Non-critical system - This is any information system that is not a critical system.

Non-repudiation - A message has a *receiver-nonrepudiation* capability if the receiver of the message cannot later deny having received the message and a *sender-nonrepudiation* capability if the sender cannot later deny having transmitted the message. Non-repudiation is provided by digital signatures.

Non-transparent - This is when firewall interfaces are the destination for traffic being sent to an internal LAN client or an external Internet host. Examples could be a NAT address or an HTTP proxy service.

No Object reuse - This is a security policy (stipulated in the Orange Book) that states that no temporarily used classified or sensitive information should ever be left in the system available for perusal by later users after a user logs off the workstation or after an application process is completed. This term has nothing to do with object-oriented programming.

Normative process - This is a description of an optimum method of how a person or machine should perform a task or set of tasks in an ideal situation.

Novice - This is a person who has little or no experience with a particular situation and environment and must use a systematic method for resolving situations.

Object - This is an entity defined in a directory database. Each object consists of properties and the values for the properties. There are three general categories of objects: container objects, leaf objects, and the root object. *Container objects* generally represent abstract entities, including the Organization and Organizational Unit objects. *Leaf objects* generally contain information about physical network entities such as users and devices. A *root object* is created during installation as the parent directory for all other objects.

Object-oriented databases - These are databases consisting of stored objects.

Object oriented programming - This is the creation of a program based on objects using a language for dealing with objects. Programming begins with the development of a collection of objects that can achieve some objective. Object-oriented programs are particularly well suited for highly interactive programs (such as graphical operating systems, games, and customer transaction stations) and programs that imitate or reflect some dynamic part of the real world (such as simulations and command and control systems).

Object reuse - This is a security policy (stipulated in the Orange Book) that states that no sensitive information temporarily stored by a previous user can be left for viewing by a later user.

Off-hook - This has two definitions: 1. It is a change in line voltage caused when the receiver or handset is lifted from the hook-switch. A traditional PBX or local telephone company recognizes this line voltage change as a request for dial tone. 2.

It is a call condition in which transmission facilities are already in use. Also known as busy.

Office of management and budget circular A-130 (Appendix III, issued in 1996) - This mandate enforces mandatory training by requiring its completion prior to granting access to the system and through periodic refresher training for continued access. Therefore, each user must be versed in acceptable rules of behavior for the application before being allowed access to the system. The training program should also inform the user on how to get help when having difficulty using the system and procedures for reporting security incidents.

One-time passwords - These are passwords that are used once and then are discarded.

Online analytical processing (OLAP) - These are applications that have query and response time characteristics that set them apart from traditional on-line transaction processing (OLTP) applications. Specialized OLAP servers are designed to give analysts the response time and functional capabilities of sophisticated PC programs with the multi-user and large database support they require. OLAP systems are often called Decision Support Systems (DSSs).
There are fundamental differences and synergies between OLAP and OLTP models and applications. OLAP and OLTP applications exist because of the shortcomings of relational database management systems and because of the OLAP associated front-end query tools for multi-dimensional analysis applications. OLAP and OLTP systems are highly complementary and should co-exist within the same enterprise environment to solve different problems.
OLAP is primarily involved with reading and aggregating large groups of diverse sample data, usually in the form of cubes (multi-dimensional tables). Unlike OLTP applications, OLAP involves many data items (frequently many thousands or even millions) that are involved in complex relationships. The objective of OLAP is to analyze these relationships and look for patterns, trends, and exception conditions. OLTP is concerned more with executing transactions representing business activities and consecutive storage/retrieval in relational databases. The databases formed by the OLTP package are used as delivered data for viewing by appropriate individuals and for providing information for forming consolidated databases for use for OLAP.
OLAP applications enable a user to perform consolidation (aggregation), drill-down (decomposition), and slicing and dicing (viewing and analyzing trends and patterns in the data).

Online transaction processing (OLTP) - This application is concerned more with executing transactions representing business activities and consecutive storage/retrieval in relational databases. The databases formed by the OLTP package are used as delivered data for viewing by appropriate individuals and for providing information for forming consolidated databases for use for OLAP. An OLTP capability can be provided by Cognos's Impromptu package or by Microsoft's Excel.

Open architecture - This is an architecture that has the following attributes, namely, its components are interoperable, portable, and scalable. These attributes are attained by making the architecture standards-based. The components of an open architecture are non-proprietary.

Open network - This is a network that can communicate with any system component (peripherals, computers, or other networks) implemented according to international standards (i.e., without special protocol conversions, such as gateways).

Open router - This is a router that performs for external clients only.

Open software foundation (OSF) - This is a consortium, of computer hardware and software manufacturers, that has over seventy companies as members.

Open systems interconnection (OSI) - This is an ISO stack model that uses seven layers for network connections, namely, 1) Physical, 2) Link, 3) Network, 4) Transport, 5) Session, 6) Presentation, and 7) Application. See the figure on the next page for a comparison of the OSI and TCP/IP models.

Operating system (OS) - This a computer program that runs on a client system or on a network server and controls system resources for a client and information processing on the entire network for the network server.

Operational controls or operations - These are the day-to-day administrative, physical, technical, and procedural mechanisms used to protect operational systems and applications.

Optical carrier - This is a communications link that uses fiber optics for carrying information, such as SONET.

Organizational facility - This is any facility that is owned, leased, or lent to the organization, where organizational information systems, or any portion of organizational information systems, will be developed, housed, or operated, or where organizational information is collected, stored, processed, disseminated, or transmitted, using organizational or non-organizational equipment.

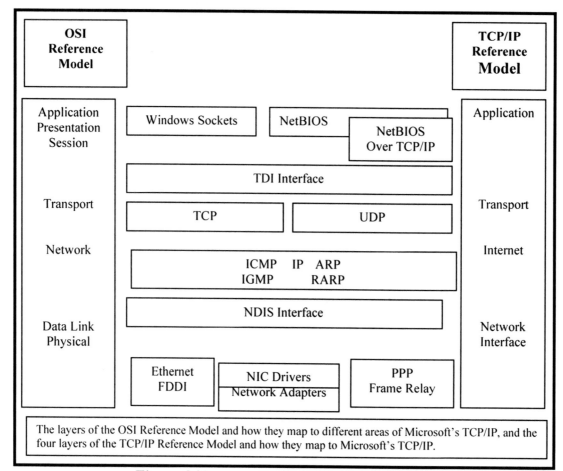

Figure OSI versus TCP/IP Reference Models

Out-of-band communications - This is a capability to provide information that does not use the inline network traffic flow. It uses other means, such as carrying information on floppy disks or providing rules with a computer that is linked only to the device that it supports (such as a firewall) but not to any other device.

Output - This is the communication from your computer. Output includes data that is printed, sent to disk, shown on your screen, or sent across a network. It can also mean data that the computer sends to the console, disk, or some other device.

Owner - In this book it represents the person who owns an organization or the lead person in an organization. In software terms it is either: 1) a file manager option used to set access permissions for the owner of a file or directory, or 2) the user who owns and controls a file or folder.

Packet - This is a unit of data transmitted over a network. It is a group of bits transmitted as a unit of information on a network. These bits include data and control elements. The control elements include the addresses of the packets source and destination and, in some cases, error-control information. In packet-switching networks, a packet is a transmission unit of a fixed maximum size that consists of binary digits representing both data and a header.

Packet filter - This is a security technology typically found on routers and other networking devices. A device with a packet filter will accept or deny traffic based on simple header information, such as source and destination IP addresses or source and destination port numbers.

Packet frame - This is a set of information added to a packet to ensure its proper transmission across the network. The format of frame information depends upon the physical medium on which the data travels.

Packet size - This is the size of an incoming or outgoing data packet. Packets contain 53 bytes. A message usually comprises many packets.

Packet-switched network - This is a group of interconnected, individually controlled computers that use packets to transmit information to each other.

Page - This is a unit of a data file. A page is the smallest unit of storage that moves between main memory and disk. Pages contain a multiple of 512 bytes (up to 4,096 bytes). A data page contains fixed-length records (or the fixed-length portion of variable length records.) An index page contains key values and pointers to the associated records for those values (which reside on a data page.) A variable page contains variable-length portions of records.

Partition - This is a logical unit into which hard disks can be divided.

Passive hub - This is a device used in some network topologies to split a transmission signal, allowing additional workstations to be added.

Password - This is a combination of characters that allows users to log on to a system or to access a program or file. Password attacks can most easily be countered by never relying on plaintext passwords.

PC card - This was formerly known as the Personal Computer Memory Card International Association (PCMCIA) card. A cryptographic PC Card contains the processor, algorithms, and cryptographic material necessary to support personalized security services.

Peer-to-peer - This is a style of networking in which a group of computers communicate directly with each other, rather than through a central server. This is often used for multiplayer situations to avoid the expense and delay of handling large amounts of traffic at a server. Unfortunately, peer-to-peer networking often has problems dealing with Network Address Translators (NATs). Some peer-to-peer networking products now work properly through several commercial NATs.

Penetration - This is the successful violation of an information system.

Penetration tests - These are the portions of security testing in which the evaluators attempt to circumvent the security features of an information system. It is used to identify the vulnerabilities in the targeted information system and is performed using a scanner. *Host and service scanning* allows you to document what systems are active on the network and what ports are open on each system by: 1) finding every system on the network, 2) finding every service running on each system, and 3) finding out which exploits each service is vulnerable to.

Performance data - This is the data that can be collected and evaluated to infer the performance (e.g., speed, response time, confidentiality, etc.) of an information system network. This data is used to instantiate (give value to) the system's performance metrics.

Performance management - This is the handling of the quality and effectiveness of the infrastructure, such as network communications. It allows for performance data to be collected and used so that performance metrics can be quantified, monitored, and used to tune (i.e., configure) the system for optimal performance.

Performance measurement - This is the collection and use of data during actual operations to determine the system's performance.

Performance metrics - These are the parameters which when quantified or qualified can be used to determine the performance of the system.

Performance values - These are the values of the performance metrics.

Perimeter - The perimeter of a network is the entrance/exit point for the network, that is, it is the point where information enters or leaves the organizational system at a particular location.

Permission or Privilege - This is a setting that determines how entities may access an object, such as a file, directory, or printer.

Personal identification number (PIN) - This is the numeric value (i.e., password) given to, or selected by, the user of a charge card of smart card that permits the user's acceptance to the computer.

Personnel security program - This is the program that provides a basis for security determinations for sensitive positions, clearances for access to classified material, and suitability for Federal employment. The authority for this program is the same as that for the investigations program.

Physical address - This is the OSI data-link layer address of a network device.

Physical layer - This is the first of seven layers of the OSI model; the physical layer details the protocols that govern transmission media and signals. In X.25, the physical layer is the layer of the interface that defines the physical interface between data terminal equipment (DTE) and data circuit-terminating equipment (DCE). It specifies the procedures used to establish, maintain, and release the physical connections or data circuits between network end points.

Physical memory - This is any memory that can be physically addressed by the processor.

Physical security - This is the combination of security controls that bar, detect, monitor, restrict, or other control access to physical sensitive areas. Physical security also refers to the measures for protecting a facility that houses Information System Security assets and its contents from damage by accident, malicious intent, fire, loss of utilities, environmental hazards, and unauthorized access.

Ping - This is a signal used to test the connectivity of a particular network node by transmitting a diagnostic packet that requires a response from that node.

Ping attack - This is a malicious saturation of a server with messages that overwhelm the server's capacity to respond. Ping attacks can also use up available bandwidth.

Ping of death - This is a command ICMP ping command that consists of a packet greater than 65,536 bytes (as opposed to the default 64 bytes). Many systems will crash if they receive a ping command like this.

Ping scanner - This scanner can send an ICMP request to each sequential IP address on a subnet and waits for a reply. A *port scanner* allows you to sequentially probe a number of ports on a target system in order to see if there is a service that is listening. You can install a network analyzer on your network to monitor traffic flow. The less information you unknowingly hand out about your network, the harder it will be for an attacker to compromise your resources. An *automated vulnerability scanner* is a software program that automatically performs all the probing and scanning steps that an attacker would normally do manually.

Plaintext - A message is in plaintext if it has not been encrypted and hence can be easily read.

Platform - This is a physical structure within an information system network that hosts a subsystem or a set of subsystem components (e.g., a mainframe computer).

Point-to-point protocol (PPP) - This is an industry-standard protocol that enables point-to-point transmissions of routed data. The data is sent across transmission facilities between interconnected LANs by using a synchronous or an asynchronous serial interface.

Policy - This is a general principal stipulated by an organization on which the organization or its employees base their behavior.

Polling - This function is performed when information collected by an SNMP management station based on queries at predetermined intervals is used to check the status of each network device.

Polyinstantiation - Polyinstantiation is used in multilevel security contexts, not considered in this book. A client requests the creation of a new object from an object manager, and the micro-kernel supplies the object manager with the subject identification (SID) of the client. The object manager sends a request for a SID for the member object to the security server, with the SID of the client, the SID of the polyinstantiated object, and the object type as parameters. The security server consults the polyinstantiation rules in the policy logic, determines a security context for the member, and returns a SID that corresponds to that security context. Finally, the object manager selects a member based on the returned SID, and creates the object as a child of the member.

Port - This is the point of contact between two hardware devices or two software elements. A hardware port can be the physical connection point between a printer cable and the computer. A software port, represented by a memory address, can be the logical contact point between a LAN driver and the protocol bound to it, or the point of access to a service on a TCP/IP host computer.

Portable - This means that system features can be easily applied to different serivces without regard for the underlying protocol. Data translation and encryption are considered to be portable features. A component is portable if it can be operated at any place in a network without any additional effort, such as modifying a software system.

Portable software - This is software that is developed for a variety of hardware platforms. It can be installed and run on different computers and operating systems.

Portal - This is a doorway (i.e., portal) to an organization's Web users where external users can access internal data and applications with a minimum of activity. In essence, it is a doorway to internal information and applications. Portals can provide billing and customer management systems software for businesses that do customer responses over the Internet.

Port driver - This is a driver that routes print jobs through the proper port (for example, LPT1, LPT2, COM1) to the printer that will handle the job.

Porting - This is the process by which a software application is made operational on a computer architecture or system different from the one on which it was originally created.

Port redirection - This is the use of a software capability that sends information addressed to a particular port to a different port. For example, if you run a production Web server (Port 80), you could then redirect telnet (Port 23) and SMTP (Port 25) to a honey pot. Because these services should not be accessed on the production system, the honey pot should immediately send an alert and/or log the activity. This requires an upstream router or firewall capable of performing port redirection.

Post office protocol (POP) - This is typically used when retrieving mail from a UNIX shell account. One of POP3's biggest drawbacks is that it does not support the automatic creation of global address books.

Preemption - This is an operating system scheduling technique that allows the operating system to take control of the processor at any instant, regardless of the state of the currently running application. Preemption guarantees better response to the user and higher data throughput. Most operating systems are not preemptive multitasking, meaning that task switching occurs asynchronously and only when an executing task relinquishes control of the processor.

Premise router - This is a router that is located on the customer's premises and is sometimes called a customer premise equipment (CPE) router.

Prescriptive process - This is a description of a recommended method of how a person or machine should perform a task or set of tasks in a real situation.

Primary logical unit (PLU) - This is the logical unit (LU) that sends the BIND command to activate a session with its partner LU, which means it contains the primary half-session for a particular LU-LU session. A logical unit can contain secondary and primary half-sessions for different active LU-LU sessions.

Primary server - This is the server that has been operating longer than its partners and is currently servicing workstations. The primary server handles workstation requests for network services. It also handles routing packets from routers on the network. If the primary server fails, the secondary server becomes the new primary server.

Print queue - This is a sequence of print job requests stored and waiting to be processed.

Print server - This is a host computer to which one or more printers are connected, or the UNIX process that manages those printers.

Private-circuit WAN topologies - These are leased lines or T1 connections that are good at ensuring privacy but introduce a single-point-of-failure. *Frame relay* is also capable of providing WAN connectivity, but it does so across a shared public network.

Private key - This is an n-bits long word used with an encryption mathematical formula that belongs to a subject (user, application, or computer) and is never revealed to anyone. The subject uses the private key to decrypt messages that it receives, and that are encrypted with the subject's public key. A private key can also encrypt a message digest sent by the subject to anyone else. Using the subject's public key, anyone can decrypt the digest and be assured that the message originated from that subject.

Process - This is a program running on the computer. In most computers, it is one execution of an application, tool, or program that uses the computer's services.

Processor - This is the data processing unit of a computer. Computers can be uniprocessing or multiprocessing. A uniprocessor system has only one run queue from which the processor can pick up threads for execution. In a multiprocessing system, however, more than one processor is available for the distribution of threads.

Promiscuous mode - A computer on a network, operating in this mode, is constantly monitoring the transmissions of all other stations on the network. A network analyzer or sniffer is a computer operating in the promiscuous mode. An issue with the Ethernet is that all systems on the network are operating in the promiscuous mode.

Property - This is an item of information about a network object, such as a name, network address, or password. Some properties can have multiple values. For example, the Telephone property, found in many objects, can contain several telephone numbers.

Protection profile (PP) - This is a combination of security requirements, including assurance and functional requirements, with the associated rationale and target environment to meet identified security needs. A protection profile is included in the SCAP.

Protection profile certification - This is a certification issued by an accredited Common Criteria evaluation facility that the PP contains requirements that are: justifiably included to counter stated threats and meet realistic security objectives, internally consistent, and coherent and technology sound.

Protocol - This is a set of rules for how information is exchanged over a communications network. The protocol dictates the formats and the sequences of the messages passed between the sender and the receiver. It establishes the rules for sending and receiving messages and for handling errors. The protocol does not need to know the details of the hardware being used or the particular communications method.

Proxy servers - A proxy server or host is a dual-homed host that is dedicated to a particular service or set of services, such as mail. One of the functions of the proxy hosts is to protect the internal users from advertising their IP addresses.

Public domain - This is the total absence of copyright protection. If something is "in the public domain" then anyone can copy it or use it in any way they wish. The author has none of the exclusive rights that apply to a copyright work. The phrase "public domain" is often used incorrectly to refer to freeware or shareware (i.e., software that is copyrighted but is distributed without (advance) payment). Public domain means no copyright, which means no exclusive rights.

Public key - This is an n-bits long word used with an encryption mathematical formula that belongs to a subject (user, application, or computer) and is revealed to everyone. It is used to encrypt messages that are sent to the subject as well as to verify the signature of the subject. To ensure that the public key really belongs to the subject, it is embedded in a digital certificate.

Public key certificate - This is a digital document verifying that a public key belongs to an individual or entity (such as a server). A public key certificate prevents unauthorized users from using phony keys to impersonate legitimate users.

Public key cryptography standard #7 (PKCS #7) - This is an RSA Data Security, Inc., standard for encapsulating signed data such as a certificate chain.

Public key cryptography standard #12 (PKCS #12) - This is an RSA Data Security, Inc., standard that specifies a portable format for storing or transporting a users private keys, certificates, miscellaneous secrets, etc.

Public key infrastructure (PKI) - This encryption method encompasses certificate management, registration functions, and public key enabled applications. PKI refers to the framework and services that provide the following:

1) Generation, production, distribution, control, revocation, archive, and tracking of public key certificates;
2) Management of entity keys;
3) Support to applications providing confidentiality and authentication of network transactions;
4) Data integrity; and
5) Non-repudiation.

Public Switched Telephone Network (PSTN) - This is the network that allows any person in the world to talk to any other person in the world provided both have a telephone and are connected to some subnetwork of the PSTN. Like the Internet, it is a collection of regional and backbone networks with associated switching devices and interconnecting links.

Public trust position - This is a position that has the potential for action or inaction by an incumbent to affect the integrity, efficiency, or effectiveness of assigned Government activities.

Publish/subscribe - A user *subscribes* to information (accesses data) in a database and a *publisher* publishes (sends) information to a database. So subscribe and publish in a publish/subscribe network are similar in nature to pull (access data from a source) and push (send data to sources) in a push/pull network, respectively.

Push/pull - *Push information* is sent to specified files by the generating system, whereas *pull information* must be accessed in order to acquire it.

Push/pull technology - *Push technology* is the transmission of information from a source or initiator without the source being requested to send that information. An example of a push technology is an *SNMP trap*. *Pull technology* is the transmission of information in response to a request for that information. An example of a pull technology is *polling*.

Quality of service (QoS) - This is the characteristic performance properties of a network service, including throughput, transit delay, and priority. Some protocols allow packets or streams to include QoS requirements. In broad terms, the QoS of a wide area network is a measure of how well it does its job, namely, how quickly and reliably it transfers various kinds of data including digitized voice and video traffic, from source to destination.

Questionnaire - This is a method of gathering information from an identified set of persons that consists of a set of questions (concerning some particular area of interest) that should be asked in the same order each time they are presented to an interviewee. The interviewees should be knowledgeable persons who understand the subject in question and they should be selected on a random basis. The answers should be recorded on the questionnaire. For best results, the questions should be multiple choice. Open-ended questions should be avoided.

R commands (viz., rcp, rlogin, and rsh) - These commands use two daemons, *login* and *exec*, to act as trusted hosts without requiring password authentication. Trust is based on security equivalency. When one system trusts another, it believes that all users will be properly authenticated and that an attack will never originate from the trusted system. Unfortunately, this can create a domino effect where an attacker need only comprimose one UNIX machine and then use the trusted host equivalency to compromise additional machines. Trusted hosts are determined by the contents of the */etc/hosts.equiv* file. This file contains a list of trusted systems, as you can see in the following example:

> *Loki.foobar.com*
> *Skylar.foobar.com*
> *Pheonix.foobar.com*

If this *host.equiv* file is located on the system named *thor.fobar.com*, then thor will accept *login* and *exec* service requests from each of these systems without requiring password authentication. If any other system attempts to gain access, the connection request is rejected. It is far too easy to exploit the minor level of security provided by the R commands. An attacker can launch a spoof attack or possibly corrupt Domain Name System in order to exploit the lack of password security. Both *login* and *exec* run as daemons under *ietd*. It is highly recommended that you disable these services.

Realm - This organizes security information and defines its scope of operations. A WebLogic Server Realm combines Users, Groups (i.e., functional roles), Permissions, and Access Control Lists. It determines who a user is authenticated and contains access control lists for given names.

Redundancy - This is the condition of excess capability in a computer system that enables it to operate properly even when the demand for services exceeds nominal.

Redundant array of inexpensive disks (RAID) - These devices provide fault tolerance against hard disk crashes but also will improve system performance. Some RAID systems are *hot swappable* and can also provide *data striping* where data is broken up across multiple disks. There are several levels of RAID capabilities, RAID 0 through 5. Server redundancy uses the concept of RAID and applies it to the entire server and is called *server fault tolerance (SFT)*. Novell's SFT-III and Microsoft's Cluster Server are examples of support at the operating-system level for running redundant servers.

Referential integrity - A database has referential integrity if all foreign keys reference existing primary keys.

Relational databases - These are databases that store data in a tabular way. E. F. Codd, of IBM, in 1970, published a paper on the many ways to manipulate tabular data. Changes to a datum in one table will be automatically relayed to the same datum element in other tables.

Remote access control (RAC) - This is the control of access within a network from a node that is not a fixed element of the network (e.g., a temporary telephone modem link).

Remote access service (RAS) - This is the support provided to remote users (those who are not fixed members of the network) so that they can have access to the network.

Remote access VPN - This is the access that VPN networks offer remote users so that the service is affordable and easier to implement and maintain, than traditional solutions such as direct dial-up access.

Remote authentication dial in user service (RADIUS) - This is a protocol that is used for Authentication and Authorization. The RADIUS protocol can be extended to cover delivery of accounting information from the Network Access Server to a RADIUS accounting server. Key features of RADIUS accounting are:

1) It uses a client/server model.
2) A Network Access Server can operate as a client of the RADIUS accounting server. The client is responsible for passing user accounting information to a designated RADIUS accounting server.

3) The RADIUS accounting server is responsible for receiving accounting requests and returning a response to the client indicating that it has successfully received the request.

4) The RADIUS accounting server can act as a proxy client to other kinds of accounting servers.

Remote method invocation (RMI) - This is the part of the Java programming language library that enables a Java program running on one computer to access the objects and methods of another Java program running on a different computer.

Remote monitoring (RMON) - This is a Cisco option that identifies activity on individual nodes and allows the user to monitor all nodes and their interaction on a LAN segment. Used in association with the SNMP agent in a router, RMON allows the user to view traffic that flows through the router as well as segment traffic not necessarily destined for the router.

Repeaters - These are simple two-port signal amplifiers used in a bus topology to extend the maximum distance that can be spanned on a cable run.

Residual risk - This is the remaining potential risk after all information technology security measures are applied. There is a residual risk associated with each threat.

Ring - This is a communications network where information is sent from client to client around a loop in the same direction.

Risk - This is the probabilistic-cost that a particular threat will be successfully mounted against the information system times a resulting potential loss in dollars.

Risk acceptance - This is the process concerned with the identification, measurement, control, and minimization of security risks in information systems to a level commensurate with the sensitivity and criticality of the information protected.

Risk analysis - This is the process of identifying: the assets you wish to protect, potential threats against them, the countermeasures to cost-effectively minimize the effects of these threats, and the implementation and testing of these countermeasures.

Risk assessment - This is the process of analyzing and interpreting risk so as to identify the various cost-effective countermeasures that are needed to protect the system and its information.

Risk management - Risk management is the process of assessing risk, taking steps to reduce risk to an acceptable level, and maintaining that level of risk. This is the total process of identifying, controlling, and mitigating information system related risk. It includes risk analysis; identification of threats, identification of the vulnerabilities, determination of the countermeasures; cost-benefit analysis (countermeasures versus vulnerabilities); and the selection, implementation, security test, and security evaluation of countermeasures. This overall system security review considers both effectiveness and efficiency, including effects on the organization and constraints due to policy, regulations, and laws. It is the maintenance of security throughout the operational phase of a system by: analyzing risk, reducing risk, and monitoring countermeasures. Risk assessment is a critical element in designing the security of systems and in re-accreditations.

rlogin - Telnet and rlogin (TCP/IP protocols) are available for making connections to a host. Telnet, a virtual terminal protocol that is part of the TCP/IP protocol suite, allows for connections to hosts. Telnet is the more widely used protocol. The *rlogin* protocol is a remote login service developed for the BSD UNIX system. It provides better control and output suppression than Telnet, but can only be used when the host (typically, a UNIX system) supports rlogin. Our implementation of rlogin does not subscribe to the trusted host model. That is, a user cannot automatically log on to a UNIX system from the remote access server, but must provide a user ID and a password for each connection.

Roaming - This is the capability of wireless users to move seamlessly among a cluster of access points.

Robot - (1) A device that responds to sensory input. (2) A program that runs automatically without human intervention. Typically, a robot is endowed with some artificial intelligence so that it can react to different situations it may encounter. Two common types of robots are agents and spiders.

Role - This is the particular job function (or sets of job functions) performed by some entity that requires access to an information system network. It is desirable that there be many more entities than roles, so that each role will contain many entities, on the average.

Role-based access control (RBAC) - This is the access control process where a user or application has access to particular information that is granted based on their appearance in a list of members of a role (e.g., a job function) and also on an access control list.

Router - This is a device that connects two networks using the same networking protocol for managing the exchange of data packets. It operates at the network layer (Layer 3) of the OSI model. Routers are multiport devices that decide how to handle the contents of a frame, based on protocol and network information. *Routers* are used to connect logical networks. The act of traversing a router from one logical network to another is referred to as a *hop*. Routers can be either statically programmed (*static routing*) with the information describing the path to follow or they can use a special type of maintenance frame (*dynamic routing*)such as the routing information protocol to relay information about known networks. Routers use these frames and static entires to create a blueprint of the network known as a *routing table. Distance* *routing* is based on a distance vector that is computed using the distance to the next nearest router between this node and the destination node.

Rule - This is a standard method or procedure and is a statement that clearly defines or constrains some aspect of an organization's (enterprise's) process.

Safeguards (Countermeasures or security mechanisms and methods) - These are the protective measures for an information system, pertaining to the technical, physical, or personnel aspects of the system.

SAMBA - This is a suite of tools that allows a UNIX machine to act as a session message block for a client or server.

Sandboxing - This is the process of executing Java code (or other applet) in an environment where the code cannot interact with any element of the software system, such as databases. This process greatly limits the effects of most applets, thus a method of signing the code can be used to guarantee that the code is okay. A trusted user can incorporate their applet code within their digital signature (hashed and encrypted) so that the receiver can then identify the sender. If the sender is trusted, then the applet can be downloaded and executed within the system environment. For code that is not signed or is from entities that are not trusted, these applets can go into the sandbox where they can be executed in a harmless manner. *Code signing* provides a method for code accountability and for authenticating the code's source. In essence, code-signing software received over the Internet is the same as shrink-wrapped software purchased from a local store.

Scalable - This is the capability (e.g., storage capacity or input/output bandwidth) of a system to have more (or less) capability without having to modify the system (hardware or software).

Scanner - This is a device for scanning a network to identify any vulnerabilities in the system.

Scanning - See **Penetration testing.**

Scenario - A scenario is a verbal description of a situation in which some particular event results from some particular cause. For example, a threat scenario is a situation where an information system suffers an attack from some specific threat that causes an identifiable damage to the system.

Schema - This is the set of rules that define how the domain tree is constructed. These rules define specific types of information that dictate the way information is stored in the Network Operating System database. A directory needs a set of rules for how its data should be stored in order to ensure that data is stored in an orderly and logical fashion. These rules are collectively called *schema* and a schema defines the following types of information: syntaxes, matching rules, attribute types, object classes, and naming and containment rules.

Screening - This represents the investigation of a potential employee to ensure that the person is trustworthy.

Seamless - A product (hardware or software) is said to be *seamless* if it can be installed with a minimum or no extra effort to ensure that it can be absorbed into the total network and capable of interoperating with all other subsystems in the network.

Search engines - These software products, such as Alta Vista, Google, or Yahoo, are an excellent method of collecting information about any object located on the Web. Usually there are several "hits" (i.e., locations of relevant information) and the user must select one or more of these. A **traceroute** command is used to trace the network path from one host to another. Probably the best search engine is at www.google.com.

Secure hash algorithm - This is an algorithm that produces a 160-bit message digest from any message of less than 264 bits in length. This is slightly slower than MD5 but considered more secure against brute-force collision and inversion attacks.

Secure multimedia Internet mail extensions (S/MIME) - This is a protocol for encrypting e-mail traffic. MIME allows the server to inform the Web browser as to what type of data it is about to receive.

Secure remote procedure call (RPC) - This is the service used to implement the network file system. It uses the data encryption standard to provide a more secure identification of the client host.

Secure shell (SSH) - This is a powerful method for performing client authentication and safeguarding multiple service sessions between two systems. When two systems are using SSH establish a connection they validate one another by performing digital certificate exchanges. *Brute force and playback* attacks are ineffective against SSH because encryption keys are periodically changed. Triple DES is used for message encryption.

Secure sockets layer (SSL) - This is the set of rules (i.e., handshake protocol) governing the exchange of information between two devices (e.g., client and server) using a public key or private key encryption system. SSL establishes and maintains secure communication between SSL-enabled servers and clients across the Internet.

Security administrator's tool for analyzing networks (SATAN) - This is a tool designed to assist system administrators in recognizing several common networking-related security problems.

Security appliance - This is a combination of hardware, software, and networking technologies with a primary purpose of providing a single security function from a simple-to-use device that operates in a plug-and-play mode. Examples of the components of a security appliance are: firewalls, VPNs, IDSs, security management, policy management, and bandwidth management. As organizations seek to implement e-business strategies, they also seek solutions that are easy to implement, use, and maintain. These same organizations are increasingly finding that security appliances are effectively meeting all of these demands. Security appliances are for those who want to have everything 'in a box' - allowing them to take it out of the shipping carton and simply plug it in, just as one plugs in a microwave oven.

Security architecture - This is a sub-architecture that minimizes the vulnerabilities of assets and resources in a reference architecture. An asset or resource is anything of value in an information system. Security refers to a complex of procedural, logical, and physical measures aimed at prevention, detection, and correction of certain kinds of misuse, together with the tools to install, operate, and maintain these measures. Most, if not all, of the newer operating systems and network communications systems are offering security and privacy controls.

Security certification and authorization package (SCAP) - This is the report that includes the results of a security test and evaluation of an information system. It is presented to the DAA who uses it to certify the information system. The SCAP includes the ISS plan, vulnerability assessment report, risk assessment, security test plan, and security test results, disaster recovery and contingency measures, and ISS certification and authorization statements.

Security compliance review - This consists of the assessments at organizational facilities, which are coordinated with a DAA, facility management and, if applicable, the Contracting Officer, that examine operational assurance as to whether a system is meeting stated or implied security requirements, including system and organizational policies.

Security identifier (SID) - This is a unique identification number that is assigned to every user or group. A good tactic ***to combat hacker attempts to find the administrator's account*** is by 1) using a strong password and 2) ensuring that all failed logon attempts are logged. For Windows NT, the *security account manager (SAM)* is the database where all user account information is stored. For Windows NT, *User Rights* are set through the User Manager by selecting Policies → User Rights. ***The Backup Files and Directories*** *right* is dangerous since it would allow a user to copy the SAM where the copy could then be subjected to a password cracker.

Security kernel - This is the hardware, firmware, and software elements of a Trusted Computing Base that implements the reference monitor concept. It must mediate all access, be protected from modification, and be verifiable as correct.

Security labels - These identify the sensitivity level of a piece of information and they can be used to limit access in a mandatory access control context.

Security mechanisms (or countermeasures) - These are the methods or devices for protecting an information system and its information. There are three types of security mechanisms: 1) Physical, 2) Behavioral, and 3) Technical.

Physical mechanisms provide protection of the system and its facilities against unauthorized physical entry.

Behavioral mechanisms are the training programs to ensure that users, managers, and administrators are knowledgeable of the importance of system security and the principal human activities required for maximizing system security assurance.

Technical network mechanisms are the various devices for computing, storing, and transmitting information, such as

Computing devices (e.g., client and server computers, mainframes)

Network devices (e.g., cables, switches, routers, firewalls, intrusion detection systems), and

Software (viz., operating systems, firmware, and applications).

Technical Non-Network Mechanisms are the devices needed to ensure that the systems is properly operating or that damage is minimal, such as

Sprinklers and other fire control devices,

Fences, gates, locks, and guards,

Isolation of certain devices,

Generators and uninterruptible power supplies (UPSs), and

Backup and recovery systems

Security policy - This is the set of laws, rules and practices that regulate how an organization manages, protects, and distributes sensitive information. It is the framework in which a system provides trust. It is a formal statement of the rules providing which people are given what access to an organization's technology and information system assets. It usually is stated in terms of subjects (e.g., users, processes, or servers) and objects (e.g., files, directories, or sockets). It is created for an organization or system and is a set of statements regarding an organization's principles or plans that guide the actions taken by a person or group or information system development. It should be a succinct statement that can easily learned and remembered.

Security target (ST) - This is a set of security functional and assurance requirements and specifications to be sued as the basis for evaluation of an identified product or system in response to a chosen protection profile.

Security test - This is the process of testing and evaluating a system's operating activities to ensure that the system is protected from security violations in accordance with the security policy formulated for the system.

Security test and evaluation (ST&E) - This is the process to determine that the system administrative, technical, and physical security measures are adequate; to document and report test findings to appropriate authorities; and to make recommendations based on test results. ST&E may be an integral part of other tests and evaluations.

Self-healing Ethernet - This means there are two Ethernet switches (intelligent hubs) where the two switches are connected via an Ethernet cable that completes the ring. The Ethernet switches use what is called a "spanning tree protocol." With this protocol, the switches make decisions based on communication path availability. That is, the switches go out and look for different paths to the same piece of equipment. In the event that the preferred path fails (due to a fiber break or hardware failure), the switch simply switches the data flow to the other path. It will then continue to test the failed (i.e., preferred) path, so that when it is repaired, it will automatically become a viable path again.

Self-monitoring analysis and reporting technology (SMART) - This is a standard for predicting the likelihood of impending failure for a hard disk. The software originated at Western Digital and was later integrated into the Advanced Technology Attachment standard. You enable SMART support in the system Basic Input-Ouput System. It's used by drives of the ATA-3 standard and above.

Sensitive security information - This is any information that, if compromised, could pose a risk to the security posture of a system or organization.

Sensitivity - This is the degree to which a system requires protection to ensure confidentiality, integrity, and availability.

Separation of duties - This is the condition that a user cannot access a piece of data for which that user may have a conflict due to access to other pieces of data.

Service - This is a process or application that runs on a server and provides some benefit to a network entity. One of the best ways to secure a UNIX system is *to shut down all unneeded services.* The more services you run the more likely one of them has a bug or is incorrectly configured and can be used by an attacker to gain entrance to your network.

Service ticket - This is trusted information used to authenticate a client to a service.

Servlet - This is a small, server-based routine written in Java programming code.

Session key - This is a key shared by at least two parties, usually a client and a server.

Session key - This is a key shared by at least two parties, usually a client and a server.

Session layer - This is the fifth of seven layers in the OSI model. The session layer allows dialog control between end systems.

Simple key management for Internet protocol (SKIP) - This is a protocol that is similar to SSL protocol because it operates at the session level. SKIP requires no prior communication to establish or exchange keys on a session-by-session basis. A shared secret is generated using the public key encryption method for IP packet-based encryption and authentication. While SKIP is efficient at encrypting data, which improves VPN performance, it relies on the long-term protection of this shared secret to maintain the integrity of each session.

Simple mail transfer protocol (SMTP) - This protocl is used to transfer mail messages between systems. SMTP is capable of transferring ASCII text only.

Simple network management protocol (SNMP) - This is a TCP/IP protocol used for communicating between a network management console (SNMP Manager) and the devices the console manages. The protocol allows the SNMP Manager to gather information about the configuration and status of the TCP/IP protocol stacks of network nodes.

Simple object access protocol (SOAP) - This is a standard that provides the format for an electronic addressing system that allows applications to communicate with one another. SOAP is a lightweight protocol exchanging information in a decentralized, distributed environment. It is an XML-based protocol consisting of three parts:

1) an envelope that defines a framework that describes what is in a message and how to process it,
2) a set of encoding rules that expresses instances of application-defined datatypes, and
3) a convention that represents remote procedure calls and responses.

Single-point-of-failure - This is a situation where the failure of a single device will cause the entire system to shut down. Single point of failure is the most common mistake made in network design. These can occur at any of the following points: 1) firewall, 2) router, 3) Channel Service Unit/Data Service Unit (CSU/DSU), and 4) a single leased line or T1 connection.

Single sign-on (SSO) - Once logged onto the SSO server, users can access all those—and only those—resources for which they are authorized, as determined by system security policies, without further logons to those individual systems, applications, resources, and networks.

Slip - This is an error that occurs due to an entity inappropriately performing some action, e.g., hitting the wrong key or misspelling a word. Normally, errors performed by an application should be found and mitigated during security testing of the application. However, this may not always be the case.

Small computer systems interface (SCSI) - This is an industry standard that defines both hardware and software communications between a host computer and any peripheral devices (such as hard drives or tape backup systems). Computers and peripheral devices designed to meet SCSI specifications have a large degree of compatibility.

Smart cards - These are plastic media that contain an embedded integrated chip with up to 32 Kbytes of data storage for storing security certificates (and other relevant information) of a user. They are activated by insertion into a reader (attached to a computer with the appropriate software) with a required additional input by the user of a Personal Identification Number for authentication.

Smurf attacks - These attacks use a combination of IP spoofing and ICMP replies in order to saturate a host with traffic, causing a denial of service. *You can stop this kind of attack* at its source by using the *standard access list* described elsewhere. Smurf attacks can also be blocked at the bounce site. By using reflexive access lists or some other firewalling device that can maintain state, you can prevent smurf packets from entering your system. Since your state table would be aware that the attack session did not originate on the local network (it would not have a table entry showing the original echo request), this attack would be handled like any other spoof attack and promptly dropped.

Sniffing - This is the action of sampling network traffic as if flows across a link. The device used for this activity is called a "sniffer."

Socket - This is a method for establishing a communications link between a client program and a server program across a LAN, WAN, or the Internet, and sometimes between processes within a computer. A socket may be considered the endpoint in a connection. If a client socket (dialer) in a PC uses a known network address to identify a server socket (receiver) on another computer, then once a connection is established both computers can exchange data (or services). Sockets are the computer programming world's equivalent of a telephone. It is a software structure that acts as a communications end point. In an IPX network, it is the part of an IPX network address, within a network node, that represents the destination of an IPX packet. Some sockets are reserved for specific applications.

Social engineering - This is obtaining security information from users under false pretenses such as by pretending to be another user or security person and asking for security information, such as a password.

SOCKS v5 - This is a circuit-level gateway that operates at the OSI transport layer and relays connections without looking into the corresponding application data. A SOCKS server may perform authentication, authorization, message security-level negotiation, and other activities also.

Spamming - This is the process of sending out unsolicited e-mail advertising, usually to many receivers and usually a lengthy message.

Spider - This is a program that automatically fetches Web pages. Spiders are used to feed pages to search engines. It is called a spider because it *crawls* over the Web, therefore another term for these programs is *webcrawler* or a *crawler*. Because most Web pages contain links to other pages, a spider can start almost anywhere. As soon as it sees a link to another page, it goes off and fetches it. Large search engines, like *Alta Vista* or *Google*, have many spiders working in parallel.

Spoofing - This is the action of an entity providing a fictitious IP address to the recipient. Strong two-factor authentication using one-time passwords is effective in combating spoofing attacks but the use of cryptographic authentication is the best form of additional authentication thus making IP spoofing irrelevant.

Star topology - This is a LAN topology in which nodes on a network are connected to a common central switch by point-to-point links.

Stateful/stateless - *Stateful* means that the computer or program keeps track of the state of message interaction (by examining the header and the message body), whereas *stateless* means that it does not (by examining the header only).

Stateful inspection - This is a firewall technology that transparently intercepts and examines data inside the packet, as well as a packet's header to accept or deny communications. Stateful inspection technology is considered to be between packet filter technology and application gateway technology.

Steganography - This is the hiding of a secret message within a larger plaintext message in such a way that others cannot discern the presence or contents of the hidden message. For example, a message might be hidden within an image by changing the least significant bits to be the message bits.

Storage-area network (SAN) - This is a separate network dedicated to storage, that supplies storage, backup, and management services. Unlike NAS, it supports block-level transfers necessary for high-performance database and transaction environments. A SAN is a high-speed, special-purpose network that connects servers to storage devices by enabling storage devices and servers from different vendors to work together. A SAN is a process whereas a Network-Attached Storage (NAS) is a thing. It is an external disk system that is connected via fiber channel or Small Computer System Interface hardware to each of the servers in a cluster. The external disk system is accessible by each of the cluster servers connected to it. Unlike stand-alone servers, a storage-area network distributes resources over a scalable, redundant network of servers and storage devices.

   1) A SAN is a network that has a primary purpose of transferring data between computer systems and storage elements and among storage elements. A SAN consists of a communication infrastructure, which provides physical connections, and a management layer, which organizes the connections, storage elements, and

computer systems so that data transfer is secure and robust. The term SAN is usually (but not necessarily) identified with block I/O services rather than file access services.

2) SAN can also mean a storage system consisting of storage elements, storage devices, computer systems, and/or appliances, plus all control software, communicating over a network.

Note: The definition specifically does not identify the term SAN with Fiber Channel technology. When the term SAN is used in connection with Fiber Channel technology, use of a qualified phrase such as "Fiber Channel SAN" is encouraged. According to this definition an Ethernet-based network whose primary purpose is to provide access to storage elements would be considered a SAN. SANs are sometimes also used for system interconnection in clusters.

Storage device - This is a device, such as a tape drive or optical disk, used to store the contents of a server's or workstation's memory or other temporary storage device. An example of a storage device is an external tape drive that backs up data from hard disk to magnetic tape. Storage can be archive (long term) or temporary.

Stored procedures - These are the coded programs (e.g., PL/SQL, Java, or C++) that are stored in databases and are executed to provide fine-grained access control.

Strategy - A plan of action.

Stream encryption - This is when a message is encyrpted a bit at a time until the entire message is encrypted.

Streaming - This is the process of playing sound or video in real-time as it is downloaded over the Internet as opposed to storing it in a local file first. A plugin to a Web browser decompresses and plays the data as it is transferred to your computer over the Web. Streaming audio or video avoids the delay entailed in downloading an entire file and then playing it with a helper application. Streaming requires a fast connection and a computer powerful enough to execute the decompression algorithm in real-time.

String - This is a contiguous sequence of symbols or values, such as a sequence of characters or a sequence of binary values.

Strong authentication - This means that the method of authentication cannot be easily circumvented, such as the use of one-time passwords, personal identification numbers, or one or more of the many forms of biometric identifiers.

Stub - This is a *dummy procedure* used when linking a program with a run-time library. The stub routine need not contain any code and is only present to prevent undefined label errors at link time. Or a stub is a *local procedure* in a remote procedure call (RPC). The client calls the stub to perform some task and need not necessarily be aware that the RPC is involved. The stub transmits parameters over the network to the server and returns the results to the client.

Supply chain management (SCM) - This is the practice of coordinating the flow of goods, services, information, and finances as they move from raw materials to parts supplier to manufacturer to wholesaler to retailer to consumer. The SCM process includes the following activities: order generation, order taking, information feedback, and the cost-effective, timely delivery of goods and/or services.

Switch - This is either a network device or a general term, namely it is a:

1) *Network device* that filters, forwards, and floods frames based on the destination address of each frame. The switch operates at the data link layer of the OSI model.
2) *General term* applied to an electronic or mechanical device that allows a connection to be established as necessary and terminated when there is no longer a session to support.

An *intelligent storage switch* centralizes management of storage devices.

Synchronous - This term refers to processes in which data streams can be delivered only at specific intervals.

Synchronous data link control (SDLC) - This is a bit-oriented, full-duplex serial protocol that has spawned numerous similar protocols, including High-Level Data Link Control and Link Access Procedure, Balanced.

Synchronous optical network (SONET) - This is an optical multiplexing hierarchy that specifies a new series of optical carrier transmission rates ranging from 51.4Mbps to 40Gbps and a set of standard interfaces and signaling formats.

SYN/ACK flags - These are flags in a TCP packet (set at the transport layer) that represent synchronize (SYN) or acknowledge (ACK) by setting them to either 0 or 1 for transferring data and acknowledging that it was properly received. A *TCP SYN attack* occurs when a sender transmits a volume of connections that cannot be completed. This causes the connection queues to fill up, thereby denying service to legitimate TCP users.

System - This is a composite of people, procedures, materials, tools, equipment, facilities, hardware, and software operating in a specific environment to perform a specific task or achieve a specific purpose, support, or mission requirement.

System authorization - This is a formal declaration by the DAA who has fiscal and operational responsibility that an information system is approved to operate in a particular security mode using a prescribed set of safeguards. This is the official management authorization for operational and is based on information provided in the Security Certification and Authorization Package (SCAP) as well as other management considerations. The authorization statement affixes security responsibility with the DAA and shows that due care has been taken for security.

System high - An architecture is referred to as system high when the different classes of information (e.g., unclassified and sensitive) are routed such that the classification of the network on which the information is routed is at least as high as the highest classification of the information being routed.

System integrity - This is a requirement that a system perform its intended function in an unimpaired manner, free from deliberate or inadvertent unauthorized manipulation of the system.

System owner - This is the manager responsible for the organization that sets policy, direction, and controls funding for an information system. Systems under development are owned by the developing organization until accepted and authorized by the operating organization.

Tape systems - These are a favorite of network administrators for providing information backup because tapes safeguard the information stored on the server. *Backup software supports three methods* of selecting which files that should be archived to tape: 1) full backup, 2) incremental backup, and 3) differential backup. *Full backup* means that there is a complete archive of every file on the server. *Incremental backup* means that there is an archive of those files on the server that have been recently changed or added. *Differential backup* means that there is an archive of those files on the server that have been changed or added since a full backup was performed.

Target of evaluation (TOE) - This is another name for an information technology product or system described in a Protection Profile or Security Target. The TOE is the entity that is the subject of a security evaluation.

Target system - This is the desired or objective system.

Taxonomy - This is a method for organizing information so that the information is structured and can be easily read. It usually organizes information so that it is classified into various logical categories.

TCP wrapper - This is an excellent way to fine tune access to your system. Even if all your UNIX systems are sitting behind a firewall, it cannot hurt to take preventative measures to lock them down further. This helps to ensure that anyone who manages to sneak past the firewall will still be denied access to your UNIX system.

Teardrop attack - This attack starts by sending a normal packet of data with a normal-size payload and a fragmentation offset of 0. From the initial packet of data, a teardrop attack is indistinguishable from a normal data transfer. Subsequent packets, however, have modified fragmentation offset and length fields. This subsequent traffic is responsible for crashing the target system. When the second packet of data is received, the fragmentation offset is consulted to see where within the datagram this information should be placed. In a teardrop attack, the offset on the second packet claims that this information should be placed somewhere within the first fragment. When the payload field is checked, the receiving system finds that this data is not even large enough to extend past the end of the first fragment. In other words, this second fragment does not overlap the first fragment; it is actually fully contained inside of it. Since this was not an error condition that anyone expected, there is no routine to handle it and this information causes a buffer overflow—crashing the receiving system. For some OSs, only one malformed packet is required. For others, the system will not crash unless multiple malformed packets are received. The IDS should be able to recognize this type of attack and report it. However, at present, it appears that an IDS *cannot prevent teardrop attacks*. Moreover, if the attacker uses IP spoofing, the IDS may not be capable of determining the actual source of the teardrop attack. However, gap appliances do seem to be capable of stopping teardrop attacks.

Telephony - This is a communications method, often two-way, for spoken information, by means of electrical signals carried by wires or radio waves. The term was used to indicate transmission of the voice, as distinguished from telegraphy (done in Morse code and usually called continuous wave transmission), radio teletypewriter transmission (also called frequency shift keying, the modulation scheme used by such machines), and later, facsimile.

Telnet - The Internet Protocol (IP) suite includes the simple remote terminal protocol called Telnet. Telnet allows a user at one site (sender) to establish a Transmission Control Protocol (TCP) connection to a login server at another site (receiver); it then passes the keystrokes from the sender system to the receiver system. Telnet accepts either an IP address or a domain name as the remote system address. Telnet offers three main services: 1) Network Virtual Terminal Connection, 2) Option Negotiation, and 3) Symmetric Connection.

Template - A *template* is a formatted system, sometimes a single page or a full document, that describes some object (based on its characteristics) or how to prepare some product.

Terminal access controller access control system (TACACS) - This is an authentication protocol, developed by the DDN community, that provides remote access authentication and related services, such as event logging. User passwords are administered in a central database rather than in individual routers, providing an easily scalable network security solution. (Also called TACACS+.)

Terminal logging - The easiest way to save the configuration information is by using *terminal logging*. The disadvantage to terminal logging is that it only works for the your specific system configuraton; you cannot save the operating system.

Thin clients - In a client/server network, the PCs are called *thin clients* if most of the software resides on servers and the client's software consists of applications such as a browser.

Thick clients - In a client/server network, the PCs are called *thick clients* if most of the software resides on clients and the server's software consists of applications such as a database management system.

Thread - In computer processing, a sequence of instructions executed as an independent entity and scheduled by system software. A thread is also known as an executable object. Within an Internet discussion group, a thread is an ongoing discussion about a particular topic and is sometimes called a conversation.

Threat - This is a circumstance or event with the potential to harm an information system through unauthorized access, destruction, disclosure, modification of data, and/or non-availability of information or systems. It is a potential for a "threat source" to exploit (intentional) or trigger (accidental) a specific vulnerability.

Throughput - This is the total amount of useful data that is processed or communicated during a specific time period between a source and a destination.

Ticket - This is a record that helps a client authenticate to a service.

Time-out condition - This is an error condition indicating that a specified amount of waiting time has elapsed without the occurrence of an expected event.

Time stamp - This is code reporting the identity of an event and the time of its occurrence. Time stamps establish the order of events (such as object creation and partition replication), record "real world" time values, and set expiration dates.

Token ring network - This is a type of LAN that uses a ring topology and token passing, as in the IEEE 802.5 standard for media access control (MAC) or the IBM token ring network. *Token Ring* was designed to be fault tolerant and is a wonderful topology when all systems operate as intended. Since Token Ring requires that each system successively pass a token to the next, a single network interface card set to the wrong speed can bring down the entire ring. Token Ring switches are less popular and far more expensive than their Ethernet counterparts. In an Ethernet network, if a single system were set to the wrong speed, only that one system would be affected. Also, Token Rings require that all media access control numbers be unique.

Tokens (authentication tokens) - These are portable devices used for authenticating a user. Authentication tokens operate by challenge/response, time-based code sequences, or other techniques. Tokens may include paper-based lists of one-time passwords or smart cards.

Tool - In the desktop environment, it is a method of doing a specific task, for example, check spelling, or a small application such as Clock. In this sense *tool* is a synonym for *utility*. In a broader sense, *tool* may be a synonym for *application*.

Topologies - There are three primary architecture topologies, ring, star, and bus: A *ring topology* is one in which each host is connected to contiguous hosts (two) around the periphery of a ring and if one host should go down, the entire ring goes down. A *star topology* is one in which each peripheral host is connected to a center host and must go through the center host in order to communicate with any other peripheral host. A *bus topology* is the linear LAN used by Ethernet networks.

TRACERT - This is a diagnostic utility that determines the route a packet has taken to a destination.

Traffic - This is an activity over a network communications channel.

Transmission control protocol (TCP) - The protocol is built upon the IP layer suite, it is a *connection-oriented protocol* that specifies the format of data and acknowledgments that two computer-systems exchange to transfer data. TCP also specifies the procedures the computers use to ensure that the data arrives correctly. TCP allows multiple applications on a system to communicate concurrently, as it handles all demultiplexing of the incoming traffic among the application programs.

Transmission rates - Four alternatives are identified.

1. **T-carrier system -** This is a series of wideband digital data transmission formats originally developed by the Bell System and used in North America and Japan.
   The basic unit of the T-carrier system is the DS0, which has a transmission rate of 64Kbps, and is commonly used for one voice circuit. Originally the 1.544Mbps T1 format carried 24 pulse-code modulated, time-division multiplexed speech signals each encoded in 64Kbps streams, leaving 8Kbps of framing information which facilitates the synchronization and demultiplexing at the receiver. T2 and T3 circuit channels carry multiple T1 channels multiplexed, resulting in transmission rates of up to 44.736Mbps.
   Asynchronous signals can be transmitted via a standard, which encodes each change of level into three bits; two of which indicate the time (within the current synchronous frame) at which the transition occurred, and the third, which indicates the direction of the transition. Although wasteful of line bandwidth, such implementation is typically used only over small distances.

2. **DS level (Digital Signal or Data Service level) -** This was originally an AT&T classification of transmitting one or more voice conversations in one digital data stream. The most popular DS levels are DS0 (a single conversation), DS1 (24 conversations multiplexed), DS1C, DS2, and DS3. By extension, the DS level can refer to the raw data rate necessary for transmission as shown in the table below.

429

DS Level	Bandwidth
DS0	64 .000Kbps
DS1	1.544Mbps
DS1C	3.150Mbps
DS2	6.310Mbps
DS3	44.736Mbps
DS4	274.100Mbps

In this sense, it can be used to measure data service rates classifying the user access rates for various point-to-point WAN technologies or standards (e.g., X.25, SMDS, ISDN, ATM, and PDH). Japan uses the U.S. standards for DS0 through DS2 but Japanese DS5 has roughly the circuit capacity of U.S. DS4, while the European standards are rather different. In the U.S. all of the transmission rates are integral multiples of 8Kbps but rates above DS1 are not necessarily integral multiples of 1.544Mbps.

3. The European standards:

E1 - This is a European framing specification for the transmission of 32 DS0 data streams. By extension, it can also denote the transmission rate required (2.048Mbps).

E2 - This is a European framing specification for the transmission of four multiplexed E1 data streams, resulting in a transmission rate of 8.448Mbps.

E3 - This is a European framing specification for the transmission of 16 E1 data streams, resulting in a transmission rate of 34.368Mbps.

E4 - This is a European framing specification for the transmission of 64 multiplexed E1 data streams, resulting in a transmission rate of 139.264Mbps.

E5 - This is a European framing specification for the transmission of 256 multiplexed E1 data streams, resulting in a transmission rate of 565.148Mbps.

4. Optical Carriers (OC):

Optical Carrier Level	Bandwidth
OC-1	51.840Mbps
OC-3	155.520Mbps
OC-12	622.080Mbps
OC-24	1.244Gbps
OC-48	2.488Gbps
OC-192	10.000Gbps
OC-256	13.271Gbps
OC-768	40.000Gbps

Service can be unprotected or protected (i.e., a redundant link is provided to be used in case the primary link fails).

Transparent - This occurs when a firewall, interface, or other network node is not the destination of a packet sent beyond it. Routers are a classic example of a transparent network device - clients send packets through a router to a destination beyond the router. In general, transparent describes a function that operates without being evident to the user.

Trap - This is an SNMP message that allows network devices to report critical events immediately back to the management station. Traps are sent when an event occurs that is important enough to not wait until the device is again polled. *Polling* is performed when information collected by an SNMP management station based on queries at predetermined intervals is used to check the status of each network device. The commands used in a trap are the **get** and **set** commands.

Trap door - This is an undocumented way to gain access to an information system. It can also be a program (such as a Trojan horse) that has been altered to allow someone to gain privileged access to a system or process.

Tree - This is a data structure accessed beginning at the *root node*. Each node is either a *leaf* or an *interior node*. An interior node has one or more *child* nodes and is called the *parent* of its child nodes. There may be one or more child nodes for each parent node. Contrary to a physical tree, the root is usually depicted at the top of the structure, and the leaves are depicted at the bottom.

Trivial file transfer protocol (TFTP) - This protocol is similar to FTP except that it uses UDP as a transport and does not use any type of authentication. Given the lack of authentication, TFTP is not something a customer wants coming through their firewall.

Trojan horse - This is a program or code fragment that hides inside an innocuous-seeming program and performs a disguised function.

Trusted computer system - This is a system that employs sufficient hardware and software assurance measures to allow its use for simultaneous processing of a range of sensitive or classified information. Such a system is often achieved by employing a trusted computing base.

Trusted computing base (TCB) - A TCB is the totality of protection mechanisms within a computer system, including hardware, firmware, and software, the combination that is responsible for enforcing a security policy. A TCB consists of one or more components that together enforce a unified security policy over a product or system. The ability of a TCB to correctly enforce a unified security policy depends solely on the mechanisms within the TCB and on the correct input by system administrative personnel of parameters (e.g., a user's clearance level) related to the security policy.

431

Trusted package - This is a PL/SQL or Java stored procedure that does not allow a client to directly access the package's data sources but will perform an activity that it should perform.

Trusted path - This is a mechanism whereby the user is assured that a direct link or path with the system's trusted software exists. The user can rely on this path to perform security relevant operations (e.g., entering a password) without fear of some application interfering with data entry or presenting misinformation.

Trusted root - This is an entity, usually a certification authority (CA), that a particular system recognizes and trusts to verify a public key. Any public key certificate signed by a trusted root is considered valid.

Tunneling - This is the process of encapsulating a packet within a packet of a different protocol.

Tunneling router - This is a router or system capable of routing traffic by encrypting it and encapsulating it for transmission across an untrusted network, for eventual de-encapsulation and decryption.

Turnkey operation - This is a situation where an organization can set up and begin operations with a minimal effort.

Uniform resource locator (URL) - A URL is the address of a file (resource) accessible on the Internet. The type of resource depends on the Internet application protocol. Using the World Wide Web's protocol, the Hypertext Transfer Protocol (Hypertext Transfer Protocol), the resource can be an HTML page, an image file, a program such as a common gateway interface application or Java applet, or any other file supported by HTTP. The URL contains the name of the protocol required to access the resource, a domain name that identifies a specific computer on the Internet, and a hierarchical description of a file location on the computer.

Uninterruptible power supply (UPS) - This is a backup power unit that supplies power if commercial power fails. A UPS can be either online or offline. Some are battery powered; others include generators that can provide standby power indefinitely.

Universal description, discovery, and integration (UDDI) - This consists of the following four components:

1) The Extensible Markup Language (XML) that tags information so that other systems know how to interpret it.
2) The proposed Simple Object Access Protocol (SOAP), a messaging specification that defines ways to package and transmit XML messages.
3) HTTP, which defines how a Web browser or server will respond to messages they receive.

432

4) The domain name system (DNS) for specifying Web addresses.

UNIX - This is a nonproprietary operating system provided by several vendors, such as the Santa Cruz Operations (SCO), which is called SCO UNIX. Another popular form of UNIX is the operating system Linux from Red Hat. A version of UNIX offered by IBM is called AIX.

Usage - This is information that can be used to determine, through an audit process, how an entity (user or application) made use of the information system.

User - This is an employee, contractor, subcontractor, Federal, state, and local government agency employees, authorized domestic and internal aviation industry partners and authorized foreign governments having access and use of information or information systems, nationally or internationally.

User datagram protocol (UDP) - This is a transport protocol in the Internet suite of protocols. UDP, like Transmission Control Protocol (TCP), uses IP for delivery; however, unlike TCP, UDP provides for exchange of datagrams without acknowledgement or guaranteed delivery.

User-level file permissions - Setting these permissions ensures that executable files do not become infected. If all applications are launched from servers (i.e., thin clients), then you can *decrease the likelihood of virus infection* by setting the minimum level of required permissions. CRC checksums can be used to detect replacement of Telnet or FTP applications, even if the replacement has the same size and timestamp.

User requirement - This is any requirement as stated by and for a user.

Utility - This is a program that adds functionality to the operating system. Examples include middleware programs, DOS-based command line utilities, Java utilities, menu utilities, and server console utilities.

Vampire tap - This is a connection to a coaxial cable in which a hole is drilled through the outer shield of the cable so that a clamp can be connected to the inner conductor of the cable. A vampire tap is used to connect each device to a Thick Ethernet coaxial cable in the bus topology of an Ethernet 10Base-T LAN. A different connection approach, the British Naval Connector, is used for the thinner coaxial cable known as a Thin Ethernet.

Verifier - This is a person or application seeking to authenticate a claimant.

Virtual local area network (VLAN) - A VLAN is a logical group of LAN segments, independent of physical location, with a common set of requirements. For example, several workstations might be grouped as a department, such as engineering or accounting. If the workstations are located close to one another, they can be grouped into a LAN segment. If any of the workstations are on a different LAN segment, such as different buildings or locations, they can be grouped into a VLAN that has the same attributes as a LAN even though the workstations are not all on the same physical segment. The information identifying a packet as being part of a specific VLAN is preserved across a switch connection to a router or another switch if they are connected via trunk ports, such as in the case of Asynchronous Transfer Mode (ATM) switching. Any VLAN can participate in the Spanning-Tree Protocol. The protocol used depends on the type of VLAN and the type of bridging function used.

Virtual memory - This allows the server to utilize more random access memory (RAM) space that is physically installed in the system. The system makes use of the hard disk as a temporary area for RAM. Hard disk memory is much slower than RAM so the use of virtual memory causes the system to perform at a much slower rate.

Virtual private database (VPD) - This is a secure data cache that provides server-enforced, fine-grained access control. It provides server-enforced security and cannot be bypassed by users accessing data directly or by using another application. Attributes, such as external name, proxy user and user ID, protocol, port number, and full distinguished name (DN) can be used for providing security. For example, by limiting users to viewing their own organization's records only, or even some part of the organization's records only (i.e., by using the DN).

Virtual private network (VPN) - A VPN is a private data network that makes use of the public telecommunication infrastructure, maintaining privacy through the use of a tunneling protocol and security procedures. A virtual private network can be contrasted with a system of owned or leased lines that can only be used by one company. The idea of the VPN is to give the company the same capabilities at much lower cost by using the shared public infrastructure rather than a private one. Phone companies have provided secure shared resources for voice messages. A virtual private network makes it possible to have the same secure sharing of public resources for data. Companies today are looking at using a private virtual network for both extranet and wide-area intranet. There are three main types of VPNs:

1) *Access VPNs* that provide access to an enterprise customer's intranet or extranet over a shared infrastructure. Access VPNs use analog, dial, ISDN, DSL, mobile IP, and cable technologies to securely connect mobile users, telcommuters, and branch offices.
2) *Intranet VPNs* that connect link enterprise customer headquarters, remote offices, and branch offices to an internal network over a shared infrastructure using dedicated connections. Intranet VPNs differ from extranet VPNs by allowing VPN access only to the enterprise customer's employees.
3) *Extranet VPNs* link outside customers, suppliers, partners, or communities of interest to an enterprise customer's network over a shared infrastructure using dedicated connections. Extranet VPNs differ from intranet VPNs by allowing users outside the enterprise to use the VPN.

Virus scanners - These scanners use signature files to locate viruses within infected files. A *signature file* is simply a database that lists all known viruses, along with their specific attributes. There are two basic types of virus scanners: 1) On demand, and 2) Memory resident. *On-demand virus scanners* must be initialized through some manual or automatic process. *Memory-resident virus scanners* are programs that run in the background of a system.

Vision - This is a succinct statement of an organization's reason for being. It is a simple statement of the high-level goals, or the whys, of an organization.

Vulnerability - This is a weakness in the physical layout, organization procedures, personnel, management, administration, hardware, or software that may be exploited to cause harm to an information system or activity. The presence of a vulnerability does not itself cause harm; a vulnerability is a condition that may allow an information system or activity to be harmed by some malicious event.

Wallet - This is a storage mechanism used to manage security credentials for a user. A wallet implements the storage and retrieval of credentials used with various cryptographic services

War dialer - This is a piece of software that dials a series of phone numbers and identifies which numbers were answered by a computer.

Weak authentication - This is any form of authentication that can be easily broken by an intruder, such as storing the passwords in plaintext or in a file that can be readily obtained by an intruder.

Web - This is an abbreviation for the World Wide Web.

Web browser - This is a client program (like Netscape's Navigator or Microsoft's Internet Explorer) that allows one to explore the Web by clicking on hot links.

Web listener - This is a lightweight HTTP Web server that enables one:

a) To build and deploy PL/SQL-based Web applications and
b) To serve static files.

Web services description language (WDSL) - This is an XML-based standard that allows applications to exchange content.

WHOIS - This is a utility used to gather information about a specific domain.

Wide area network (WAN) - *Wide area network (WAN) topologies* are network configurations that are designed to carry data over a great distance. *Point-to-point* means that the technology was developed to support only two nodes sending and receiving data. If multiple nodes need access to the WAN, then a LAN will be placed behind it to accommodate this functionality. A *T1 link* (1.544Mbps) is a full-duplex signal over two-pair wire cabling. T1s use time division multiplexing to break two wire pairs into 24 separate channels. There are two common ways to deploy leased lines or T1s:

a) The circuit constitutes the entire length of the connection between the two organizational facilities (most secure but more costly), or
b) The leased line is used for the connection from each location to its local exchange carrier, such as frame relay.

While it is possible to sniff one of these circuits, an attacker would need to gain physical access to some point along its path. The attacker would also need to be able to identify the specific circuit to monitor. For large WAN environemnts, frame relay can be far more cost effective than dedicated circuits. The WAN connection point is defined through the use of a unique *data link connection identifier (DCLI)*. An attacker can divert traffic to their network by using the same local exchange carrier and the same physical switch, and know your DCLI.

Window Internet name service (WINS) - This is a server that allows NetBIOS systems to communicated across a router using IP encapsulation of NetBIOS. The WINS server acts as a NetBIOS Name Server (NBNS) for p-node and h-node systems located on the NT Server's local subnet. WINS can store the system's NetBIOS name as well as its IP address. When a p-node system needs the address of another Net BIOS system, it sends a discovery packet to its local WINS server. If the system in question happens to be located on a remote subnet, the WINS server returns the remote system's IP address. This allows the remote system to be discovered without propagating broadcast frames throughout the network. When h-nodes are used, the functionality is identical to the p-node, except that an h-node can fall back to broadcast discovery if the WINS server does not have an entry for a specific host.

Wired equivalent privacy (WEP) - This is the 802.11 standard describes the communication that occurs in wireless LANs. A part of the 802.11 standard is the WEP algorithm, which is used to protect wireless communications from eavesdropping.

Wireless local area network (LAN) - This is a flexible data communications system implemented as an extension to, or as an alternative for, a wired LAN.

Wiring closets and server rooms - These facilities tend to be junction points for many comunications sessions and are pirme targets for malicious persons.

Wizard - A wizard is an agent-like program that is used to guide a user though a complex task and answer questions when issues arise.

Workstation - This is a high-performance, single-user microcomputer that is used for graphics, CAD, CAE, simulation, or scientific applications, but is often synonymous with personal computers. It is typically a RISC-based computer that runs under some variation of UNIX, but may run under the Mac OS or some version of Windows or other OS.

World Wide Web (Web) - This is a communications network that allows users or servers to communicate with other users or servers anywhere in the world provided each is connected to the Web. The Web port on the gateway server is port 80.

Wrapper - This is code that is combined with an original piece of code to determine how the original code is executed. The wrapper acts as an interface between its caller and the wrapped code (original code). This may be done for compatibility, e.g. if the wrapped code is in a different programming language or uses different calling conventions, or for security (such as preventing the calling program from executing certain functions). The implication is that the wrapped code can only be accessed via the wrapper.

Charles L. Smith, Sr.

APPENDIX B

ACRONYMS AND ABBREVIATIONS

3DES	Triple Data Encryption Standard
A&E	Ammunition and Explosives
Abilene	packet-over SONET network (very high-speed (IPv4 at OC-48))
ACAT	Acquisition Categories
ACC	Application Cryptographic Command
ACI	Access Control Information (or Instructions)
ACK	Acknowledge
ACL	Access Control List
ACO	Administrative Contracting Officer
ACRN	Accounting Classification Reference Number
AD	Applications Development
ADM	Add/Drop Multiplexer
AES	Advanced Encryption Standard
AH	Authentication Header
AIS	Automated Information System (outdated term for IS)
AISSO	AIS Security Officer
AIX	Advanced Interactive Executive (IBM version of UNIX)
ALE	Annualized Loss Expectancy
ALNP	Aggregated Link and Node Protection
ANK	Alphanumeric Keyboard
AP	Access Point
APB	Automated Program Baseline
API	Application Programming Interface
ARP	Address Resolution Protocol
ASACS	Advanced Smart Access Control System
ASAP	As soon as possible
ASBR	Autonomous System Border Router
ASCII	American Standard Code for Information Interchange
ASIC	Application-Specific Integrated Circuit
ASO	Advanced Security Option
ASP	Application Service Provider
ATA	Advanced Technology Attachment
ATM	Asynchronous Transfer Mode
AV	Anti-Virus
B&C	Builds and Controls
B2B	Business-to-Business
B2C	Business-to-Customer
BA	Business Area

BAPI	Biometric API
BDC	Backup Domain Controller
BDK	Beans Development Kit
BER	Bit Error Rate
BGP	Border Gateway Protocol
BI	Business Intelligence
BIC	Business Information Center
BIND	Berkeley Internet Name Domain (Internet server for domain names)
BOOTP	Bootstrap Protocol (an alternative to DHCP)
BSA	Business Software Alliance
C&A	Certification and Accreditation
CA	Certification Authority
CAD/CAM	Computer-Aided Design/Computer-Aided Manufacturing
CAM	Certificate Arbitrator Module
CAP	Common Authentication Protocol
CASE	Computer-Assisted Software Engineering
CAT	Contract Administration Team
CATV	Cable Television
CAW	Certificate Authority Workstation
CBC	Cipher Block Chaining
CBI	Computer-Based Instruction
CBK	Common Body of Knowledge
CBOM	Component Bill Of Materials
CBT	Computer Based Training
CC	Common Criteria
CCB	Configuration Control Board
CCC or C^3	Command, Control, and Communications
C^3I	Command, Control, Communications, and Intelligence
CCD	Configuration Control Decision
CCdB	Closed Contract Database
CCI	Common Client Interface
CCITT	Consulting Committee, International Telephone and Telegraph
CCK	Complementary Code Keying
C-Commerce	Collaborative Commerce
CD	Compact Disk
CDN	Content Delivery Network
CDP	Cisco Discovery Protocol
CDPD	Cellular Digital Packet Data
CDR	Critical Design Review
CDRL	Contract Data Requirements List
CD-ROM	Compact Disk-Read Only Memory
CDSA	Common Data Security Architecture
CERT	Computer Emergency Response Team
CFB	Cipher Feedback
CGI	Common Gateway Interface

CHAP	Challenge Handshake Authentication Protocol
CIAO	Critical Information Assurance Office
CICS	Customer Information Control System
CID	Computer Identification (or Card Image Dataset)
CIFS	Common Internet File System
CIM	Common Information Model
CINC	Commander-In-Chief
CIP	Critical Infrastructure Protection or Communications Interface Processor (Processing)
CIPR	Contractor Insurance and Pension Reviews
CIPSO	Commercial Internet Protocol Security Option
CIR	Committed Information Rate
CIRT	Computer Incident Response Team
CISO	Chief Information Security Officer
CISSP	Certified Information Systems Security Professional
CKL	Compromised Key List
CLEC	Competitive Local Exchange Carrier
CLIN	Contract Line Item Number
CM	Configuration Management
CMA	Contract Management Assistant
CMCP	Change Management and Control Process
CMIP	Common Management Information Protocol
CMM	Capability Maturity Model
CMS	Common Message Set
CMVC	Configuration Management Version Control
COMPUSEC	Computer Security
COMSEC	Communications Security
CONOPS	Concept of Operations
CONUS	Continental or Coterminous United States
COOP	Continuity of Operations
COPS	Common Open Policy Standard, Computer Oracle and Password System
CORBA	Common Object Request Broker Architecture
COS	Cache Object Store (Novell iChain architecture)
COTS	Commercial Off-The-Shelf
CPC	Collaborative Product Commerce
CPE	Customer Premises Equipment
CPI	Control Processor Interface
CPL	Customer Priority List
CPM	Contractor Performance Measurement
CPS	Certification Practice Statement
CPSS	Contractor Priority Surveillance System
CPU	Central Processing Unit
CR	Change Request
CRL	Certificate Revocation List
CRLCMP	Computer Resources Life-Cycle Management Plan

CRM	Customer Relationship Management
CRMP	Computer Resource Management Plan
CRT	Cathode Ray Tube
CSC	Cryptographic Service Call
CSCI	Computer Software Configuration Item
CSIRC	Computer Security Incident Response Center
CSL	Computer System Laboratory (NIST)
CSM	Customer Service Management
CSMA/CA	Carrier Sense Multiple Access/Collision Avoidance
CSP	Commerce Service Provider, Cryptographic Services Provider
CSR	Certificate Signing Request
CSS	Cascading Style Sheets, Customer Service & Support
CSU/DSU	Channel Service Unit/Data Service Unit
CTAK	Ciphertext Autokey
CTO	Chief Technology Officer
CVE	Common Vulnerabilities and Exposures
CWDM	Coarse Wavelength Division Multiplexing
D&T	Deloitte and Touche
DAA	Designated Approving Authority, Data Authentication Algorithm
DAC	Discretionary Access Control, Data Authentication Code
DAS	Data Acquisition Subsystem or Direct Access Storage
DASD	Direct Access Storage Device
DAT	Digital Automatic Tape
DB	Database
DBA	Database Administrator
DBCA	Database Configuration Assistant
DBMS	Database Management System
DBPS	Differential Binary Phase Shifting
DCD	DFAS Corporate Database
DCE	Distributed Computing Environment
DCID	Director of Central Intelligence Directive
DCLI	Data Link Connection Identifier
DCMC	Defense Contract Management Command
DCMD	Defense Contract Management District
DCOM	Distributed Component Object Model
DCU	Data Control Unit
DCW	DFAS Corporate Warehouse
DDDS	Defense Data Dictionary System
DDOS	Distributed Denial-of-Service
DDE	Distributed Development Environment
DEN	Directory Enabled Networking
DES	Data Encryption Standard
DFAS	Defense Financial and Accounting System
DGSA	Defense Goal Security Architecture
DHCP	Dynamic Host Configuration Protocol

DHSS	Direct-sequence spread-spectrum
DHTML	Dynamic Hypertext Markup Language
DID	Data Item Description
Diff-Serv	Differentiated Services
DII	Dynamic Invocation Interface or Defense Information Infrastructure
DIRAMS	DCMC Information Repository Automated Metrics System
DITSCAP	DOD Information Technology Security Certification and Accreditation Process
DISA	Defense Information Systems Agency
DISN	Defense Information System Network
DISSP	Defense Wide Information Systems Security Program
DIT	Directory Information Tree
DITC	DCMC Information Technology Center (formerly DLA Systems Design Center (DSCD))
DITSCAP	DOD IT Security Certification and Accreditation Process
DLA	Defense Logistics Agency
DLAR	DLA Regulation
DLC	Data Link Control
DLCI	Data Link Connection Identifier
DLS	Distributed Link Service
DMC	Defense MegaCenter, see RPC
DMS	Defense Message System
DMTF	Distributed Management Task Force
DMX	Documentrix
DMZ	Demilitarized Zone
DNA	Distributed Internet Application
DNS	Domain Name Service (or System or Server)
DOD	Department of Defense
DODAAC	DOD Activity Address Code
DODIIS	Department of Defense Intelligence Information System
DOORS	Dynamic Object-Oriented Requirements System
DOS	Disk Operating System
DoS	Denial of Service
DPACS	DLA Pre-Award Contracting System
DPAS	Defense Priority Allocation System
DPPS	Defense Procurement Payment System
DQPSK	Differential Quadrature Phase Shift Keying
DRAM	Dynamic Random Access Memory
DSA	Digital Signature Algorithm
DSCC	Defense Supply Center Columbus (Ohio)
DSDC	DLA Systems Design Center
DSL	Digital Subscriber Line
DSML	Directory Services Markup Language
DSN	Defense Switching Network
DSP	Digital Signal Processor
DSS	Digital Signature Standard, Decision Support System

DSSS	Direct Sequence Spread Spectrum
DT	Developmental Testing
DTD	Document Type Definition
DT&E	Developmental Test and Evaluation
DUAL	Diffusing Update Algorithm
DUNS	Data Universal Numbering System
DVD	Digital Video Disk
DWDM	Dense Wavelength Division Multiplexing
E&Y	Ernst and Young
EAI	Enterprise Application Integration
EAP	Enterprise Application Portals or Extensible Authentication Protocol
EBCDIC	Extended Binary Coded Decimal Interchange Code
ebXML	Electronic Business Extensible Markup Language
EC	Electronic Commerce
ECA	External Certification Authority
ECB	Electronic Codebook
E-Commerce	Electronic Commerce
ECON	Electronic Commerce Processing Node
EDA	Electronic Document Access
EDI	Electronic Data Interchange
EDIFACT	Electronic Data Interchange For Administration, Commerce, And Transport
EDW	Electronic Document Workflow
EEPP	End-to-End Path Protection
EES	Escrowed Encryption Standard
EGP	Exterior Gateway Protocol
EIP	Enterprise Information Portal
EIS	Executive Information System
EJB	Enterprise JavaBeans
ELIN	Exhibit Line Identification Number
EMSEC	Emissions Security
EOM	End-Of-Message
EOS	End-Of-Service
EP	Extension Point
EPIC	Explicitly Parallel Instruction Computing
EPL	Evaluated Product List
EPROM	Erasable Programmable Read Only Memory
ERP	Enterprise Resource Planning
ERTZ	Electromagnetic Radiation TEMPEST Zone
ESCON	Enterprise Systems Connection
ESI	Extremely Sensitive Information
ESP	Encapsulating Security Payload, External Service Provider
ESS	Electronic Signature System
ETH	Ethernet
ETL	(Data) Extraction, Transformation, and Load

444

EVPN	Enterprise VPN
FC	Fiber Channel
FDDI	Fiber Digital Data Interface
FEP	Front End Processor`
FGAC	Fine Grained Access Control
FHSS	Frequency-hopping spread-spectrum
FIAT	Facility Information Analysis Tool
FICON	Fiber Connection
FICS	File Inventory Control System
FIDNET	Federal Intrusion Detection Network
FIPS	Federal Information Processing Standard
FIRMR	Federal Information Resources Management Regulation
FITL	Fiber In The Loop
FLA	Field Level Activity
FME	Firmware Maintenance Environment
FR	Frame Relay
FRAP	Facilitated Risk Analysis Process
FR/ATM	Frame Relay to Asynchronous Transfer Mode
FRF	Frame Relay Forum
FSO	Free Space Optics
FT	Functional Testing
FTP	File Transfer Protocol
FTS	Federal Technology Service
FY	Fiscal Year
GAO	General Accounting Office
GB	Gigabyte
GbE	Gigabit Ethernet
GEOMAP	Geographic Map
GFE	Government-Furnished Equipment
GFI	Government-Furnished Information
GFP	Government-Furnished Property
GIF	Graphics Image Format
GMS	Global Monitor Support
GMT	Greenwich Mean Time
GNSS	Global Navigation Satellite System
GOES	Geostationary Operational Environment Satellite
GOTS	Government Off-The-Shelf
GPS	Global Positioning System
GRE	Generic Routing Encapsulation
GSA	General Services Administration
GSM	Global System for Mobile Communication
GSR	Gigabit Switch Router
GTE	General Telephone and Electronics or Government Transition Evaluation

445

GUI	Graphical User Interface
GW	Gateway
HA-API	Human Authentication API
HCI	Human-Computer Interface
HEP	Horizontal Enterprise Portals
HIDS	Host-based Intrusion Detection System
HIPAA	Health Insurance Portability and Accountability Act
HMI	Human-Machine Interface
HP	Hewlett Packard
HPC	High Performance Computing
HPFS	High Performance File System
HQ	Headquarters
HRMS	Human Resource Management System
HSM	Hierarchical Storage Management
HSP	High Speed Printer
HTML	HyperText Markup Language
HTMP	HyperText Markup Protocol
HTTP	Hypertext Transport Protocol
HTTPS	Secure HTTP (sometimes written SHTTP)
HVAC	Heating, Ventilation, and Air Conditioning
HW	Hardware
HWCI	Hardware Configuration Item
I&A	Identification and Authentication
I&C	Installation and Checkout
I&I	Installation and Integration
I&T	Integration and Test
IA	Information Assurance
IAD	Integrated Access Device
IANA	Internet Address and Numbering Authority
iAS	iPlanet Application Server, Internet Application Server
IASE	Information Assurance Support Environment
IATO	Interim Approval to Operate
IAW	In Accordance With
IBM	International Business Machines
ice	Import/Conversion/Export
ICMP	Internet Control Message Protocol
ICS	Internet Caching Server
ID	Identification or Identifier
IDE	Integrated Development Environment
IDEA	International Data Encryption Algorithm
IDL	Interface Definition Language (for Java)
IDSL	ISDN Digital Subscriber Line
IECA	Interim External Certification Authority
IESG	Internet Engineering Steering Group

IETF	Internet Engineering Task Force
I/F	Interface
IGP	Interior Gateway Protocol
IGRP	Interior Gateway Routing Protocol
IIOP	Internet Inter-ORB (Object Request Broker) Protocol
IKE	Internet Key Exchange formerly called ISAKMP/Oakley
ILEC	Incumbent Local Exchange Carrier
IM	Instant Messaging
IMAP	Internet Messaging Access Protocol
IMS	Information Management System
IMUX	Inverse Multiplexing (= DSL Bonding)
INFOSEC	(National) Information Systems Security
INTI/INTO	Interfacility Input/Interfacility Output
I/O	Input/Output
IOS	Internetworking Operating System (Cisco's OS for routers and firewalls)
IOT&E	Initial Operational Test and Evaluation
IP	Internet Protocol
IPFR	IP-enabled Frame Relay
IPRA	Internet Policy Registration Authority
IPSec	Internet Protocol Security
IPSO	Internet Protocol Security Option
IPv4	IP version 4
IPv6	IP version 6 (sometimes called IPng, for next generation)
IPX	Internetwork Packet Exchange
IPX/SPX	An Internet Protocol that is less robust than TCP/IP
IR	Infrared
IRC	Incident Response Capability
IRD	Interface Requirements Document
IRDP	Internet Control Message Protocol (ICMP) Router Discovery Protocol
IRM	Information Resource Management
IS	Information System, Industrial Specialist
ISA	Interagency Service Agreement
ISAC	Information Sharing and Analysis Center
iSCSI	Internet Small Computer System Interface
ISDN	Integrated Services Digital Network
ISAKMP	Internet Security Association and Key Management Protocol
ISO	International Organization for Standards
ISP	Internet Service Provider
ISS	Internet Security Systems (a company), Information Security Strategies
ISSC	Information System Security Certifier
ISSM	Information Systems Security Manager
ISSO	Information Systems Security Officer
ISV	Independent Software Vendor
IT	Information Technology
ITL	Information Technology Laboratory (NIST)

ITSEC	Information Technology Security
ITU	International Telecommunications Union
IV&V	Independent Verification and Validation
IVR	Interactive Voice Response
JAAS	Java Authentication and Authorization Service
JAIN	Java APIs for Intelligent Networking
JAR	Java Archive
JAVA	Object-Oriented Programming Language that can be downloaded over the Internet
JCL	Job Control Language
JTA	Joint Technical Architecture
JDBC	Java Database Connectivity
JDK	Java Development Kit
JMS	Java Message Service
JNDI	Java Naming and Directory Interface
JNI	Java Native Interface
JOVIAL	Jules Own Version of an Intermediate Assembly Language
JRC	Joint Resources Council
JSP	Java Server Page
JVM	Java Virtual Machine
J2EE	Java 2 Platform, Enterprise Edition
Kbps	Kilobits per second
KDC	Key Distribution Center
KEA	Key Exchange Algorithm
KEK	Key Encryption Key
Km	Kilometer
KMF	Key Management Facility
KTC	Key Translation Center
L2F	Layer 2 Forwarding (Protocol)
L2TP	Layer 2 Tunneling Protocol
LAN	Local Area Network
LANE	Local Area Network Emulation
LATA	Local Access and Transport Area
Lat/Lon	Latitude/Longitude
LCM	Life-Cycle Management
LCN	Local Communications Network
LCS	Local Communications System
LDAP	Lightweight Directory Access Protocol
LDIF	LDAP Data Interchange Format
LDM	Logical Data Model
LEAP	not an acronym, it is an authentication system from Cisco
LEC	Local Exchange Carrier
LED	Light Emitting Diode

Linux	A new version of UNIX
LLC	Limited Liability Corporation
LLP	Limited Liability Partnership
LO	Log Out
LOA	Letter of Agreement
LOB	Line Of Business
LOE	Level Of Effort
LRA	Local Registration Authority
LSA	Link-State Advertisement
LSP	Label Switched Path
LSR	Label Switch Router
MA	Maintenance Authority
MAC	Media Access Control, Message Authentication Code
MAISRC	Military Automated Information Systems Resource Council
MAN	Metropolitan Area Network
MB	Megabyte
MD5	Message Digest #5
MDA	Milestone Decision Authority
MHz	Megahertz
MI	Middle Initial
MIB	Management Information Base
MID	Message Identifier
MILDEP	Military Deputy
MIL-STD	Military Standard
MILSTRIP	Military Standard Requisitioning and Issue Procedures
MIS	Management Information System
MFR	Multi-Link Frame Relay
MIM	Man In the Middle
MIME	Multipurpose Internet Mail Extension
MISSI	Multilevel Information System Security Initiative
MLPPP	Multi-Link PPP
MLS	Multi-Level Security
MMDS	Multi-channel Multipoint Distribution System
MMI	Man-Machine Interface
MNS	Mission Need Statement
MOA	Memorandum of Agreement
MOLAP	Multidimensional Online Analytical Processing
MPEG	Moving Pictures Expert Group
MPLS	Multi-Protocol Label Switching
MPOA	Multi-Protocol Over ATM
MPP	Massively Parallel Processing (Systems)
MPTS	Multiple Program Transport Stream
MRP	Materials Replacement Program
MS EAP/TLS	Microsoft Extensible Authentication Protocol/Transport Level Security
MSP	Message Security Protocol

MTBF	Mean Time Between Failures
MTTF	Mean Time To Failure
MTTR	Mean Time To Recover
MTS	Multi-Threaded Server
MTTR	Mean Time To Repair
MTU	Multi-tenant Unit
MVS	Multiple Virtual System
MVS/XA	MVS/Extended Architecture
NAS	Network-Attached Storage
NAT	Network Address Translation
NATO	North Atlantic Treaty Organization
NCS	Non-Critical Sensitive
NCSC	National Computer Security Center
NDAP	Novell Directory Access Protocol
NDI	Non-Developmental Item
NDS	Network Directory Service or Novell Directory Server
NetBEUI	NetBIOS Extended User Interface
NetBIOS	Network Basic Input Output System
NFS	Network File Service
NIC	Network Interface Card
NIDS	Network-based Intrusion Detection System
NII	National Information Infrastructure
NIPRNET	Unclassified (but sensitive) Internet Protocol Routing Network
NIS	Network Information Service
NIST	National Institute for Standards and Technology
NLSP	Network Layer Security Protocol, NetWare Link Services Protocol
nm	Nautical Mile
NMAS	Novell Modular Authentication Service
NMI	Native Method Invocation
NMS	Network Management System
NNI	Network-to-Network Interface
NNTP	Network News Transfer Protocol
NOAA	National Oceanic and Atmospheric Administration
NOC	Network Operating Center
NOFORN	No Foreign Dissemination
NORAD	North American Air Defense
NOS	Network Operating System
NSA	National Security Agency
NSC	Network Security Center
NSM	Network and Systems Management
NSN	National Stock Number
NSP	Network Service Provider
NSTISSC	National Security Telecommunications and Information Systems Security Committee
NSTISSI	National Security Telecommunications and Information Systems

	Security Instruction
NSTISSP	National Security Telecommunications and Information System Security Policy
NUMA	Non-Uniform Memory Access
NVRAM	Nonvolatile Random-Access Memory
OADM	Optical Add/Drop Multiplexer
OASYS	Over and Above System
ODBC	Open Database Connectivity
OEM	Original Equipment Manufacturer
OID	Oracle Internet Directory
OLAC	Object Level Access Control (Novell iChain architecture)
OLAP	Online Analytical Processing
OLE	Online linking and embedding
OLTP	Online Transaction Processing
OMB	Office of Management and Budget
OMI	OPSEC Management Interface
OO	Object Orientation
OPR	Organization of Primary Responsibility
OPSEC	Open Platform for Security
ORB	Object Request Broker or Operational Review Board
ORD	Operational Requirements Document
OS	Operating System
OS/2	IBM's OS
OSD	Office of the Secretary of Defense
OSF	Open System Foundation
OSI	Open Systems Interconnection
OSPF	Open Shortest Path First
OSS	Oracle Security Server
OT	Operational Testing
OTP	One-Time Password
OU	Organizational Unit
P3P	Platform for Privacy Preferences Project
PAC	Privilege Attribute Certificate
PAT	Process Action Team
P&L	Profit and Loss
PBX	Phone Base Exchange (telephone switchboard)
PC	Personal Computer
PCA	Physical Configuration Audit, Policy Creation Authority
PCARSS	Plant Clearance Automated Reutilization Screening System
PCI	Personal Computer Interface
PCMCIA	Personal Computer Memory Card International Association (Card)
PCS	Physical Control Space
PCT	Private Communication Technology
PDA	Personal Digital Assistant

PDC	Primary Domain Controller
PDD	Presidential Decision Directive
PDF	Portable Document Format
PDM	Product Data Management
PDN	Public Data Network, Pentagon Digital Network
PDR	Preliminary Design Review
PDU	Protocol Description Unit
PEM	Privacy Enhanced Mail
PERL	Practical Extraction and Reporting Language
PGP	Pretty Good Privacy
PHY	Physical Layer (Layer 2 of the OSI Protocol Model)
PID	Packet Identifier
PIIN	Procurement Instrument Identification Number
PIN	Personal Identification Number, Process Improvement Network
PIP	Partner Interface Process or Product Integration Plan
PKCS	Public Key Cryptography Standard (PKCS #1, 3, 5, 7, 10, 12, and 13)
PKI	Public Key Infrastructure
PKIX	Public Key Infrastructure X.509
PLAS	Performance Labor Accounting System
PM	Project Manager
PMP	PERMIT Management Protocol
POC	Point Of Contact
POP	Point of Presence, Post Office Protocol
POTS	Plain Old Telephone Service
PPP	Point-to-Point Protocol
PPPoE	Point-to-Point Protocol over Ethernet
PPTP	Point-to-Point Tunneling Protocol
PRF	Pseudo-Random Function
PRNG	Pseudorandom Number Generator
PROM	Programmable Read Only Memory
PS/2	Personal System/2
PSN	Packet Switched Network
PSTN	Public Switched Telephone Network
PUB L	Public Law
PVC	Permanent Virtual Connection (Circuit)
PwC	Price Waterhouse Coopers
QA	Quality Assurance
QAR	Quality Assurance Representative
QL	Quick Look
QoS	Quality of Service
R&M	Reliability and Maintainability
RADAR	Radio Detection and Ranging
RADIUS	Remote Authentication Dial-In User Service
RAID	Redundant Array of Independent (Inexpensive) Disks

RAM	Random Access Memory or Requirements Allocation Matrix
RAMP	Risk Assessment Management Program
RARP	Reverse Address Resolution Protocol (an alternative to DHCP)
RAS	Remote Access Server
RBAC	Role-Based Access Control
RBS	Role-Based Services
RC2	Rivest Cipher #2
RDBMS	Relational Database Management System
RDF	Revised Delivery Forecast or Reason for Delay
RDS	Remote Data Services
RFC	Request For Comment
RFP	Request For Proposal
RIP	Routing Information Protocol
RIPSO	Revised Internet Protocol Security Option
RISC	Reduced Instruction Set Computer
RMI	Remote Method Invocation
RMON	Remote Monitoring
RPM	Rotations Per Minute
ROBO	Remote Office/Branch Office
ROI	Return On Investment
ROLAP	Relational Online Analytical Processing
RPC	Remote Procedure Call or Regional Processing Center, see DMC
RQS	Recoverable Querying Server
RRAS	Routing and Remote Access Service
RS/6000	IBM RISC Server
RSA	Rivest, Shamir, and Adelman (cipher codes)
RSVP	Resource Reservation Protocol
RTDB	Real Time Database
RTM	Requirements Traceability Matrix
RTMP	Routing Table Maintenance Protocol
SA	Security Association
SAA	Systems Application Architecture
SAFE	Secure Blueprint for Enterprise Networks (Cisco, this is not an acronym)
SALT	Speech Application Language Tags
SAMMS	Standard Automated Material Management System
SAMP	Suspicious Activity Monitoring Protocol
SAN	Storage-Area Network
SANS	Systems Administration, Networking, and Security (Institute)
SAP	Special Access Program, (Systems, Applications, and Products in Data Processing defined for the SAP AG product SAP R/3)
SAS	Server Attached Storage
SASL	Simple Authentication and Security Layer
SATA	Serial Advanced Technology Attachment
SATAN	Security Administrator Tool for Analyzing Networks

SBU	Sensitive but Unclassified, Strategic Business Unity
SCAP	Security Certification and Authorization Package
SCCB	Software Configuration Control Board
SCCI	Source Code Control Interface
SCE	Software Capability Evaluation
SCI	Sensitive Compartmented Information
SCIF	Sensitive Compartmented Information Facility
SCM	Supply Chain Management
SCSI	Small Computer System Interface
SDD	Software Design Description
SDH	Synchronous Digital Hierarchy
SDK	Software Development Kit
SDM	System Decision Memorandum
SDNS	Secure Data Network System
SDSL	Symmetric Digital Subscriber Line
SDW	Shared Data Warehouse
SEI	Software Engineering Institute
SEP	Secure Exchange Protocol
SET	Secure Electronic Transaction (protocol)
SETA	Systems Engineering, Testing, and Analysis
SFA	Sales Force Automation
SFC	System File Checker
SFP	System File Protection
SFUG	Security Features Users Guide
SGML	Standard Generalized Markup Language
SHA	Secure Hash Algorithm
SHDSL	Symmetric High-bit-rate Digital Subscriber Line
SHTTP	Secure Hypertext Transfer Protocol
SI	Systems Integrator
SICM	Standard Information Contract Management
SIMS	Software Information Management System
SIOP	Single Integrated Operations Plan
SIOP-ESI	Single Integrated Operations Plan - Extremely Sensitive Information
SIP	Site Installation Plan
SIPRNET	Secret Internet Protocol Routing Network
SIR	System Implementation Review
SIS	Supplier Information Service, Secure Integrated Software (or Service)
SI/SO	Sign In/Sign Out
SIU	Service Interface Unit
SKIP	Simple Key Interchange Protocol
SLA	Service Level Agreement
SLP	Service Location Protocol
SMB	Session Message Block
SME	Session Management Exit
SMF	System Management Facility
SMI	Security Management Infrastructure

S/MIME	Secure/Multipurpose Internet Mail Extension
SMIT	System Management Interface Tool
SMP	Symmetric Multi-Processing
SMS	Short Messaging Service, System Managed Storage
SMTP	Simple Mail Transfer Protocol
SNA	Simple Network Architecture (IBM network architecture)
SNMP	Simple Network Management Protocol
SOAP	Simple Object Access Protocol
SOHO	Small Office/Home Office
SONET	Synchronous Optical Network
SOP	Standard Operating Procedure
SPECS	Software Professional Estimating and Collection System
SPF	Shortest Path First
SPI	Security Parameter Index
SPICE	Software Process Improvement and Capability Determination [sic]
SPIIN	Supplemental Procurement Instrument Identification Number
SPIRNet	Secret IP Router Network
SPS	Standard Procurement System
SPX	Sequenced Packet Exchange
SRM	Storage Resource Management
SQL	Structured Query Language
SQLJ	Structured Query Language Java
SRAM	Static Random Access Memory
SSA	Single Station Administration
SSAA	Systems Security Authorization Agreement
SSBI	Single Scope Background Investigation
SSC	Single Source of Control
SSIWG	SPS Security Integration Working Group
SSL	Secure Sockets Layer
SSO	Single Sign-On
SSP	Storage Service Provider
SSR	Secondary Surveillance Radar
SSS	System/Subsystem Specification
ST	Security Target
STACS	Storage Access Coordination System
ST&E	Security Test and Evaluation
STD	Standard
STP	shielded twisted pair, Spanning Tree Protocol
SUID	Secure User ID or Set User ID
SVC	Switched Virtual Connection (Circuit)
SVN	Secure Virtual Network
SW	Software
SysOp	System Operator
T&E	Test and Evaluation
TAFIM	Technical Architecture Framework for Information Management

TAMS	Termination Automation Management System
TASO	Terminal Area Security Officer
TCB	Trusted Computing Base
TCL	Tool Command Language
TCO	Total Cost of Ownership
TCP	Transmission Control Protocol
TCP/IP	Transmission Control Protocol/Internet Protocol
TCSC	Trusted Criteria for Secure Computers
TCSEC	Trusted Computer System Evaluation Criteria
TDM	Time Division Multiplexing
TEMP	Test and Evaluation Master Plan
TEMPEST	not an acronym
TFM	Trusted Facility Manual
TFTP	Trivial File Transfer Protocol (sometimes written "tftp")
TGT	Ticket Granting Ticket
TIFF	Tagged Image File Format
TLD	Top-Level Domain
TLS	Transport Layer Security (similar to SSLv3) or Transparent LAN Service
TNG	The Next Generation (Computer Associates security software package)
TNM	Telecommunications Network Management (architecture)
TNS	Transit Network Selection
TS	Top Secret
TSN	Technical Support to Negotiations
TTY	Teletypewriter
UDDI	Universal Description, Discovery, and Integration
UDP	User Datagram Protocol
UFP	URL Filtering Protocol
UHF	Ultra-High Frequency
UID	User Identification
ULO	Unliquidated Obligations
UNIX	Universal Operating System
UPS	Uninterruptible Power Supply
URL	Uniform Resource Locator
USA	United States Army or United States of America
USAF	United States Air Force
USGS	United States Geological Survey
USN	United States Navy
UTC	Coordinated Universal Time
UTP	unshielded twisted pair
VAN	Value-Added Network
VAR	Value-Added Reseller
VB	Visual Basic
vBNS	very high-performance Backbone Network Service

VC	Virtual Circuit
VEP	Vertical Enterprise Portals
VLAN	Virtual Local Area Network
VLSM	variable-length subnetwork mask
VM	Virtual Machine (Operating System)
VoFR	Voice over Frame Relay
VoIP	Voice over IP
VPN	Virtual Private Network
VRML	Virtual Reality Markup Language
VSAM	Virtual Storage Access Method
VSAT	Very Small Aperture Terminal
VTAM	Virtual Telecommunications Access Method
VTP	VLAN Trunk Protocol
W3C	World Wide Web Consortium
WAN	Wide Area Network
WAP	Wireless Application Protocol
WBEM	Web-Based Enterprise Management
WCDMA	Wideband Code Division Multiple Access
WDM	Wave Division Multiplexing
WDSL	Web Services Description Language
Web	World Wide Web
WEP	Wired Equivalent Privacy
WLAN	Wireless Local Area Network
WS	Workstation
WWW (= W3)	World Wide Web
X.25	A packet-switched protocol.
X.500	A directory standard
X.509	A certificate standard
XDR	XML Data Reduced (Schema)
xDSL	any of the variations in DSL
XIWT	Cross-Industry Working Team
XML	Extensible Markup Language

Charles L. Smith, Sr.

APPENDIX C

QUESTIONNAIRE: A SAMPLE

1.0   Introduction

A questionnaire is a document that contains a set of questions that should be addressed to a knowledgeable person in some particular area of interest. For best results, the questions should be multiple choice, binary if at all possible (e.g., Yes-No, On-Off, etc.), but no more than four or five options, which should be included on the questionnaire after each question. Unfortunately, many questions that seem to be binary actually have more answers than Yes-or-No, for example, you may get "Sometimes Yes, sometimes No," or "We don't have that option" or some other answer. The actual answer given should be recorded in the space provided for the multiple choices, even if one of the choices is not selected.

The persons chosen as the source of answers for the questions should be selected randomly to ensure that the results are statistically acceptable, if that is what is desired. In some cases, the interviewee is the only person that is interviewed for a particular questionnaire, so there is no need to worry about randomness. The randomness of the interviewees is important only if some sort of statistical compendium of the results is desired. Many questionnaires are for a specific set of persons and therefore randomness is not required.

2.0   Sample Questionnaire

A sample questionnaire is provided in Table C.1, *Sample Questionnaire for System Assets and Values*. If more than one interviewee is required, then the same exact sequence of questions should be posed to each interviewee, that is, present the questions to each person in exactly the same order as listed in the questionnaire. This avoids the possibility of bias. Record the answers by circling the answers as given by the interviewee, or if the answer must be written in, write in the answer, immediately, not later. Never try to remember what an interviewee said, always write down, or note, the answer right after asking the question.

Some of the security issues can be addressed using a questionnaire since they require knowledge from some particular person, such as a security administrator or manager who has detailed information, either in their memory or in a handy reference book, about the relevant system.

There are a variety of reasons for using a questionnaire. You might be interested in performing a risk assessment for a local area network, or a set of local area networks. Or you might be interested in determining if a system has a contingency plan, and if not, determining the kind of plan required. Or you might be interested in identifying the system assets and their values by addressing upper management, system administrators, and important users. This is the example given. Similarly, you might be interested in identifying what management thinks are the most important threats to the system.

Table C.1 Sample Questionnaire (For System Assets and Values)

Interviewee: _____ Interviewer: _____
Date: _____ Location: _____

No	Question	Answers
1	What organizational functions are performed using your information system?	a) personnel, b) financial, c) business transactions, d) other
2	What is the most important product or service performed by your organization?	a) named product, b) named service, c) other
3	What proportion of your business transactions are done through your information system?	a) 100%, b) 75%, c) 50%, d) 25%, e) less than 25%
4	What is the most important aspect of your information system?	(a) system access by customers, b) system access by users, c) generated database information, d) information about customers, e) other
5	What is the value of this item?	Answer
6	What is the next most important aspect of your information system?	(a) system access by customers, b) system access by users, c) data information, d) information about customers, e) other
7	What is the value of this item?	Answer
8	Repeat questions 6 and 7 until most of the assets of the system are identified.	
9	What is the probability that an ex-employee might perform a malicious action on the organization's information system?	a) high, b) likely, c) medium, d) fair, e) small, f) very small
10	What is the probability that someone else might perform a malicious action on the organization's information system?	a) high, b) likely, c) medium, d) fair, e) small, f) very small
11	Are your applications backed up at a remote site?	a) Yes, b) No, c) Other
12	Is your middleware backed up at a remote site?	a) Yes, b) No, c) Other
13	Are your database information backed up at a remote site?	a) Yes, b) No, c) Other
14	Do you have replacement equipment at a remote location?	a) Yes, b) No, c) Other
15	Do you have a contingency plan?	a) Yes, b) No, c) Other
16	Do you ever practice the software recovery procedures?	a) Yes, b) No, c) Other
17	If so, how frequently?	Answer
18	Do you ever practice the hardware recovery procedures?	a) Yes, b) No, c) Other
19	If so, how frequently?	Answer
20	Do you have an uninterruptible power supply?	a) Yes, b) No, c) Other

21	Do you have backup heating, ventilation, and air conditioning?	a) Yes, b) No, c) Other
22	Do you have limited access for authorized personnel to server rooms?	a) Yes, b) No, c) Other

Charles L. Smith, Sr.

APPENDIX D.1

SECURITY POLICY: AN EXAMPLE

"A security policy is the set of rules and practices that regulate how an organization manages, protects, and distributes sensitive information. It's the framework in which a system provides trust," (Russell 1991).

By categorizing the security policy into the three types of controls:

1) Physical controls,
2) Behavioral controls, and
3) Technical controls,

the process of creating the security policy is simplified.

Some elements of a security architecture cannot be easily placed into a single category, for example, firewalls are computer devices (hardware) yet depend on the rapid and simple operating system and a set of rules (software) in order to operate. Similarly, routers are also hardware devices depending on software to operate. In both these cases, firewalls and routers, the device is placed in the hardware category.

The security perimeter is the boundary between the organization's primary information system and all other systems that provide sources of inputs or outputs to the system. If the organization's system is local, then the boundary is the point of attachment of the communications systems (e.g., the Internet and perhaps a Wide Area Network (WAN) link such as those provided by Internet Service Providers (ISPs)) to the local system. If the organization's system is distributed, then the perimeter (or demarcation points) is the set of connection points that act as the communications paths among the various intranets that compose the entire system. In these cases, the links also provide for communications paths to and from extranets or remote users or even the untrusted public (some of whom are malicious hackers). If the system has private communications networks, such as leased T1 links (1.544Mbps), then the link can also be considered inside the boundary (or perimeter) of the organization's information system.

This appendix comprises two parts, the first part is an example of a security policy and the second part is an example of a security rules base. At a minimum, a good security policy should:

1) Be readily accessible to all members of the organization.
2) Be readily accessible to all customers of the organization.
3) Define a clear set of security goals.
4) Accurately define each issue discussed in the policy.
5) Clearly show the organization's position on each issue.
6) Describe the justification of the policy regarding each issue.
7) Define under what circumstances the issue is applicable.
8) State the roles and responsibilities of organizational members with regard to the described issue.

9) Spell out the consequences of noncompliance with the described policy.
10) Provide contract information for further details or clarification regarding the described issue.
11) Define the user's expected level of privacy.
12) Include the organization's stance on issues not specifically defined.

An organization's security policy should be aligned with the goals of the organization. Some rules for ensuring that an organization's security policy is properly worded are (Andress 2001):

1) The security policy should be created so that its rules can be easily monitored and enforced.
2) The security policy should clearly state the penalties for any violations of the policy rules.
3) The security policy should be disseminated to all employees and each employee should be expected to sign a statement that they have read and understand the policy rules.

The security policy should be created so that the procedures and required detailed instructions for implementing and enforcing the rules are clearly delineated.

5) The security policy rules should be continuously monitored using audit systems to ensure that all those affected by the rules properly adhere to them.
6) The security policy should be reviewed periodically so that any changes due to modified organizational goals, improvements in security technology, or changes in threats can be immediately addressed. Subsequent to any changes, the updated policy should be made immediately available to all employees and others affected by the changed policy.
7) The security policy should clearly state who is responsible for what parts of the policy.

A security policy for an organization might look like the following.

An Example of an Organization's Security Policy

This organization's information system network policy for its system components (hardware and software) and information, is as follows:

1. Physical access to the information system facility shall be controlled to limit only authorized personnel.
2. Hardware devices (e.g., routers, cables, firewalls, servers, intrusion detection systems, etc.) that handle sensitive but unclassified information shall be housed in protected areas so that only authorized personnel can access them.
3. Sensitive but unclassified information shall be safeguarded at all times whether stored or transmitted. Countermeasures shall be applied so that such information is accessed only by authorized persons, is used only for its intended purpose, retains its content integrity, and is printed on pages marked "For Official Use Only" or "For Organizational Use Only" or similar marking, as required.

4. Sensitive but unclassified information shall be safeguarded against tampering, loss, and destruction and shall be available when needed. This is necessary to protect the organization's investment in obtaining and using information and to prevent fraud, waste, and abuse. Suggested countermeasures can be categorized as physical (e.g., badges and guards), behavioral (e.g., administration and training), or technical (e.g., firewalls, clustered servers, intrusion detection systems, and anti-virus software).

5. System countermeasures against sabotage, tampering, denial-of-service, organizational espionage, fraud, misappropriation, misuse, access to forbidden Web sites (e.g., pornographic sites), or release to unauthorized persons, shall be accomplished through the continuous employment of auditing countermeasures as well as other techniques, as required. The mix of countermeasures selected shall achieve the requisite level of security or protection.

6. Any employee who is found guilty of participating in any violation of a security rule will be dismissed. Any non-employee found guilty of a violation will be prosecuted to the full extent of any laws that are broken.

7. The selected mix of countermeasures shall ensure that the system meets the minimum requirements as set forth in the organization's *Minimum Security Requirements Statement*. These minimum requirements shall be met through automated and manual means in a cost-effective and integrated manner.

8. A risk assessment shall be performed to identify cost-effective countermeasures and any additional security requirements over and above the set of minimum requirements.

9. Countermeasure products that are commercial off-the-shelf (COTS) or existing software (no matter how it was developed or obtained (but not pirated software)) are preferred but other available products (such as from the Government or from other organizations) shall be evaluated (as requested) for designation as trusted computer products for inclusion on an Evaluated Products List (EPL). Evaluated products shall be designated as meeting security criteria maintained by a respected security organizations (such as the National Computer Security Center (NCSC) at the National Security Agency (NSA)) regarding its capabilities and features (e.g., C2 controlled access protection, see Appendix I).

10. The following timetable (schedule) shall be adhered to:

   a) All systems that process or handle sensitive unclassified information and that require at least controlled access protection (i.e., class C2 security), based on the risk assessment procedure requested above, shall implement the required security features within one year subsequent to the date of acceptance of the organization's original or updated *Minimum Security Requirements Statement*.

   b) If security features above class C2 are required, based on the risk assessment results, a timetable for meeting these more stringent requirements shall be determined on an individual system basis and submitted to the Designated Approving Authority (DAA) for approval. These requirements shall be met either by implementing trusted computer products listed on the EPL or by using a product not on the EPL that has security features that meet the level of trust required for the system. In either case, a certification and accreditation (C&A) must be accomplished and approved by the appropriate DAA to assess whether adequate security measures have been taken to permit the system to be used operationally.

11. If the introduction of additional computer-based security features, according to the schedule described above, for an existing system or a system already under development, is prohibitively expensive, time-consuming, technically unsound, or may have unacceptable effects on the system's operations, then the following shall apply:

   a) Other countermeasures (e.g., physical controls, administrative controls, etc.) may be substituted as long as the requisite level of system security or protection, as determined by the DAA, is attained.
   b) Only the organization's owners or high-level managers or the appropriate DAA may authorize exceptions. Such authorization shall be based on a written determination that one or more of the exceptional conditions exists. Exceptions shall be reviewed at each reaccreditation.

12. System C&A shall be performed at least every three years or subsequent to any system upgrade that affects the security aspects of the system, or if the security technology has been greatly improved, or if increases in the threat causes the security requirements to become more stringent. The C&A shall follow the planning, testing, and analysis and evaluation of test data collected during the security test and evaluation (ST&E) process.

13. When systems managed by different DAAs are interfaced or networked, a memorandum of agreement (MOA) is required that addresses the accreditation requirements for each subsystem involved. The MOA should include the following:

   a) Descriptions of the data;
   b) Access levels of the users;
   c) Designation of the DAA who shall resolve conflicts among the DAAs; and
   d) Countermeasures to be implemented before interfacing the subsystems.

14. MOAs are required when one Organizational Component's system interfaces with another system within the same Organizational Component or in another Organizational Component and when a contractor's system interfaces with an Organizational Component's system or to another contractor's system.

15. For a multi-user telecommunications network (e.g., the Internet or a public Wide Area Network (WAN)), a DAA shall be designated as responsible for the overall security of the network and shall determine the security requirements for connection of systems to the primary network.

16. Necessary countermeasures shall be agreed to and implemented and the systems accredited for interconnection before they are connected to the network.

17. The security of each system connected to the network remains the responsibility of its DAA.

18. The DAA responsible for the overall security of the network shall have the authority and responsibility to remove from the network any system not adhering to the security requirements of the network.

19. It is permissible to categorize network interfaces and system boundaries into manageable subnetworks based upon physical or logical boundaries, when there is a need to do so. Cryptographic separation and/or equivalent computer security measures, as defined by recognized security authorities where applicable, shall be a basis for defining such network and/or subnetwork interfaces or boundaries.

20. Networks, including all connected subnetworks, shall be accredited for the highest security level required based on the concepts and procedures in the risk assessment and risk mitigation.

21. Security policy shall be considered throughout the lifecycle of the system from security policy development, security requirements development, concept development, security architecture development, through design, development, operation, and maintenance until replacement or disposal. A DAA shall be designated as responsible for the overall security of the information system. The following conditions shall be met:

    a) The system developer is responsible for ensuring the early and continuous involvement of the users, information system security officers, data owners, and DAA(s) in defining and implementing security requirements of the system. There shall be an evaluation plan for the system showing progress towards meeting full compliance with stated security requirements through the use of necessary computer and network countermeasures.

    b) Mandatory statements of security requirements shall be included, as applicable in the acquisition and procurement specifications for system. The statements shall be the result of an initial risk assessment, and shall specify the level of trust required under any primary reference document, such as OMB Circular A-130, Appendix III.

    c) No sensitive unclassified data shall be introduced into the system without designation of the sensitivity of the data. Approval to enter the data shall be obtained from the data owner where applicable. The accreditation of the system shall be supported by a certification plan, a risk analysis of the system in its operational environment, an evaluation of the countermeasures, and a certification report, all approved by the DAA. Accreditation of computers embedded in a system may be at the system level.

    d) A program for conducting periodic reviews of the adequacy of the countermeasures for operational, accredited systems shall be established. To the extent possible, reviews are to be conducted by persons who are independent of the user organization and of the system operation or facility.

    e) Where required, as specified in OMB Circular No. A-130, a program for developing and testing contingency plans shall be established. The objective of contingency planning is to provide reasonable continuity of system support if events occur that prevent normal operations. The plans should be tested periodically under realistic operational conditions.

f) Changes affecting the security of an information system must be anticipated. Any changes to the system or associated environment that affect the accredited safeguards or result in changes to the prescribed security requirements shall require reaccreditation. Reaccreditation shall take place before the revised system is declared operational. Minimally, the system shall be reaccredited every three years, regardless of changes.

22. Access to the system by anyone outside the organization may be authorized only by the System or Security Administrator, and shall be consistent with this policy.

a) An information system accredited for processing and/or storing sensitive but unclassified information may use automated means (software, firmware, or hardware) to permit both nonsensitive unclassified and sensitive unclassified information to be accessed and stored.

b) No employee is allowed to place a sniffer on the network without explicit permission from an organizational owner. Any employee found taking this action without proper permission shall be immediately dismissed.

c) Malicious and non-malicious actions will be identified (identification of user, event, and results of event) and the user shall be held accountable. Employees will be immediately dismissed for confirmed malicious actions. Employees may be disciplined for unintentional actions that do damage to the system. Employees found making consistent errors (i.e., more than two occurrences within a single year), even though unintentional, that cause damage to the system may be reassigned or dismissed.

d) Sensitive information shall be protected from unauthorized viewing or access during storage or transmission.

References:

Russell, Deborah and G. T. Gangemi, Sr. (1991). "Computer Security Basics," O'Reilly, 1991.

Andress, Mandy (2001). "Effective Security Starts With Policies," InfoWorld, p. 56, November 19, 2001.

APPENDIX D.2

SECURITY RULES BASE: AN EXAMPLE

A sample set of rules for a rules base is presented in the second column of Table D.2, *An Example of a Security Rules Base*, below. The Open Systems Interconnection (OSI) level for each rule is provided in the first column. The various categories for the rules are also presented in the table prior to the listing of the rules. The security mechanisms for obeying the rules are not included since the table is not vendor specific. However, the owner may decide to select those products that are verified by the NIST as being evaluated by an independent organization using the Common Criteria. The contents of this table are a modified version of the material that appeared in the following reference (Smith 1999).

Table D.2 An Example of a Security Rules Base

OSI Levels	Security Rules
Identification and Authentication (The "A" category)	
5, 6, or 7	A.1 Identification and authentication should be by user identification (User ID) and password for regular users and include an additional authentication mechanism, such as a smart card or fingerprint biometric for system and security administrators.
5, 6, or 7	A.2 All entities (persons, applications, or services/servers) should have a secure identification represented by a certificate.
5, 6, or 7	A.3 Identification should be provided on the basis of the user's name.
5, 6, or 7	A.4 Each user ID shall be unique and each password shall be unique.
5, 6, or 7	A.5 Passwords shall not be "guessable," for example, do not use words from a dictionary.
5, 6, or 7	A.6 Passwords should never be written down and posted where other users can see them.
5, 6, or 7	A.7 Passwords should be at least eight characters long.
5, 6, or 7	A.8 Passwords should include at least one "numerical character" and at least one "special character."
5, 6, or 7	A.9 Passwords should have a lifetime of 90 days (or whatever is deemed appropriate for your organization) and be enforced by a software mechanism.
5, 6, or 7	A.10 Users should be notified by the system if their remaining password lifetime is less than or equal to five days.
5, 6, or 7	A.11 If an incorrect password is provided by an entity three times in succession, then the entity is forbidden from access for 24 hours (or whatever is deemed appropriate for your organization).
5, 6, or 7	A.12 Each entity shall have a certificate
5, 6, or 7	A.13 Expired certificates shall be entered in a expired access control list
5, 6, or 7	A.14 Infrastructure access by external entities shall be based the policy that "all entering entities that are not explicitly allowed (by some rule) are disallowed."

OSI Levels	Security Rules
5, 6, or 7	A.15 All users (except administrators) shall have a time-of-day limitation for their authentication.
5, 6, or 7	A.16 A single sign-on (SSO) capability shall be provided.
5, 6, or 7	A.17 Remote internal users shall be identified and authenticated by either a RADIUS or TACACS subsystem.
5, 6, or 7	A.18 All employees with access to the information system shall have badges or be escorted and are required to enter the facility through a guard-protected gate.
5, 6, or 7	A.19 All information facilities shall be enclosed with fences and gates and be regularly patrolled by guards.
5, 6, or 7	A.20 Identification and authentication shall be provided by the operating system and the single sign-on (SSO) servers, not each application.
Authorization (The "B" category)	
5, 6, or 7	B.1 All authorized entities shall require a distinguished name for determining their permissions or privileges for application and database access.
5, 6, or 7	B.2 Database access shall be determined to a specific row of a specific database, if need be, to ensure that the appropriate granularity is in place.
5, 6, or 7	B.3 Separation of duties shall be enforced. That is, no entity shall be allowed access to any database information that could be used in a malicious manner.
Access Control (The "C" category)	
5, 6, or 7	C.1 All authorized entities shall use a role-based access control (RBAC) mechanism for access control.
5, 6, or 7	C.2 The RBAC capabilities shall include: identification of the entity, roles to which each entity belongs, and privileges (or permissions) for each role.
5, 6, or 7	C.3 All job functions (i.e., roles) shall be defined for the organization.
5, 6, or 7	C.4 All privileges (or permissions) shall be defined for each role.
5, 6, or 7	C.5 The system administrator shall be responsible for immediately removing from the Access Control List (ACL) the user's name for those who have left the organization and placing their user identification in a Certificate Revocation List (CRL).
5, 6, or 7	C.6Database and application access shall be based on a need-to-know or need-to-use.
5, 6, or 7	C.7 No user shall have access to information that they do not need in order to perform their job functions (i.e., least privilege).
Availability (The "D" category)	
2-7	D.1 The information system shall have at least a 0.999 (or whatever is deemed appropriate for your organization) availability performance.
2-7	D.2 All single-points-of-failure shall have at least a dual capability so that there are no single-points-of-failure.

OSI Levels	Security Rules
2-7	D.3 All servers shall be designed as a clustered server with concomitant software for load balancing and automatic failover.
5, 6, or 7	D.4 The system shall be protected from denial-of-service (DoS) attacks by using anti-DoS software and anti-Distributed DoS software.
Accountability (The "E" category)	
5, 6, or 7	E.1 All activities on the system shall be audited and the audit trail shall include the following: identification of the performing entity, the type of event, the success or failure of the event, and the date and time of the event.
5, 6, or 7	E.2 All intruders shall be detected and the intrusion and intruder shall be identified and properly recorded for immediate or later response.
5, 6, or 7	E.3 All auditing and IDS information shall be stored in a database that its contents cannot be modified or deleted.
Confidentiality (The "F" category)	
5, 6, or 7	F.1 All sensitive information (such as user passwords) stored in memory shall be encrypted.
2-7	F.2 All sensitive information (such as user passwords) transmitted from any server to another server shall be encrypted.
2-7	F.3 All sensitive information transmitted over a public network (e.g., the Internet) shall be encrypted.
2-7	F.4 Encryption shall be asymmetric.
5, 6, or 7	F.5 Each entity shall have a certificate.
5, 6, or 7	F.6 Session keys shall be used for symmetric encryption.
3	F.7 All transmitted information shall be encrypted using a Virtual Private Network (VPN) service of the firewall.
Data Integrity (The "G" category)	
5, 6, or 7	G.1 Data integrity shall be provided using a hashing function or by encryption.
5, 6, or 7	G.2 All transmitted sensitive information shall be protected with data integrity.
Non-Repudiation (The "H" category)	
5, 6, or 7	H.1 All entities shall use digital signatures (based on asymmetric encryption) to ensure that neither the sender nor the receiver can later deny having sent or received a transmitted message.
No Object Reuse (The "I" category)	
5, 6, or 7	I.1 All sensitive temporary information produced by any user shall be erased prior to a later user logging on.
Firewalls, routers, computers, and gateways and cabling (The "J" category)	
2-7	J.1 Infrastructure access by external entities shall be based the policy that "all entering entities that are not explicitly allowed (by some rule) are disallowed."
2-5	J.2 All internal user addresses shall be protected from external viewing.
2-7	J.3 All rules for any firewall, router, or gateway shall be input to the device by an out-of-band computer.

OSI Levels	Security Rules
2-7	J.4 All routers shall have dynamic rules.
2-7	J.5 All firewalls shall be placed between the public wide area network (WAN) and internal users (an intranet).
2-7	J.6 All firewalls between a WAN and an intranet shall have a demilitarized zone (DMZ) for connecting Web servers, FTP servers, and e-mail servers, so that these servers are accessible to the public but no internal device is accessible to the public (i.e., an untrusted user).
1	J.7 All cabling shall be protected against monitoring of electronic emanations.
1	J.8 The information system shall routinely be inspected for sniffers.
3	J.9 All traffic passing through a firewall, route, or gateway shall be logged.
3-7	J.10 All traffic through the firewall shall be subjected to stateful inspection.
3-7	J.11 All physical assets of the information system shall be protected against theft, modification, or damage.
3-7	J.12 All traffic through the firewall between the Internet and an intranet shall be audited (e.g., source and destination addresses, connection start/stop time, etc.).
3-7	J.13 All network products shall be COTS, unless there is an overriding rebuttal reason.
3-7	J.14 Firewall auditing procedures shall include the capability to mitigate operational risks and detect malicious intrusions.
Not applicable	J.15 Procedures for analyzing, evaluating, and potentially responding to hacking attacks shall be defined and improved as available and implemented in the appropriate mechanism to minimize delays in packet handling.
Not applicable	J.16 Tactics, such as using "honeypots," shall be implemented to ensure that intruders are caught and identified.
Not applicable	J.17 Access to server platforms, network components, cabling, and sensitive workstations shall be controlled.
Not applicable	J.18 All network subsystems shall have an expanded performance capability over that which is immediately required so that the system can grow without requiring expensive upgrades.
Not applicable	J.19 All network subsystems shall be scalable, interoperable, and portable so that they naturally are a part of an open architecture.
Not applicable	J.20 The capability to detect the failure of a subsystem shall be provided.
Not applicable	J.21 In case of detected failed subsystems, the capability to redirect communications to alternative processing nodes and/or across alternative transmission paths shall be provided.
Not applicable	J.22 All subsystem components shall be accredited/certified, according to some recognized authority.

OSI Levels	Security Rules
Not applicable	J.23 All operational facilities shall establish configuration management over all database and file structures to include: 1) database/table descriptions as well as the purpose and use of the information; and 2) access control criteria for all entities authorized for information system access with proofs of authorization required for acceptance.
Accreditation (The "K" category)	
Not applicable	K.1 All subsystems shall be accredited by a certification authority (CA) as being security worthy.
1-3	K.2 All communications connections via modems shall be subjected to controls (e.g., encryption, call back, strong authentication, etc.). Modems shall not be allowed to bypass the accredited network traffic control points (e.g., firewalls or routers).
Not applicable	K.3 All servers shall be housed in a secure room.
4-7	K.4 Any non-accredited LAN subsystem (i.e., extranet) shall be connected to this organization's information system through a firewall with an appropriately defined set of access rules.
Behavioral Security (The "L" category)	
Not applicable	L.1 All users shall be required to attend a training session every twelve months.
Not applicable	L.2 All managers shall be required to attend a training session every twelve months.
Not applicable	L.3 All system and security administrators shall be required to attend a training session every twelve months.
Not applicable	L.4 All users shall be required to attend an educational (i.e., awareness) session every twelve months.
Not applicable	L.5 All managers shall be required to attend an educational (i.e., awareness) session every twelve months.
Not applicable	L.6 All system and security administrators shall be required to attend an educational (i.e., awareness) session every twelve months.
Not applicable	L.7 All commercial off-the-shelf (COTS) software shall be upgraded according to patches received from the appropriate vendor within a reasonable time (e.g., no more than three weeks) after the patch arrives.
Not applicable	L.8 All physical entries of the facility shall be protected by accredited Security Guards.
Not applicable	L.9 All persons seeking to become an authorized user must get the required permissions from the System and Security Administrators.
Not applicable	L.10 No user shall record (on floppies, compact disks, zip disks, or super disks, or on removable hard disks) any information that is intended to leave the facilities without explicit written and signed permission.
Not applicable	L.11 No user shall place a sniffer on the network without explicit written and signed permission.
Not applicable	L.12 No employee shall give out sensitive information (such as passwords) to callers whom they cannot identify with certainty.

OSI Levels	Security Rules
Not applicable	L.13 The information system shall be subjected to a comprehensive security test and evaluation of its security mechanisms at least once every three years or whenever the system is upgraded with new subsystems that affect its security posture.
Not applicable	L.14 The ST&E process shall be well defined.
Not applicable	L.15 The ST&E process shall include ST&E for each organizational intranet and the network end-to-end situations.
Not applicable	L.16 The security administrator shall periodically perform a scanning of the network to identify any holes or other vulnerabilities, and then to determine and implement the appropriate actions to counter the vulnerabilities.
Not applicable	L.17 Subsequent to any "security incident," the security administrator shall immediately disseminate the incident to relevant personnel and identify a counter to future such incidents.
Not applicable	L.18 All persons with operations and maintenance responsibilities shall be properly trained.
Not applicable	L.19 All persons with network communications responsibilities shall be properly trained.
Not applicable	L.20 The effectiveness of all training and awareness programs shall be determined and used as feedback to improve the programs.
Not applicable	L.21 A comprehensive risk analysis shall be required for the information system.
Not applicable	L.22 All users shall be trained not to open any e-mail attachments if the mail is from someone that they do not know or do not trust.
Backup and Recovery (The "M" category)	
4, 5, 6, or 7	M.1 All sensitive information, other information, applications, operating systems, and middleware shall be replicated at a survivable location.
4, 5, 6, or 7	M.2 All sensitive information shall be replicated to a survivable location where it can be easily retrieved to quickly bring the system back up in case of a catastrophic event that erases the system's capabilities. Replications shall be performed at the end of each workday (e.g., 9 PM till Midnight) or more frequently, as required, for example, every time an update is required, then the updated information is also sent to a remote site for backup and recovery purposes.
4, 5, 6, or 7	M.3 All software items shall be subject to configuration management as defined for this organization.
Not applicable	M.4 Each LAN and the overall system shall have a contingency plan prepared and shall practice recovery once every 24 months (or more frequently, as required).
4-7	M.5 All information that may be required at some later time shall be archived on tape.

OSI Levels	Security Rules
4-7	M.6 Configuration control shall be maintained over all archived files. Configuration Management shall include: date file archived, the contents of the operational facility/platform/file archived, each file's name/size/other descriptors contained in the archived file, and tape/file integrity verification parameters.

No doubt there are many possibilities for rules that are not in this table, however, the contents of this table should provide an excellent stimulus for identifying the specific rules for any organization (e.g., as an input for beginning a brainstorming session). In most cases, these rules will actually be an overstatement of the rules base and the desired rules will likely be a subset of these.

Security control can be grouped into three categories, namely, 1) management, 2) operational, and 3) technical (NIST 1995).

1) *Management Controls* address the managerial aspects of security. These controls are processes for managers of the organization's computer security program. In general, these controls are applied by managers to regulate the computer security program and to properly handle risk within the organization.
2) *Operational Controls* address those processes that are implemented and executed by people. These controls are put in place to improve the security of a particular system (or group of systems). Operational controls often require technical or specialized expertise, and may rely upon management and technical controls.
3) *Technical Controls* address the computerized processes. These controls depend upon the proper functioning of the system for their effectiveness. Implementation of technical controls requires significant operational considerations and should be consistent with the management of security within the organization.

References:

Smith, Sr., Charles L. (1999). "Draft of Security Rules Base for the Pentagon Digital Network," Single Agency Manager (SAM), 1999.

NIST (1995). "An Introduction to Computer Security: The NIST Handbook," Special Publication 800-12, National Institute of Standards and Technology (NIST), October 1995.

Charles L. Smith, Sr.

APPENDIX E

SECURITY THREATS

Table E.1, *Summary of Threats*, presents the potential threats that could be used to take advantage of information system vulnerabilities. The use of Tier I, Tier II, and Tier III are references to advanced client/server architecture hardware and software components residing on the following three platform levels:

Tier I - Client Workstations,
Tier II - Application Servers, and
Tier III - Database Servers.

Table E.1 Summary of Threats

Threat	Implementation	Targets

Transmitted information monitoring	Communications sniffing
	  - Induction loops
	  - Vampire taps (fiber)
	  - Splices
	  - Offices with active Ethernet ports
	  - Cables to open platform ports
	    (e.g. serial port)
	  - Impostor platforms/inserted
	    components
	  - Radio transmission interceptions
		- LAN/MAN/WAN
		  transmissions
		- Radio transmissions
		- All infrastructure
		  components generating
		  electromagnetic
		  emanations
		- All
		  network/communications
		  components with
		  physically open (i.e., active
		  daemons) or unused ports

	Unauthorized monitoring
	  - Sniffers (Software only based or
	    Hardware/software based)
	  - Unauthorized platform activity as
	    recorded by auditing logs
	  - Retrieval of cached memory
	    printer/screen memory buffers
		- LAN/MAN/WAN network
		  transmissions
		- Cross-tier traffic
		  monitoring
		    - Tier I to Tier II
		    - Tier I to Tier III
		    - Tier II to Tier II
		    - Tier II to Tier III
		    - Tier III to Tier III
		- All
		  network/communications
		  components with buffered
		  memory or storage

Threat	Implementation	Targets

Transmitted information tampering	Radio signal interception/jamming
	Backdoors for communications
	   devices
	   - Routers, Switches, or Firewalls
	Message Routing
	Redirection of ICMP router
	Message or data integrity
	   - Replay attacks
	   - Cut-and-Paste attacks
	   - Man-in-the-middle attacks
	   - Spoofing
		- Radio based transmissions
		- All transmissions employing transmission protocols with exploitable flaws
		- Tier I, II, and III platforms offering direct access to services with exploitable flaws

	Limitations in communications protocols
	   - Sequence number attacks/spoofing
	   - Internet message modifications to TCP or UDP
	   - Corruption of RPC-based protocols
	   - Modifications to Sendmail
	   - Portmapper service
	   - Contamination of Domain Name Services info	

Decrypt attacks	Scanning of networks and brute force or dictionary attacks
	
	Practical cryptanalysis
	   - Known-plaintext attack
	   - Chosen plaintext
	   - Exhaustive search
	   - Birthday attack
	
	Universal keys (e.g., cryptographic backdoors)
		- Tier I, II, & III encrypted data repositories/transmissions
		- Local and wide area network component-encrypted transmissions
		- Encrypted OS and application files
		   - e.g., Password files
		- Off-line system/data archives
		- Recorded transmissions and/or captured data files

Threat	Implementation	Targets

Unauthorized entry (electronic)	Abuse of authorized info services to Identify potential security targets
	  - via Whois instruction
	  - via Ping instruction
	  - via finger instruction
	  - via Traceroute instruction
	  - via Inverse DNS tree
	  - using network mappers:
	    - Inverse DNS tree
	    - UDP based
	    - ICMP based
	Exploitation of known application and network service deficiencies
	  - Automated Scanning of Hosts:
	    - War dialing
	    - SATAN (or equivalent, e.g., nmap)
	  - Automated or manual attacks
	Bypassing of security mechanisms
	  - Exploitation of improper router level filtering
	  - Use of unmonitored or uncontrolled firewall holes
	  - Use of covert channels
	Improper Identification/Authentication
	  - Insecure Key Distribution servers
	  - Spoofing
	  - Social engineering
	  - Stolen/decrypted system passwords
	  - Physically stolen passwords (theft/trickery)
	  - Guessable or reused passwords
	  - Poor password dissemination techniques
	  - Stolen passwords
	  - Sniffing plaintext passwords
	  - Untrustworthy trusted hosts
	  - Connection laundering
	  - Use of stolen smart cards and concomitant PINs
	Internal attacks by authorized users
		Role-based access controls
		  - Tier I client access RBACs
		  - Tier II application server RBACs
		  - Tier III database server RBACs
		  - Tier I, II, III operating system RBACs
		- Programmable network component RBACs
		Threats to information can be accomplished as follows:
		a. steal the information (ostensibly for use by a competitor of some sort),
		b. damage or erase of information so that it cannot be used, or
		c. modify the information so that it cannot be used properly.
		Modified information can possibly cause great harm.

Threat	Implementation	Targets
Malicious code or actions	Programmed threats - Viruses, Worms, Trojan horses, Logic bombs - Modified platform services - Back doors to firmware (e.g., through OS holes) Attacks on OSs - filename attacks - Startup file attacks - Path attacks - War dialing - Spoofing - Off-line platform attacks Physical intrusions - Climb fences or gates - Access unprotected platform resources - Access unattended workstations	- Tier I, II, III platforms - Network components which use modifiable or stored code - Operating systems anti-malicious software (client and network) - Secured facility - Secured server room - Unattended workstation automatic logoff
Data integrity attack	Threats introduced by applications - Unrestricted/unmonitored user queries or modifications - Application running with root or improperly segmented file system access Unauthorized electronic access as stated above Platform corruption as stated above	Client and server systems Tier I, II, and III hosts or disks
Denial-of-service attacks	Message flooding Service overloading (e.g., syslog) Signal grounding (Physical) Physical destruction of communications or platform infrastructure components	- Specifically targeted applications on specifically targeted platforms - WAN network segments - LAN network segments - Physically targeted platforms and communications components

Threat	Implementation	Targets
Unauthorized removal of information	Storage and removal of sensitive information via floppy disks, super disks, zip disks, and compact disks Corruption of data/system archives Archival of long term, corrupted data Archival of corrupted system applications Archival of data that is not encrypted nor sanitized Use of improperly secured physical media/archives Copied and stolen storage media Physical destruction of information	Online archives - Stored electronic media - Remote archives

Charles L. Smith, Sr.

APPENDIX F

VULNERABILITIES

A *vulnerability* is any existing situation in your information system facility that makes the system susceptible to some threat. All information systems are vulnerable to some threats and one of the tasks for a security analyst is to identify these vulnerabilities and determine the seriousness of these vulnerabilities. Vulnerabilities can be categorized as physical, behavioral, or technical.

Physical and behavioral vulnerabilities can be identified through interviews with owners and users, sometimes using questionnaires, and through observations of the information system facilities. *Technical* vulnerabilities can be identified by: 1) scanning the network, 2) performing an analysis of the system using representations and descriptions of the system, or 3) comparing this system's architecture and operational environment with other similar systems and environments that have known vulnerabilities. The extent of the identified vulnerabilities depends on:

1) The countermeasures currently implemented,
2) The size of the system,
3) The volume of business done by the system,
4) Value of information stored in the system,
5) Notoriety of the system,
6) Perceived susceptibility of the system to non-catastrophic threats, namely, power outages, minor fires, minor water damage, random server failures, and other such events, and
7) Other factors such as system familiarity among potential malicious hackers.

The potential damage that can be done to a system with a particular set of vulnerabilities depends on the assets and the values of these assets to the organization. Thus, the identification of the information system assets and the determination of the values of these assets are important tasks for any security analyst.

Vulnerabilities are often divided into three categories:

1) *Critical* - functions or services that, if lost, would prevent the organization from exercising safe control over its assets.
2) *Essential* - functions or services that, if lost, would reduce the capability of the organization to exercise safe control over its assets.
3) *Routine* - functions or services that, if lost, would not significantly degrade the capability of the organization to exercise safe control over its assets.

These categories can be of use for deciding which vulnerabilities are the most important and for performing a risk analysis of the vulnerabilities that are not countered because it was not cost-effective to do so.

Vulnerabilities

Two methods of determining system vulnerabilities are through the use of architectural analyses or through use of a scanner. If a detailed architecture of the system is available, then it is possible to perform an analysis to identify some of the system's vulnerabilities.

To identify the system *technical vulnerabilities*, a scanner can be used.

The *physical and behavioral vulnerabilities* in your system must be identified by security analysts who, at least, may wish to:

1) Review relevant documentation (such as the policy statement, the security policy statement, the concept of operations, and the information system architecture model),
2) Question some of the owners, users, or other employees,
3) Inspect the information system facilities, and
4) Scrutinize other areas, as required.

Other sources of information may come from owners and named representatives who are aware of certain physical and technical intrusions that have occurred in the past.

The scanning description should identify the:

1) Specific products used,
2) Where they are located in the network,
3) What their capabilities are,
4) What their vulnerabilities might be, and
5) Any other attributes or disadvantages that the vendor might freely admit.

If an architecture model exists, then this model could be used. A knowledgeable analyst can identify many of the system's vulnerabilities using this method, though probably not as comprehensively as would be the case using a scanning tool.

List All Known Vulnerabilities

Sources of the knowledge required to answer questions about system vulnerabilities include:

1) Scanner output and analysis,
2) User testimonials, and
3) Analyses of your system using knowledge of threats and system vulnerabilities that usually exist in other similar systems, or analyses using the identified threats and a description of the system.

The primary objective of this task is to identify the vulnerabilities, that either have occurred or might occur, that could cause harm to this system should a particular type of threat attack be launched against the system by some attacker.

The two primary situations for a vulnerability assessment are:

1) The situation of an existing information system and
2) The situation of an information system to be built for which there is only an architecture model.

For the first case, the owner can have a scanning of the information system performed, or the owner can have a consultant or technical analyst perform an analysis of the architecture and existing documentation to determine abstractly what the vulnerabilities might be.

For the second case, the owner can only determine abstractly what the vulnerabilities might be, based on an architecture of the new system and a sample list of vulnerabilities, such as Table F.1, *List of Potential Vulnerabilities*. Any user should begin with this table and then expand it to include any other vulnerabilities that come to mind. When requirements are developed, they should be placed in a table that allows for tracing back to any previous requirement from a lower level requirement. The reason for this is to incorporate hyperlinking of requirements so that a user can always view the upper level requirements from which a lower level requirement was derived.

Identifying vulnerabilities can be accomplished in many different ways. A process for identifying the vulnerabilities is (NSA 1998):

1) Search through the graphical model of the system architecture for the absence of a particular type of countermeasure that will defeat some identified threat (and scenario). Do this for all of the identified threats.

2) Again, using the architecture model, examine the flow of information to see if there are any weaknesses in the information flow design by theorizing attacks that might exploit these weaknesses.

3) Perform a code analysis of listings of legacy software to see if there are any weaknesses that could be exploited. (This step would be extremely laborious and is not recommended.)

4) Perform a security test and evaluation of the legacy system to identify any weaknesses that could be exploited. (This step is not recommended since a scanning process is the better method for identifying vulnerabilities.)

Table F.1 List of Potential Vulnerabilities

Threats	Potential Vulnerabilities
Unauthorized Physical Entry	Physical entry by unauthorized person and subsequent theft or destruction of some portion of the premises and/or network equipment by that person
Non-catastrophic Event	Loss of power, minor water damage, minor fire damage, random server failure, and loss of heating or air conditioning
Catastrophic Event	Total loss of system due to terrorist bomb or aircraft crash, earthquake, volcanic eruption, major mud flow, flood or tsunami, wind damage from tornado or hurricane, or major fire
Social Engineering Attack	Divulging by employee of privileged information to malicious individual
Internal User Action	1) Compromise of sensitive information recorded on portable device (e.g., floppy disk, compact disk, zip drive disk, super disk, etc.) and then illegally removed from premises by employee 2) Intentional damage to sensitive information Unintentional damage to sensitive information by user
Sneaker Attack	Loss of information that has been illegally recorded by attacker
Virus (Malicious Code) Attack	Loss of system due to damage from virus or other malicious code sent as an attachment through the e-mail system and opened by the victim
Denial-of-Service Attack	Loss of system or information availability due to a denial-of-service attack
Sniffer Attack	Loss of unencrypted information through use of sniffer by an attacker
Scanner Attack	Loss of information concerning system vulnerabilities to attacker using a scanning device and/or software
Hacker Attack	Unauthorized system entry (electronic) to perform malicious actions, such as stealing sensitive information, damage to applications or middleware or databases, deface Web page, loss of transmissions, and loss of system control (availability)
Man-in-the-middle attack	Loss of sensitive information, including user IDs and passwords, to attacker who intercepts communications between two users and acts as the source and destination of the messages
Brute Force Attack	Loss of encrypted information to attacker with computational capability to decrypt encrypted information by trying every key possibility
Ex-Employee Attack	Unauthorized entry and possible malicious attacks on, or loss of, assets to former employee who is knowledgeable of the organization's information system

Threats	Potential Vulnerabilities
Employee Accident	Damage to applications or database information due to accidents
Untrusted Connection	Connecting to an untrusted network without any protection in between such as a firewall followed by hacker entry through the untrusted network
Spoofing	Intentionally using incorrect IP addresses to fool the system

5) Perform a security test and evaluation of the upgraded system to identify any weaknesses that could be exploited.

6) Build a replication of the system in a laboratory where various tests can be performed to identify any weaknesses. (This is recommended for those who can afford it since it allows the client to investigate the capabilities of the implemented security mechanisms prior to a formal implementation.)

Unfortunately, many of these methods are quite time consuming and expensive. The use of scanning tools will perform essentially the same actions as the labor-intensive methods shown above. For these reasons, the first four are not recommended, but the fifth (that proposes an ST&E) is highly recommended. The sixth (system mock-up in a lab) can be too expensive for most organizations, but can lead to a most rigorous security evaluation without any potential of crashing the primary online system.

List Relevant Vulnerabilities

Some of the vulnerabilities that will be discovered may not be pertinent to this particular system because the threats that would take advantage of a particular vulnerability are not considered credible.

The primary inputs to the creation of a list of vulnerabilities, for any particular system, are the results of an analysis based on the findings using the scanner mechanism, and from an analysis of the model of the system architecture, provided the architecture model has sufficient detail.

Charles L. Smith, Sr.

APPENDIX G

SECURITY MECHANISMS

A list of the recommended security mechanisms to counter the various threats discussed in this book are contained in the Table G.1, *Technical Threats Versus Security Mechanisms.* In addition, the table also contains the URLs for some of the vendors that manufacture the security mechanism products. Although threats are divided into the three categories: 1) physical, 2) behavioral, and 3) technical, the table presents just the security mechanisms for the *technical threats.* Of course, *physical security mechanisms* provide security against physical incursions by unauthorized persons and *behavioral security methods* are those aspects of security that provide training and education of user, managerial, technical, and administrative personnel.

A recommended method for finding a complete set of vendors for any technical security mechanism is to go to www.google.com and type in the name of the security mechanism, such as "firewall" or "intrusion detection system." The list of hits that appears will include the names of most or all of the vendors that sell this type of mechanism. By clicking on any hit the customer will obtain the Web page of the vendor and an explanation of the relevant security mechanisms that the vendor manufactures.

Table G.1 Technical Threats Versus Security Mechanisms

Threat	Security Mechanism	Some Vendor URLs
Unauthorized entry (electronic)	**Identification and authentication software (SW), RADIUS and TACACS$^+$ servers, Smart cards, Biometrics**	**User ID and Password** - In-house generated code, www.devshed.com, www.microsoft.com **RADIUS** - http://www.radius.cistron.nl/, http://web.blastradius.com/ **TACACS$^+$** - www.cisco.com, **Smart cards** - http://www1.slb.com/smartcards/, www.gemplus.com/, www.microsoft.com/ **Biometrics** - http://www.acsysbiometricscorp.com/, http://www.appliedbiometrics.net/ie.htm
Authorization	**Access control lists**	http://java.sun.com/products/jaas/, http://mail.cc.umanitoba.ca/drac/

Threat	Security Mechanism	Some Vendor URLs
Access control	**Role-based access control (RBAC)**[75]	www.oracle.com/, http://www.entrust.com/entrustcygnacom/labs/act021.htm, http://www.opennetwork.com/products/info/features/rbpm.php
External intruders (i.e., hackers), Malicious code (i.e., e-mail), Denial-of-service attacks, Sniffers and man-in-the-middle attacks	**Firewalls** (with blocking/passing rules, VPN SW, intrusion detection SW, anti-DoS SW, anti-malicious code SW)	www.cisco.com, www.checkpoint.com, www.zonelabs.com/, http://www.symantec.com/sabu/nis/npf/, http://www.ifi.unizh.ch/ikm/SINUS/firewall/, http://www.consealfirewall.com/, http://www.sygate.com/swat/default.htm
DoS attacks	**Gap appliance** and **anti-DoS** SW	**Gap application** - Spearhead, Inc. bcarmeli@spearheadsecurity.com. **Anti-DoS SW** - www.mcafee.com, www.microsoft.com, www.zeus.com, www.baytsp.com
External intruders (i.e., hackers) and Internal intruders	**Intrusion detection systems**	www.cisco.com, www.checkpoint.com, www.snort.org, www.dshield.org, www.uac.com
Sniffers and man-in-the-middle attacks	**Virtual private networks and other methods of message encryption (e.g., PKI or PGP)**	www.checkpoint.com, www.cisco.com, www.snapgear.com, www.nokia.com/vpn, www.rad.com
Spoofing	**Anti-spoofing**	www.checkpoint.com, www.cisco.com, http://msgs.securepoint.com/cgi-bin/get/netfilter-0112/243.html
Spamming	**Anti-spamming**	**Anti-spam ISPs** can be found at: http://spam.abuse.net/goodsites/a.shtml

---

[75] Although there are other access controls, the only one recommended for unclassified networks is RBAC.

Threat	Security Mechanism	Some Vendor URLs
Catastrophic events	**Backup and recovery systems**	A list of **backup and recovery** providers is at: http://www.win2000mag.com/Techware/InteractiveProduct/Backup2001/
Non-catastrophic events	**UPSs, clustered servers**	**UPS** - www.apcc.com/, www.mgeups.com/, www.gamatronic.com/ **Clustered servers and SW** - www.dell.com, www.gateway.com, www.ibm.com, www.compaq.com **Heating, air conditioning, and ventilation** - Local contractors

Charles L. Smith, Sr.

APPENDIX H

MECHANISMS VERSUS THREATS

Definitions of the security functions, namely, identification and authentication, authorization, accountability, assurance, confidentiality, data integrity, non-repudiation, and security management can be found in Appendix A, *Glossary*. The correlation of these security functions with defensive measures and relevant threats is presented in Table H.1, *Security Functions versus Security Measures and Relevant Threats*.

Table H.1 Security Functions versus Security Measures and Relevant Threats

Security Functions	Security Measures	Relevant Threats

Accountability	1) Access Control
	  a) Identification/Authentication
	  b) Authorization
	2) Detection of Unauthorized Activity
	  a) Auditing
	  b) Intrusion detection
		1) Internal (electronic) users
		2) External (electronic) users
		3) Unauthorized physical access
		4) Sniffing
		5) Spoofing

Assurance	1) All security mechanisms 2) Configuration management	All Threat Groups
Availability	1) Clustered servers 2) Anti-denial-of-service software 3) Identification and authentication 4) Authorization 5) Detection of Unauthorized Activity 6) Backup and recovery 7) Non-catastrophic events	1) System failure 2) Denial of service attacks 3) Unauthorized (electronic) access 4) Unauthorized (physical) access 5) Lack of Role-based access 6) Catastrophic and non-catastrophic events
Confidentiality	1) VPN 2) Encryption 3) Access control 4) Authorization	1) Use of hacker sniffers 2) Hacker and other unauthorized electronic entries 3) Improper access to applications and databases
Data Integrity	1) Checksums 2) Encryption (symmetric and asymmetric)	1) Sniffers 2) Man-in-the-middle attacks
Non-repudiation	1) Use of digital signatures	1) Denial of actual transmissions 2) Denial of actual receptions

Security Functions	Security Measures	Relevant Threats
Security Management	1) Role-based access control 2) Configuration management 3) Identification and authentication 4) Auditing and IDS 5) Physical protection 6) Training and education	1) Unauthorized (electronic) access 2) User or hacker damages 3) Unauthorized (physical) access 4) Lack of user knowledge 5) Social engineering

APPENDIX I

C2 REQUIREMENTS: CONTROLLED ACCESS PROTECTION

Overview

Controlled access protection (often called C2 access control or a C2 Level of Security) provides the needed capabilities for handling sensitive but unclassified information. These requirements were obtained from the DoD Orange Book, titled "DoD Trusted Computer System Evaluation Criteria," DoD 5200.28-STD, 1984. This level of protection requires the various capabilities described below. For commercial organizations, the C2 requirements are somewhat different from those of government organizations. For example, a commercial organization may not wish to create the documentation required for government organizations and create its own required documentation nor use the Discrimination Access Control mechanism and instead use the Role-Based Access Control mechanism.

1.0   C2 Security Policy

1.1   Discretionary Access Control

The Trusted Computing Base (TCB)[76] shall define and control access between named users and named objects (e.g., files, and applications) in the information system. The enforcement mechanisms (e.g., self/group/public controls and access control lists) shall allow users to specify and control sharing of those objects by named individuals, or both, and shall provide controls to limit propagation of access rights. The discretionary access control mechanism shall, either by explicit user action or by default, provide that objects are protected from unauthorized access. These access controls shall be capable of including or excluding access to the granularity of a single user. Access permission to an object for users not already possessing access permission shall only be assigned by authorized users.

1.2   Object Reuse

No information, including encrypted representations of information, produced by a prior subject's actions is to be available to any subject that obtains access to an object that has been released back to the system. All authorizations to the information contained within a storage object shall be revoked prior to initial assignment, allocation, or reallocation to a subject from the pool of unused storage objects.

---

[76] The TCB (sometimes called the security architecture) is the totality of protection mechanisms within a computer system, namely, hardware, firmware, and software and physical and training measures that form the security devices and methods for protecting the computer system. The combination of these is responsible for enforcing a security policy. A TCB consists of the components that together enforce a unified security policy for a computer system. The capability of a TCB to correctly enforce a security policy depends on the countermeasures within the TCB and on the correct actions by system administrative personnel (e.g., identifying the proper roles for a specific user) and users related to the security policy.

2.0 Accountability

Accountability means identifying what an entity did and who that entity is. So accountability is a combination of both identification and authentication as well as the results of an auditing process.

2.1 Identification and Authentication

The TCB shall require users to identify themselves to it before beginning to perform any other actions that the TCB is expected to mediate. Furthermore the TCB shall use a protection mechanism (e.g., passwords) to authenticate the user's identity. The TCB shall protect authentication data so that it cannot be accessed by any unauthorized user. The TCB shall be able to enforce individual accountability by providing the capability to uniquely identity each individual information system user or application. The TCB shall also provide the capability of associating this identity with all auditable actions taken by that user or application.

2.2 Audit

The TCB shall be able to create, maintain, and protect from modification or unauthorized access or destruction an audit trail of accesses to the objects it protects. The audit data shall be protected by the TCB so that read access to it is limited to those who are authorized to access and modify audit data. The TCB shall be able to record the following types of events: use of identification and authentication mechanisms, introduction of objects into a user's or application's address space (e.g., file open or program initiation), deletion of objects, actions taken by computer operators and system administrators and/or system security officers, and other security relevant events.

For each recorded event, the audit record shall identify:

1) Date and time of the event,
2) User or application performing the event,
3) Type of event, and
4) Success or failure of the event.

For identification and authentication events, the origin of request (e.g., personal computer identification number) shall be included in the audit record. For events that introduce an object into a user's address space and for object deletion events, the audit record shall include the name of the object. The information system administrator shall be able to selectively audit the actions of any one or more users or applications based on individual identity.

3.0   Assurance

3.1   Operational Assurance

3.1.1   System Architecture

The TCB shall maintain a domain for its own execution that protects it from external interference or tampering (e.g., by modification of its code or data structures). Resources controlled by the TCB may be a defined subset of the subjects and objects in the information system. The TCB shall isolate the resources to be protected so that they are subject to the access control and auditing requirements.

3.1.2   System Integrity

Hardware and/or software features shall be provided that can be used to periodically validate the correct operation of the on-site hardware and firmware[77] elements of the TCB.

3.2   Life-Cycle Assurance, Security Testing

The security mechanisms of the information system shall be tested and found to work as claimed in the system documentation. Testing shall be done to assure that there are no obvious ways for an unauthorized user to bypass or otherwise defeat the security protection mechanisms of the TCB. Testing shall also include a search for obvious flaws that would allow violation of resource isolation, or that would permit unauthorized access to the audit or authentication data.

4.0   Documentation

4.1   Security Features User's Guide

A single summary, chapter, or manual in user documentation shall describe the protection mechanisms provided by the TCB, guidelines on their use, and how they interact with one another.

4.2   Trusted Facility Manual

A manual addressed to the information system administrator shall present cautions about functions and privileges that should be controlled when running a secure facility. The procedures for examining and maintaining the audit files as well as the detailed audit record structure for each type of audit shall be given.

---

[77] Firmware is the set of computer programs that provide the interfacing of the hardware and software subsystems and consists of client operating systems, network operating systems, database management systems, and application programming interfaces.

4.3   Test Documentation

The system developer shall provide to the evaluators a document that describes the test plan, test procedures that show how the security mechanisms were tested, and results of the security mechanisms' functional testing.

4.4   Design Documentation

Documentation shall be available that provides a description of the manufacturer's philosophy of protection and an explanation of how this philosophy is translated into the TCB. If the TCB is composed of distinct modules, the interfaces between these modules shall be described.

4.5   Documentation for Commercial Organizations

Commercial organizations likely will not wish to generate the various documents required for government organizations. However, even commercial organizations should have a set of documents that ensure that the analysts either have the needed information or have generated the proper system. This information should include the following:

1) An organizational Security Policy,
2) An organizational Security Rules Base,
3) An Information System Overall Architecture,
4) A Security Test Plan and Results,
5) A Risk Assessment, and
6) A Migration Plan (i.e., implementation roadmap).

These documents most likely will contain the same information contained in the four documents listed above and probably much more.

To conclude that you have a C2 Level of Security because the vendor who sold you the operating system says, "Our operating system has a C2 Level of Security" is not a good enough rationale. Only by testing your system and supplying evidence that your system actually meets the requirements stated above can you realistically conclude that your system has a C2 Level of Security.

APPENDIX J

MINIMUM SECURITY REQUIREMENTS

A minimum set of security requirements should contain those organizational security requirements that the organization's information system must meet. These requirements are stated in the same manner as other requirements, namely, in a "shall be" sense. A suggested minimal set of security requirements is presented below. If the organization's information system comprises several intranets, such as LANs located at distributed sites, then the security requirements shall be interpreted to be relevant to each site.

The following minimum requirements shall be met through automated or manual means in a cost-effective manner and integrated fashion.

1.0   Identification and Authentication

There shall be in place a capability to identify each entity (person, application, service, or server) that wishes to logon to the information system and to authenticate that they are who they say they are. To identify the entity, a user identification (User ID) must be used. Authentication can be done through use of a Password. IN some cases, the system may wish to institute strong authentication with the use of tokens, smart cards (and personal identification numbers (PINs)), or biometrics.

2.0   Authorization

Before any entity can access an application or database, the entity must be authorized by the system to determine if the entity has the proper access privileges. The authorization can be determined using an access control list (ACL) and role-based access control (RBAC) databases.

3.0   Access Control

There shall be in place an access control policy for each information system. It shall include features and/or procedures to enforce the access control policy of the information within the information system. The identity of each entity that is authorized access to the information system shall be established positively before authorizing access. Access control shall be enforced with an ACL and RBAC policies, that is, users are identified by their job functions.

4.0   Accountability

There shall be in place countermeasures to ensure each person having access to the information system may be held accountable for his or her actions on the system. There shall be an audit trail providing a documented history of information system use. The audit trail shall be of sufficient detail to reconstruct events in determining the cause or magnitude of

499

compromise should a security violation or malfunction occur. To fulfill this requirement, the manual and/or automated audit trail shall document the following:

1) The identity of each entity accessing the information system,
2) The date and time of the access,
3) A description of the activity sufficient to ensure that the entity's actions can be identified, including attempts to modify, bypass, or negate audit countermeasures controlled by the information system, and
4) Recording of security-relevant actions associated with periods processing or the changing of security levels or categories of information.

DAAs shall cause a review to be made of audit trails associated with the information system(s) over which the DAAs have cognizance to determine an adequate retention period for the audit information. The decision to require an audit trail of user access to a stand-alone, single-user information system (e.g., personal computer (PC)) should be left to the discretion of the DAA.

In addition, the system shall provide for the detection and a proper response to intruders by using an intrusion detection system that scans the audit trails for proof of intrusion. The intrusion detection can be performed offline (not in real time) or online (in real time).

5.0   Availability

The system and its information shall be ready for use by any authorized entity at least 99.99 percent of the time.

6.0   Data Integrity

There shall be countermeasures in place to detect and minimize inadvertent modification or destruction of data, and detect and prevent malicious destruction or modification of data.

7.0   Confidentiality

There shall be a countermeasure that ensures that sensitive information that is stored or transmitted is protected from interception or viewing by unauthorized entities. This countermeasure will be enforced using encryption.

8.0   Non-Repudiation

There shall be a countermeasure (digital signature) that ensures that all entities that receive a message cannot later say that they did not receive the message, and all entities that send a message cannot later deny that they did not transmit the message.

9.0   Least Privilege

The information system shall function so that each entity has access to all of the information to which the entity is entitled (by virtue of the entity's role or formal access approval), but to no more. In the case of "need-to-know" for sensitive information, access must be essential for accomplishment of lawful and authorized organizational purposes.

10.0   Data Continuity

Each file or data collection in the information system shall have an identifiable source throughout its life cycle. Its accessibility, maintenance, movement, and disposition shall be governed by sensitivity level, formal access approval, and need-to-know.

11.0   Marking

Sensitive unclassified output shall be marked to accurately reflect the sensitivity of the information. Requirements for the marking of sensitive information as stipulated in the organization's security policy shall be implemented.

The marking may be automated (i.e., the information system shall have a feature that produces the markings). Automated markings on output must not be relied on to be accurate, unless the sensitivity features and assurances of the information system meet the requirements for the organization's sensitive information.

All output shall be protected, as appropriate, at the sensitive level of the information handled by the information system until manually reviewed by an authorized person to ensure that the output was marked accurately with the appropriate markings and caveats. All media (and containers) shall be marked and protected commensurate with the requirements for the sensitive level and most restrictive category of the information ever stored until the media are declassified (e.g., degaussed or erased) using an approved methodology set forth in the organization's information system Security Policy statement, or unless the information has been downgraded from sensitive but unclassified to unclassified non-sensitive.

12.0   Contingency Planning

Contingency plans and mechanisms shall be developed, installed, and tested in accordance with the organization's Security Policy (or as mandated by OMB Circular No. A-130) to ensure that information system sensitivity controls function reliably and, if not, that adequate backup functions are in place to ensure that security functions are maintained continuously during interrupted service.

If data is modified or destroyed, procedures must be in place to recover the original information as stored in a remote safe location.

13.0   Security Training and Awareness

There shall be in place a security training and awareness program with training for the security needs of all persons accessing the information system. The program shall ensure that all persons responsible for the information system and its associate information, and all entities that access the information system are aware of proper operational and security-related procedures and risks.

14.0   Physical Controls

Information system hardware, software, and documentation, and all sensitive unclassified data handled by the information system shall be protected to prevent unauthorized (intentional or unintentional) disclosure, destruction, or modification (i.e., data integrity shall be maintained). The level of control and protection shall be commensurate with the maximum sensitivity of the information and shall provide the most restrictive control measures required by the data to be handled. This includes having personnel, physical, administrative, and configuration controls. Additionally, protection against denial-of-service of information system resources (e.g., hardware, software, firmware, and information) shall be consistent with the sensitivity of the information handled by the information system.

Unclassified hardware, software, or documentation of an information system shall be protected if access to such hardware, software, or documentation reveals sensitive information, or access provides information that may be used to eliminate, circumvent, or otherwise render ineffective the security countermeasures for sensitive information. Software development and related activities (e.g., systems analysis) shall be controlled by physical controls (e.g., two-person control) and protected when it is determined that the software shall be used for handling sensitive unclassified data.

15.0   Risk Management

There should be in place a risk management program to determine how much protection is required, how much exists, and the most economical way of providing the needed protection.

16.0   Accreditation

The information system (main network and all intranets) shall be accredited to operate in accordance with a DAA-approved set of security countermeasures.

APPENDIX K

RULES FOR ARCHITECTURE MODEL DEVELOPMENT

1. Introduction

Seven high-level rules for architecture model development are:

1) Identify an uncomplicated version of a Reference Architecture model,
2) Identify and/or develop a simple statement of the Security Policy,
3) Identify and/or develop a simple statement of the Security Requirements,
4) Identify all Relevant Security Mandates,
5) Develop an uncomplicated Security Infrastructure definition,
6) Develop a Security Architecture model, and
7) Modify the Reference Architecture model, as required, where a reference system architecture is that architecture into which the security architecture will be embedded.

Each of these high-level rules can be decomposed into more detailed rules as follows.

2. Identify a Reference Architecture Model

   a. Assuming that a reference architecture model exists, this rule means that the analyst must search for an existing architecture model. Otherwise, the analyst will have to create one and the rules for this procedure are not in this document.

3. Identify and/or Develop a Simple Statement of the Security Policy

   a. Using the appropriate references, identify the appropriate security policy statements.
   b. Ensure that the policy statements include the following: statements concerning policy and statements concerning the marking or labeling of objects.

4. Identify and/or Develop a Simple Statement of the Security Requirements

   a. Based on the policy statements, list the requirements statements.
   b. Ensure that the requirements include the following:
      1) *Security Policy* - there must be an explicit and well-defined policy enforced by the system (i.e., policies to be rigorously followed by people, hardware, and software);
      2) *Marking* - access control labels must be associated with objects;
      3) *Accountability* - audit information must be selectively kept and protected so that actions affecting security can be traced to the responsible party;
      4) *Identification* - individual subjects must be identified;

5) *Assurance* - the computer system must contain hardware and software mechanisms that can be independently evaluated to provide sufficient assurance that the system enforces requirements 1 through 4 above; and

6) *Continuous Protection* - the trusted mechanisms that enforce these basic requirements must be continuously protected against tampering and/or unauthorized changes.

5. Identify All Relevant Security Mandates

a. List the mandates that are concerned with the relevant security architecture model.

6. Develop An Uncomplicated Security Infrastructure Definition

a. Develop a model that provides a definition of the security infrastructure.

7. Develop A Security Architecture Model

a. Develop a model of the security architecture.

8. Modify The Reference Architecture Model, If Required

a. If it is required, modify the model of the reference architecture.

APPENDIX L

WHY IS INFORMATION SYSTEM SECURITY NEEDED?

This appendix contains a mathematical analysis approach to answering the question of the need for security for an information system. The analysis begins with the question:

QUESTION: "Why spend money on information system security mechanisms?"

There are two primary cases to be addressed here and they are:

Case I: "Why do I want to spend money for a set of security mechanisms for my information system?"

Case II: "What do I gain from the implemented security mechanisms?"

The approach is based on organizational risk. Only security risk is considered in this appendix. Other risks are not relevant to this analysis.

Case I: Why do I need security?

The need for a set of security mechanisms is addressed first. In essence, an argument will be presented that proves the following: 1) the threats to your organization's information system are great and 2) an appropriate set of security mechanisms that can diminish the threat can be identified that will provide some benefit to your organization.

1) Threats to the system

There are at least 100 million users on the Internet. If just one-thousandth of one percent are users with malicious purposes, then that means that at least 1,000 people might be attempting to gain access to the information or subsystems on your information system. Thus, there is a need for some security mechanisms to diminish the possibility of potential damage from this hoard of malicious users.

Some large corporations and certain Government agencies (e.g., General Motors or the Department of Defense) may be the objective of hundreds of thousands of attacks on their information system each year. Thus, we conclude that because of the great number of potentially malicious users, there is a significant probability of harmful attacks on the system.

2) Benefit of security mechanisms

The answer to this issue is based on security risk and is obtained mathematically. I begin with the following definitions:

1. Let $P(a)$ = probability of a particular attack on some information system.
2. Let $P(p)$ = the probability that the attack is on your particular information system,
3. Let $P(s)$ = probability that the attack is successful,
4. Let D_s = smallest possible damage from the attacker's actions,
5. Let D_g = largest possible damage from the attacker's actions,

   where $0 \leq P(\text{-}) \leq 1.0$.

Then,

6. Risk (smallest) = $P(a) * P(p) * P(s) * D_s$, and
7. Risk (greatest) = $P(a) * P(p) * P(s) * D_g$.

If we assume that $D_s = 0$ (i.e., we assume that the minimum harm done is from a non-malicious attacker who causes no damage), then the expected risk (ER) is;

8. $ER = \frac{1}{2} * P(a) * P(p) * P(s) * D_g$,

   so that ER (minimum) = 0 and ER (maximum) = $D/2$.

We cannot affect $P(a)$ except by indirect means. Whenever any malicious activities are identified by an intrusion detection system (IDS) (i.e., auditing of network activities to determine if they are malicious), then there should be remedies that include modifying the system so that this attack cannot happen again and/or collecting relevant information to support taking punitive (legal) action against the attacker. (One objective of an IDS or an auditing system is to collect information on any attacker's activities to properly support a legal case against the attacker.) Now even though $P(a)$ cannot be determined accurately, it is highly likely that if intruders are identified and severely punished, and that this punishment is widely publicized in the news media, then $P(a)$ can be reduced assuming that potential intruders (or at least some of them) will fear the potential repercussions of their malicious actions.

The value for $P(p)$ is dependent on many variables, such as the size of your organization, the publicity your organization receives in the media, the value of your information, the ego value to an attacker of having broken into your system, and the likelihood that a former employee might decide to attack your system. A fairly precise value of $P(p)$ for any organization is probably better acquired using statistics rather than creating and using some mathematical formula. The values of these variables are not easily affected by your organization's actions. But one can, by use of security mechanisms, affect $P(s)$ in a positive sense, that is, reduce its value.

If nothing is done, then $P(s)$ is virtually 1.0 (at least for some types of intrusions), but if security mechanisms are deployed to diminish intruder activities, $P(s)$[78] can be reduced to nearly 0.0, depending on how much your organization is willing to spend relative to its security requirements. In this manner, ER can be made close to 0.0.

There are two conclusions that can be deduced from the above analysis. They are:

---

[78] As mentioned before, no amount of security mechanisms can reduce the risk to zero.

1) Whenever any malicious activities are identified (say by the auditing and administration capabilities) then there should be remedies that include fixing the system so that this attack cannot happen again and taking punitive (legal) action against the attacker. Even though P (a) cannot be determined accurately, it is highly likely that if intruders are identified and severely punished, and that information about this punishment is widely publicized in the news media, then P(a) can be reduced because potential intruders will fear repercussions. However, some organizations do not wish to publicize breaches to their information system so this approach is not an alternative.

2) With the use of security mechanisms to reduce the value of P(s), the potential risk to organizations from attackers can be reduced.

Case II: What are the potential gains from implementing security mechanisms?

The potential gains from implementing the different security mechanisms are difficult to determine in a precise manner. The threats or tactics are numerous and include the following:

1) Unauthorized access - Access to the system by unauthorized persons or systems.
2) Data corruption - Diminution of data availability.
3) Platform corruption - Diminution of subsystem availability.
4) Denial of service - Use of tactics to diminish system and data availability.
5) Physical destruction or maliciousness - Physical attacks on the system.
6) Traffic monitoring or tampering - Use of electronic systems to access or modify transmitted information.

Security mechanism categories for diminishing the potential success of such tactics include the following:

1) Identification and authentication - Use of methods to accurately identify and authenticate all authorized users.
2) Access control - Assurance that all entities are properly assigned access according to an appropriate access control list (ACL).
3) Auditing - Use of monitors to gather data on all system activities.
4) Back-Up - Use of restoration methods to quickly store and retrieve all data.
5) Physical Protection - Use of locked rooms, policemen, etc. to provide protection to the physical assets of the system.
6) Encryption - Use of data encryption to ensure confidentiality, data integrity, user authentication, and non-repudiation.
7) Scanning - Use of a scanner (e.g., SATAN or nmap) to identify network security issues (e.g., open ports or software holes).
8) Other anti-attacker mechanisms - Use of firewalls, virus walls, anti-virus software, DMZs, proxy servers, and intrusion detection systems.

The needed security mechanisms should be determined relative to the security requirements for some particular application in your information system. If the security requirements for the system are not so stringent, then the need for security mechanisms is reduced.

The potential gains for your organization through the use of security mechanisms to defeat potential attackers should be ascertained through an analysis of the recommended and implemented security mechanisms versus the considered threats to the system.

The security mechanisms needed are dependent upon the value of the potential benefits gained by attacks on your information system, such as any of the following:

1) Utility of stolen information,
2) Losses incurred due to denial of access to your system,
3) Losses incurred due to damaged or erased information,
4) Losses incurred due to publication of the attacks on your system (which is difficult to quantify),
5) Losses due to unauthorized viewing of sensitive information (such as credit card numbers) in your system, and
6) Other such events.

For example, a security analyst might conclude that if your *organization's data* is not of too much value to a potential attacker, but your *system's availability* is highly critical, then security mechanisms that can assure system uptime (i.e., system and data availability) are much more valuable to your organization than security mechanisms that provide data protection (i.e., confidentiality). However, even in this case, data integrity (i.e., protection from modification by unauthorized entities) can be very important and a data integrity mechanism, such as adding a checksum to transmitted or stored information, is of great utility.

INDEX

ABOUT THE AUTHOR

Dr. Charles L. Smith, Sr. is a sometimes consultant but currently works full time for TLA Associates in charge of the security aspects and overview of the Standard Procurement System as the security representative of the Defense Contract Management Agency.

He has worked on many computer system architecture and security efforts primarily as a consultant, including the security aspects of the Federal Aviation Administration's computer program for improved airspace control, Chrysler Corporation's powertrain testing system, the Defense Contract Management Agency's worldwide computer network, the Pentagon's Digital Network, and the Internal Revenue Service's tax systems modernization program. He has a Ph.D. in Information Technology from George Mason University (1992) and is a Certified Information Systems Security Professional (2001).

Printed in the United States
68728LVS00006B/5-42